Computer Mathematics

Proceedings of the
Sixth Asian Symposium (ASCM 2003)

LECTURE NOTES SERIES ON COMPUTING
Editor-in-Chief: D T Lee (*Northwestern Univ., USA*)

Published

Vol. 1: Computing in Euclidean Geometry
Eds. D-Z Du & F Hwang

Vol. 2: Algorithmic Aspects of VLSI Layout
Eds. D T Lee & M Sarrafzadeh

Vol. 3: String Searching Algorithms
G A Stephen

Vol. 4: Computing in Euclidean Geometry (Second Edition)
Eds. D-Z Du & F Hwang

Vol. 5: Proceedings of the Conference on Parallel Symbolic Computation — PASCO '94
Ed. H Hong

Vol. 6: VLSI Physical Design Automation: Theory and Practice
Sadiq M Sait & Habib Youssef

Vol. 7: Algorithms: Design Techniques and Analysis
Ed. M H Alsuwaiyel

Vol. 8: Computer Mathematics
Proceedings of the Fourth Asian Symposium (ASCM 2000)
Eds. X-S Gao & D Wang

Vol. 9: Computer Mathematics
Proceedings of the Fifth Asian Symposium (ASCM 2001)
Eds. K Yokoyama & K Shirayanagi

Lecture Notes Series on Computing Vol. 10

Computer Mathematics

Proceedings of the Sixth Asian Symposium (ASCM 2003)

Beijing, China 17 - 19 April 2003

Editors

Ziming Li
Chinese Academy of Sciences, China
and University of Waterloo, Canada

William Sit
The City College of
The City University of New York, USA

World Scientific
New Jersey • London • Singapore • Hong Kong

Published by

World Scientific Publishing Co. Pte. Ltd.
5 Toh Tuck Link, Singapore 596224
USA office: Suite 202, 1060 Main Street, River Edge, NJ 07661
UK office: 57 Shelton Street, Covent Garden, London WC2H 9HE

British Library Cataloguing-in-Publication Data
A catalogue record for this book is available from the British Library.

COMPUTER MATHEMATICS
Proceedings of the Sixth Asian Symposium (ASCM 2003)

ISBN 981-238-220-8

Printed in Singapore by World Scientific Printers (S) Pte Ltd

PREFACE

This volume contains papers of the invited speakers and of contributing authors that were presented at the Sixth Asian Symposium on Computer Mathematics (ASCM) on April 17–19, 2003 in Beijing, China. The twenty-one contributed papers were selected by the Program Committee after a standard refereeing process from forty-one submissions, and together with two papers and an extended abstract from the invited speakers, cover some of the most recent research in computer mathematics. The articles present and survey advanced algorithms for polynomial algebra, algebraic geometry, non-commutative algebra, geometric modelling and applications, differential and difference equations, numerical methods, and pertubation analysis. This volume thus reflects the state-of-the-art and trends on research in computation methods for mathematics and their applications.

The ASCM symposia have been organized under the close collaboration between the Japan Society for Symbolic and Algebraic Computation (JSSAC) and the Mathematics Mechanization Research Center (MMRC) of the Chinese Academy of Sciences. Held in Asian countries with international participation since 1995, these symposia have become the ideal forum for the presentation of original results and networking among researchers interested in the development and application of computer software for mathematics. The previous five ASCM symposia in 1995, 1996, 1998, 2000, and 2001 were held in Beijing (China), Kobe (Japan), Lanzhou (China), Chiang Mai (Thailand), and Matsuyama (Japan) respectively.

This sixth symposium, preceded on April 16, 2003 by the Workshop on Geometric Constraint Solving, has attracted a wide international participation. The ASCM program includes three invited talks, by Christoph M. Hoffmann, Josef Schicho, and Nobuki Takayama, and twenty-one technical presentations. Together they represent fifty-two authors from thirteen countries and regions in Asia, Europe, Australia, and North America. The Sixth Symposium will also host a poster session for the latest discoveries which, unfortunately, will not be included in the proceedings due to an early publication deadline. The complete programs for this and the preceding symposia are recorded at the official website `http://www.mmrc.iss.ac.cn/~ascm/`. ASCM 2003 is hosted by MMRC with financial support from the Institute of Systems Science of the Chinese Academy of Sciences and the Chinese National Key Basic Research (973) Project *Mathematics-Mechanization and Automated Reasoning Platform.*

Many people have contributed to the organization and preparation of ASCM 2003 and in particular, to the present volume. The expertise and timely work of the Program Committee members, external reviewers (see below), and authors are responsible for the quality of the conference proceedings. The publication of the proceedings as a volume in the Lecture Notes Series on Computing by World Scientific makes ASCM more accessible to the international scientific community. We thank all of them for their generous efforts and cooperation despite a very tight schedule.

Waterloo — Ziming Li
New York — William Sit

January, 2003

External Reviewers

V. Adamchik, M. Bronstein, L. Chao, C. Chen, H. Chen, X. Chen, Y. Chen, E. W. Chionh, A. Chtcherba, R. Coulter, H. Crapo, H. Du, I. Emiris, E. Fan, Y. Feng, Y. Gao, I. Gessel, J. Giesl, B. Han, G. Han, M. Henderson, W. Hereman, J. Hietarinta, M. Hoffman, H. Kai, M. Kreuzer, B. Jones, B. Juettler, A. Khetan, I.Ṡ. Kotsireas, F. Lamnabhi-Lagarrigue, D. Lazard, Z. B. Li, Q. Liao, W. Liu, P. Luo, H. Ma, Y. Ma, D. Manocha, F. Muller-Hoissen, H. Murao, B. Mourrain, J. Rokne, K. Sakai, J. Sanders, N. Sasa, T. Sasaki, T. Satoh, H. Sekigawa, H. Stetter, J. Schicho, M. Singer, Y. B. Suris, A. Suzuki, T. Takahashi, W. Tong, J. Verschelde, B. Wang, J. Wang, J. P. Wang, J. Y. Wang, S. K. Wang, V. Weispfenning, N. White, K. Wu, W. Wu, Y. Wu, B. Xia, C. Xu, G. Xu, N. Yan, B. Yu, F. Zeilfelder, H. F. Zhang, H. Q. Zhang, J. Zheng.

Conference Organizers

General Chair

Wen-tsün Wu	Chinese Academy of Sciences, China

Program Committee

Falai Chen	University of Science and Technology of China, China
Gouting Chen	University of Lille I, France
Shang-Ching Chou	Wichita State University, USA
Ding-Zhu Du	University of Minnesota, USA
Xiaoshan Gao	Chinese Academy of Sciences, China
George Havas	University of Queensland, Australia
Hoon Hong	North Carolina State University, USA
Jie Hsiang	National Taiwan University, Taiwan
Deepak Kapur	University of New Mexico, USA
Wen-Shin Lee	University of Waterloo, Canada
Hongbo Li	Chinese Academy of Sciences, China
Ziming Li (co-chair)	Chinese Academy of Sciences, China and University of Waterloo, Canada
Matu-Tarow Noda	Ehime University, Japan
Masayuki Noro	Kobe University Japan
Hong Qin,	SUNY at Stony Brook, USA
Yosuke Sato	Ritsumeikan University, Japan
Kiyoshi Shirayanagi	NTT Communication Science Labs, Japan
William Sit (co-chair)	The City College of New York, USA
Nobuki Takayama	Kobe University, Japan
Dongming Wang	Centre National de la Recherche Scientifique, France
Paul S. Wang	Kent State University, USA
Wenping Wang	University of Hongkong, China
Lu Yang	Chinese Academy of Sciences, China
Kasuhiro Yokoyama	Kyushu University, Japan
Lihong Zhi	Chinese Academy of Sciences, China

Local Arrangements Committee

Xiaoshan Gao (chair)	Chinese Academy of Sciences, China
Dingkang Wang	Chinese Academy of Sciences, China
Dai-zhen Zhou	Chinese Academy of Sciences, China

Poster Session Committee

Dingkang Wang (chair)	Chinese Academy of Sciences, China

CONTENTS

Preface v
Conference Organizers vii

Invited Talks

Compliant Motion Constraints 1
C. M. Hoffmann (speaker), W. Yang (USA)

Parametrization of Rational Surfaces (Extended Abstract) 17
J. Schicho (Austria)

Algebraic Algorithms for *D*-Modules and Numerical Analysis 23
T. Oaku, Y. Shiraki, N. Takayama (speaker) (Japan)

Contributed Papers

On the Division of Generalized Polynomials 40
N. Aris, A. Rahman (Malasia)

On One Property of Hurwitz Determinants 52
L. Burlakova (Russia)

Interval Parametrization of Planar Algebraic Curves 64
F. Chen, L. Deng (China)

Blending Quadric Surfacecs via a Base Curve Method 77
J. Cheng (China)

Bivariate Hermite Interpolation and Linear Systems of Plane Curves with Base Fat Points 87
C. Ciliberto, F. Cioffi, R. Miranda F. Orecchia (Italy and USA)

A Series of Exact Solutions of Two Extended Coupled Ito Systems .. 103
E. Fan (China)

Corner Point Pasting and Dixon $\mathcal{A}$-resultant Quotients 114
M. Foo, E. Chionh (Singapore)

Zero Decomposition Theorems for Counting the Number of Solutions for Parametric Equation Systems 129
X. Gao, D. Wang (China)

An Exploration of Homotopy Solving in Maple 145
K. Hazaveh, D. J. Jeffrey, G. J. Reid, S. M. Watt, A. D. Wittkopf (Canada)

Densities and Fluxes of Differential-Difference Equations 163
M. Hickman, W. Hereman (New Zealand and USA)

A Complete Maple Package for Noncommutative Rational Power Series 174
V. Houseaux, G. Jacob, N.E. Oussous, M. Petitot (France)

Global Superconvergence of Biquadratic Lagrange Elements for Poisson's Equation 189
H. Huang, Z. Li, A. Zhou (Taiwan and China)

Automation of Perturbation Analysis in Computer Algebra: Progress and Difficulties 204
R. Khanin (United Kingdom)

Implicitization of Polynomial Curves 217
I. S. Kotsireas, E. S. C. Lau (Canada)

A Bracket Method for Judging the Intersection of Convex Bodies 227
H. Li, Y. Chen (China)

Discrete Comprehensive Gröbner Bases, II 240
Y. Sato, A. Suzuki, K. Nabeshima (Japan)

Computational Aspects of Hyperelliptic Curves 248
T. Shaska (USA)

Application of the Wu-Ritt Differential Elimination Method to the Painlevé Test 258
F. Xie, H. Zhang, Y. Chen, B. Li (China)

Degree Reduction of Rational Curves by μ-Bases 265
F. Zeng, F. Chen (China)

A Molecular Inverse Kinematics Problem: An Approximation Approach and Challenges 276
M. Zhang, R. A. White (USA)

Displacement Structure in Computing Approximate GCD of Univariate Polynomials 288
L. Zhi (China)

Author Index 299

COMPLIANT MOTION CONSTRAINTS*

CHRISTOPH M. HOFFMANN, WEIQIANG YANG
Department of Computer Science
Purdue University
West Lafayette, IN 47907, USA

We examine a dual-quaternion formulation for expressing the relative rigid body motion between two objects when incidence constraints are to be observed. The incidences are between points, lines and planes, of the two parts. Both parametric and implicit representations are investigated. Several examples illustrate the techniques.

Keywords: Rigid body motion, compliant motion, quaternion, dual quaternion, relative motion, incidence constraint, virtual reality, kinematics, geometric constraint.

1. Introduction

Geometric constraints are used in two different contexts. In one application area we define a set of geometric primitives and constraints upon them, and then are asked to find an arrangement of the primitives such that the constraints are satisfied. Let us call this the *construction problem*. The construction problem arises for example when defining CAD models for discrete manufacturing.

In a second application area we are given a set of (usually composite) geometric objects as well as constraints upon their spatial relationship, with the objective of constraining the relative motion of the objects with respect to each other. Let us call this the *compliance problem*. The compliance problem arises in assembly modeling, kinematic simulation of machinery, and in virtual reality, to name a few uses.

In this paper we consider the compliance problem in 3-space and investigate basic techniques to solve it. There is a wealth of prior work, and we give a few example references below. The area of kinematics to a large extent considers and solves compliance problems, for instance when

*Work supported in part by NSF Grants EIA 02-16131, DMS/CCR 01-38098, CCR 99-02025, and by ARO contract 39136-MA.

investigating linkages, or, more generally, when designing and analyzing machinery [6,7]. Other relevant research is done in robotics [1], and in some areas of geometric constructions [5]. To some extent, the compliance problem overlaps with the construction problem, as seen in [4], where a system of equations is attacked by considering the residual compliant motion of geometric primitives when restricting to a subset of the given constraints.

Much of the research into compliance is dominated by seeking elegant mathematical formalisms that would simplify expressing and analyzing compliant motion. In addition to ad-hoc techniques that are highly successful in special cases such as four-bar linkages, three main formalisms have emerged: (4×4) transforms, screws, and dual quaternions. The three formalisms offer a general description of rigid body motion in 3-space. Note that screws are essentially dual quaternions, but the reduced coordinate set may introduce ambiguities in some cases. For this reason we do not consider them further.

2. Tools and Notation

In this section, we review some basics and notations on quaternions, their relations to rotations, and dual quaternions.

2.1. *Quaternions*

The field of quaternions has elements of the form $\mathbf{a} = a_0 + a_1 i + a_2 j + a_3 k$ where the coefficients a_r are real numbers and the units i, j, and k obey the equations:

$$i^2 = j^2 = k^2 = -1,$$
$$ij = -ji = k, \qquad jk = -kj = i, \qquad ki = -ik = j$$

Complex numbers are quaternions with $a_2 = a_3 = 0$. The *length* of a quaternion $\mathbf{a}$, is defined as $\|\mathbf{a}\| = \sqrt{a_0^2 + a_1^2 + a_2^2 + a_3^2}$. Quaternions are due to Maxwell.

The *conjugate* of the quaternion $\mathbf{a} = a_0 + a_1 i + a_2 j + a_3 k$ is the quaternion $\bar{\mathbf{a}} = a_0 - a_1 i - a_2 j - a_3 k$. The *norm* of $\mathbf{a}$ is the square of the length of $\mathbf{a}$ and is equal to the quaternion product $\mathbf{a}\bar{\mathbf{a}}$. We define the *inner product* $(\mathbf{a} \cdot \mathbf{b})$ of two quaternions $\mathbf{a}$ and $\mathbf{b}$ as $(\mathbf{a} \cdot \mathbf{b}) = a_0 b_0 + a_1 b_1 + a_2 b_2 + a_3 b_3$. Note that the norm of $\mathbf{a}$ is $(\mathbf{a} \cdot \mathbf{a}) = \mathbf{a}\bar{\mathbf{a}}$.

A quaternion with $a_0 = 0$ is called a *vector*. More generally, we can call a_0 the *real part* of the quaternion $\mathbf{a}$ and call $a_1 i + a_2 j + a_3 k$ the *vector part*. We denote quaternions with lower-case bold letters. Vectors in $\mathbf{R}^3$

are denoted by bold lower-case letters with an arrow; for example, $\vec{\mathbf{p}}$. The arrow is omitted when it is clear from the context that we speak of a vector. The vector part of a quaternion is denoted in the same way. Thus, if $\mathbf{a} = a_0+a_1i+a_2j+a_3k$, then $\vec{\mathbf{a}} = a_1i+a_2j+a_3k$. We note that $\overline{\mathbf{a}} = a_0-\vec{\mathbf{a}}$.

We will use the inner product, denoted by $\cdot$ and the cross product of vectors, denoted by $\times$, to express quaternion operations more succinctly. For example, the product of two quaternions $\mathbf{a} = a_0 + \vec{\mathbf{a}}$ and $\mathbf{b} = b_0 + \vec{\mathbf{b}}$ is the quaternion $\mathbf{ab} = a_0b_0 - (\vec{\mathbf{a}} \cdot \vec{\mathbf{b}}) + a_0\vec{\mathbf{b}} + b_0\vec{\mathbf{a}} + \vec{\mathbf{a}} \times \vec{\mathbf{b}}$.

2.2. *Rotations*

With Cartesian point coordinates in 3-space, a rotation in 3-space about the origin can be represented by the orthogonal matrix

$$R = \begin{pmatrix} r_{11} & r_{12} & r_{13} \\ r_{21} & r_{22} & r_{23} \\ r_{31} & r_{32} & r_{33} \end{pmatrix},$$

where $RR^T = I$ and $\det(R) = 1$.

It is well-known that unit-length quaternions can represent rotations about the origin. Wittenburg [10] gives the following conversion formulae. For any unit-length quaternion $\mathbf{a}$, the entries of the rotation matrix are

$$\begin{aligned} r_{11} &= 2(a_0^2 + a_1^2) - 1, & r_{12} &= 2(a_1a_2 + a_0a_3), & r_{13} &= 2(a_1a_3 - a_0a_2), \\ r_{21} &= 2(a_1a_2 - a_0a_3), & r_{22} &= 2(a_0^2 + a_2^2) - 1, & r_{13} &= 2(a_2a_3 + a_0a_1), \\ r_{31} &= 2(a_1a_3 + a_0a_2), & r_{32} &= 2(a_2a_3 - a_0a_1), & r_{33} &= 2(a_0^2 + a_3^2) - 1, \end{aligned}$$

and for any rotation matrix with entries r_{pq} the quaternion coefficients are:

$$\begin{aligned} a_0^2 &= (r_{11} + r_{22} + r_{33} + 1)/4, \\ a_1^2 &= r_{11}/2 - u, \\ a_2^2 &= r_{22}/2 - u, \\ a_3^2 &= r_{33}/2 - u, \\ \text{where} \quad u &= (r_{11} + r_{22} + r_{33} - 1)/4. \end{aligned}$$

Other, equivalent conversion formulae are given in, for example, [8].

There is a well-known geometric interpretation of the quaternion representation of such rotations. Let $\mathbf{v} = (v_1, v_2, v_3)$ be the unit-length direction vector of the axis of rotation, and let 2θ be the angle of rotation. With $c = \cos(\theta)$ and $s = \sin(\theta)$, the rotation is represented by the quaternion $c + sv_1i + sv_2j + sv_3k$.

2.3. *Dual Numbers and Dual Quaternions*

A *dual number* is defined as $A = a + b\epsilon$, where a and b are from a field and $\epsilon^2 = 0$. Dual numbers form a Clifford algebra. If $A = a + b\epsilon$ is a dual number, $A_\epsilon = a - b\epsilon$ is its *conjugate*.

A *dual quaternion* is defined as $\mathbf{A} = \mathbf{a} + \mathbf{b}\epsilon$, where $\mathbf{a}$ and $\mathbf{b}$ are quaternions. Equivalently, a dual quaternion is a quaternion whose components are dual numbers (with real coefficients). Dual quaternions can represent points, lines and planes in 3-space, as well as general rigid body motions, as will be discussed in the next section.

As in [2], we define three different conjugations of a dual quaternion, according to whether the quaternion components are conjugated, the dual numbers are conjugated, or both. Let $\mathbf{A} = \mathbf{a} + \mathbf{b}\epsilon$, where $\mathbf{a}$ and $\mathbf{b}$ are the quaternions. We define

$$\overline{\mathbf{A}} = \overline{\mathbf{a}} + \overline{\mathbf{b}}\epsilon, \quad \mathbf{A}_\epsilon = \mathbf{a} - \mathbf{b}\epsilon, \quad \overline{\mathbf{A}}_\epsilon = \overline{\mathbf{a}} - \overline{\mathbf{b}}\epsilon.$$

3. Representations

We adopt the algebraic schema of [2] to represent points, lines and planes in 3-space, as well as rigid-body transformations on them.

3.1. *Points, Lines, and Planes*

In Cartesian coordinates, points are specified by their position vector (p_1, p_2, p_3) which we represent by the dual quaternion

$$\mathbf{P} = 1 + \mathbf{p}\epsilon$$

A plane has the equation $ap_1 + bp_2 + cp_3 + d = 0$, where we require that $a^2 + b^2 + c^2 = 1$. Such a plane is represented by the dual quaternion

$$\mathbf{E} = \mathbf{n} + d\epsilon.$$

The first quaternion is the plane normal vector $\mathbf{n} = ai + bj + ck$, and the second quaternion, which is real, is the constant of the implicit plane equation.

Using Plücker coordinates, lines in 3-space can be represented by two 3-vectors $\mathbf{t} = (t_1, t_2, t_3)$ and $\mathbf{m} = (m_1, m_2, m_3)$, where $\mathbf{t}$ is the line direction vector, normed to unit length, and $\mathbf{m}$ is the moment vector $\mathbf{p} \times \mathbf{t}$ of some point $\mathbf{p}$ on the line. Clearly, the inner product of the moment vector and the direction vector is zero; that is, $(\mathbf{m} \cdot \mathbf{t}) = 0$. Identifying, as before, the vector (a, b, c) with the quaternion $ai + bj + ck$, we represent the line $(\mathbf{t}, \mathbf{m})$ as the dual quaternion

$$\mathbf{L} = \mathbf{t} + \mathbf{m}\epsilon.$$

The first quaternion is the unit-length direction vector, the second quaternion is the moment vector. For lines through the origin $\mathbf{m} = 0$.

3.2. *Rigid Body Motion*

The unit quaternion $\mathbf{q}$ was noted to represent a rotation about the origin. The dual quaternion $\mathbf{Q} = \mathbf{q}$ with the zero quaternion as the ϵ coordinate is chosen to represent the same rotation. Furthermore, we represent a translation by the vector $(2s_1, 2s_2, 2s_3)$ by the dual quaternion $\mathbf{S} = 1 + \mathbf{s}\epsilon$, where $\mathbf{s} = s_1 i + s_2 j + s_3 k$. A rigid body motion in 3-space can therefore be represented by a dual quaternion $\mathbf{T} = \mathbf{SQ}$ that is the product of the rotation quaternion $\mathbf{Q}$ and the translation quaternion $\mathbf{S}$, imitating the action of 4-by-4 transforms. The representation of rigid motions by dual quaternions is due to Study [9].

Screw Motion

Chasles' theorem [3] states that every rigid motion is equivalent to a screw motion. Here, the screw with axis $(\mathbf{t}, \mathbf{m})$, angle of rotation 2θ, and a displacement $2d$ is represented as the dual quaternion

$$\mathbf{M}_{\text{screw}} = \cos(\theta) + \sin(\theta)\,\mathbf{t} + (-d\sin(\theta) + \sin(\theta)\,\mathbf{m} + d\cos(\theta)\,\mathbf{t})\,\epsilon \quad (3.1)$$

Note that for $\theta = 0$, the motion $\mathbf{M}$ simplifies to $1 + d\,\mathbf{t}\epsilon$, a translation by $2d\,\mathbf{t}$, and for $d = 0$ and $\mathbf{m} = 0$ it simplifies to $\cos(\theta) + \sin(\theta)\,\mathbf{t}$, a rotation about the origin.

Other Motion Representation

The general rigid body motion can be expressed as

$$\mathbf{M} = \mathbf{q} + \mathbf{u}\,\epsilon, \quad \|\mathbf{q}\| = 1, \quad (\mathbf{q} \cdot \mathbf{u}) = 0. \quad (3.2)$$

We prove that this is true.

The conditions of Equation (3.2) are clearly satisfied by the screw motions of Equation 3.1. Thus, all rigid motions can be represented.

Conversely, assume that the above conditions are satisfied by the dual quaternion $\mathbf{M} = \mathbf{q} + \mathbf{u}\epsilon$. Writing $\mathbf{q} = q_0 + q_1 i + q_2 j + q_3 k$, we may define θ, t_1, t_2, t_3 by setting

$$q_0 = \cos(\theta), \qquad \text{and } q_r = \sin(\theta) t_r, \qquad r = 1, 2, 3.$$

Let $\mathbf{t} = t_1 i + t_2 j + t_3 k$. Since $\|\mathbf{q}\| = 1$, we have $\|\mathbf{t}\| = 1$ in general.

If $|q_0| = 1$, then $q_1 = q_2 = q_3 = 0$ and $u_0 = 0$. In that case $\mathbf{M}$ is a translation.

If $u_0 = 0$, the vector $\vec{\mathbf{u}}$ must be perpendicular to the vector $\vec{\mathbf{q}}$. We may assume $|q_0| \neq 1$ and define the vector $\mathbf{m} = \vec{\mathbf{u}}/\sin(\theta)$. We now see that $\mathbf{M}$ represents a pure rotation about an axis with direction $\mathbf{t}$ and moment $\mathbf{m}$.

Otherwise, with $|q_0| \neq 1$ and $u_0 \neq 0$, we have $\sin(\theta) \neq 0$, and we can define the nonzero quantity d from $u_0 = -d\sin(\theta)$. Define $m_0 = 0$ and $m_r \sin(\theta) = u_r - dt_r \cos(\theta)$, $r = 1, 2, 3$. Then

$$\begin{aligned}
0 &= q_0u_0 + q_1u_1 + q_2u_2 + q_3u_3 \\
&= -d\sin(\theta)\cos(\theta) + \textstyle\sum_{r=1}^{3} \sin(\theta)t_r(m_r\sin(\theta) + dt_r\cos(\theta)) \\
&= -d\sin(\theta)\cos(\theta) + \textstyle\sum_{r=1}^{3}(d\sin(\theta)\cos(\theta)t_r^2 + t_rm_r\sin^2(\theta)) \\
&= \sin^2(\theta)(\mathbf{t}\cdot\mathbf{m})
\end{aligned}$$

Therefore the vector $\mathbf{m} = m_1i + m_2j + m_3k$ is perpendicular to the vector $\mathbf{t}$, which means that $\mathbf{M}$ is a screw motion with axis $(\mathbf{t}, \mathbf{m})$, rotation angle 2θ, and displacement $2d$.

3.3. *Motion of Points, Lines and Planes*

Let $\mathbf{P}$ be a dual quaternion representing a point or a plane. Then the dual quaternion $\mathbf{P}'$ that represents the result of a rigid body motion $\mathbf{M}$, applied to the point or plane represented by $\mathbf{P}$, is calculated as

$$\mathbf{P}' = \mathbf{MP}\overline{\mathbf{M}}_{\epsilon}. \tag{3.3}$$

Similarly, the line represented by the dual quaternion $\mathbf{L}$ is transformed into the line represented by $\mathbf{L}'$, where

$$\mathbf{L}' = \mathbf{ML}\overline{\mathbf{M}} \tag{3.4}$$

An algebraic computation verifies this definition; see also [2].

Summarizing, dual quaternions allow us to represent points, lines and planes in 3-space uniformly, and express rigid body transformations of them uniformly as well.

4. Constrained Motion

We investigate what relative motion is possible when requiring a single incidence constraint of a point, line or plane on another point, line or plane. First, we formulate incidence conditions in terms of dual quaternions. Then we investigate relative motion assuming that the incidences are currently satisfied.

4.1. *Incidence*

Six elementary incidence conditions arise when requiring points, lines and planes to be incident to each other. Among features of equal type, incidence is trivial, as it requires equal coordinates. Note, however, that for planes and lines incidence with opposite orientation should be accounted for.

Let **P** be a point, **E** a plane, and **L** a line. We require that the plane normal and the line direction vectors have unit length. The following are the incidence conditions between features of different type.

$$\begin{aligned} \mathbf{EP}_\epsilon + \mathbf{PE}_\epsilon &= 0 \qquad \text{point on plane} \\ \mathbf{LP} - \mathbf{PL}_\epsilon &= 0 \qquad \text{point on line} \\ \mathbf{LP} + \mathbf{PL}_\epsilon &= 0 \qquad \text{line on plane} \end{aligned}$$

See [2] for a proof.

4.2. *Parametric Relative Motion*

It is not difficult to express parametrically the relative motion that obeys a single elementary incidence constraint. In particular, if the elements are of the same type, we are asking for motion expressions that leave a point, a line or a plane invariant. However, a parametric representation in the presence of multiple incidence constraints between different features of two rigid bodies is not so easy. We will show that it can be done based on the parametric representation, in a number of cases. A related problem commonly investigated in robotics is to synthesize the motion of a kinematic chain, such as an articulated robotic arm. Such work typically assumes fixed common lower-pair connections between the links such as a revolute or a prismatic joint.

If we express the relative motion of a single incidence constraint parametrically, then we can combine the equations into a single system and obtain a combined parameterization using elimination computations. It is advantageous to keep the equation system as simple as possible, and this would argue for performing every algebraic simplification possible as a preprocessing step before undertaking the actual evaluation. We begin with expressing the elementary constraints. As before, we denote points, lines and planes with dual quaternions **P**, **L** and **E**, respectively.

4.2.1. *Incidences of Equal Type*

These are incidences of point on point, line on line, and plane on plane. To express the relative motions, we ask which rigid body transformations fix a point, a line or a plane.

For the point represented by the dual quaternion $\mathbf{P} = 1 + \epsilon\mathbf{p}$ we obtain

$$\mathbf{M_P} = \mathbf{q} + \epsilon\vec{\mathbf{p}} \times \vec{\mathbf{q}}, \quad \|\mathbf{q}\| = 1. \tag{4.1}$$

Note that we require that $\mathbf{q}$ has length 1, and that $\mathbf{p}$ is a vector quaternion, that is, $p_0 = 0$.

We can derive $\mathbf{M_P}$ by conjugating a general rotation about the origin by the translation of the fixed point to the origin. Let $\mathbf{T}$ be the translation from the point $\mathbf{P}$ to the origin, represented as a dual quaternion, and let its inverse be $\mathbf{T}'$. Then $\mathbf{M_P} = \mathbf{T}'\mathbf{Q}\mathbf{T}$, where $\mathbf{Q}$ represents a rotation about the origin. The representation has four parameters which reduce to three independent ones because of the unit-length requirement on $\mathbf{q}$.

A different parameterization derivation is possible by considering a screw motion that has a zero displacement along the axis. With $\mathbf{t}$ an arbitrary unit length vector, we then obtain the equivalent form

$$\begin{aligned} \mathbf{M_P} &= \cos(\theta) + \mathbf{t}\sin(\theta) + \mathbf{m}\sin(\theta)\epsilon, \\ \|\mathbf{t}\| &= 1, \qquad \mathbf{m} = \mathbf{p} \times \mathbf{t}. \end{aligned}$$

Again, there are 4 parameters reducing to three independent ones because of the unit-length requirement. The resulting parameterization is identical to (4.1).

Next, we consider the motion that leaves the plane $\mathbf{E} = \mathbf{n} + d\epsilon$ invariant. Here, $\mathbf{n}$ is the unit-length normal vector of the plane. The motion that leaves the plane invariant can be considered as a rotation about an axis through the origin in the direction $\vec{\mathbf{n}}$ plus a translation by a vector $\mathbf{t}$ in the plane, which therefore satisfies $(\mathbf{t} \cdot \mathbf{n}) = 0$. We obtain

$$\begin{gathered} \mathbf{M_E} = \cos(\theta) + \sin(\theta)\mathbf{n} + \epsilon(\cos(\theta)\mathbf{t} + \sin(\theta)(\mathbf{t} \times \mathbf{n})), \\ (\mathbf{t} \cdot \mathbf{n}) = 0. \end{gathered} \tag{4.2}$$

The four parameters reduce to three independent ones by the (linear) equation $(\mathbf{t} \cdot \mathbf{n}) = 0$.

Finally, the motion that leaves the line $\mathbf{L}$ with direction $\mathbf{t}$ and moment $\mathbf{m}$ invariant is given by

$$\begin{gathered} \mathbf{M_L} = \cos(\theta) + \sin(\theta)\mathbf{t} + \epsilon(-d\sin(\theta) + \sin(\theta)\mathbf{m} + d\cos(\theta)\mathbf{t}), \\ (\mathbf{t} \cdot \mathbf{m}) = 0. \end{gathered} \tag{4.3}$$

It represents the screw with axis $\mathbf{L}$, displacement $2d$, and angle of rotation 2θ. The two parameters d and θ are independent. Thus, this relative motion parameterization is irredundant.

4.2.2. *Incidences of Different Types*

Unequal type incidence constraints may be obtained by combining motions that include fixing one of the features, to account for symmetries, followed by displacing it within the geometry of the other feature.

The relative motion subject to requiring that the point **P** stay in the plane **E** can be obtained by composing the relative motion that fixes **P** with a subsequent translation in the plane. With 2**s** the vector of translation, we obtain

$$\begin{aligned} \mathbf{M_{PE}} &= \mathbf{TM_P} \\ &= \mathbf{q} + (-(\vec{\mathbf{s}}\cdot\vec{\mathbf{q}}) + q_0\vec{\mathbf{s}} + \vec{\mathbf{p}}\times\vec{\mathbf{q}} + \vec{\mathbf{s}}\times\vec{\mathbf{q}})\epsilon \\ &= \mathbf{q} + (\mathbf{sq} + \vec{\mathbf{p}}\times\vec{\mathbf{q}})\epsilon, \\ \|\mathbf{q}\| &= 1, \qquad (\vec{\mathbf{s}}\cdot\vec{\mathbf{n}}) = 0. \end{aligned} \tag{4.4}$$

The condition that the translation vector be perpendicular to the plane normal implies two independent parameters in the choice of **s**, bringing the total degrees of freedom of the motion to 5.

Applying the same procedure we obtain the following for keeping the point **P** on the line **L**. The translation must be along the line direction, thus we obtain

$$\mathbf{M_{PL}} = \mathbf{TM_P} = \mathbf{q} + (\mathbf{tq} + \mathbf{t}\times\vec{\mathbf{q}})\epsilon. \tag{4.5}$$

Finally, consider keeping a line $\mathbf{L} = \mathbf{t}+\mathbf{m}\epsilon$ incident to a plane $\mathbf{E} = \mathbf{n}+d\epsilon$. Geometrically, the motion can be considered a screw motion with axis **L** followed by a translation of the line in the plane which can be restricted to a displacement perpendicular to the line. Since **t** is perpendicular to the plane normal, the subsequent translation is in the direction $\mathbf{t}\times\mathbf{n}$.

We obtain the following representation:

$$\begin{aligned} \mathbf{M_{LE}} = {} & \cos(\theta) + \sin(\theta)\mathbf{t} + \epsilon(-d_1\sin(\theta) + \sin(\theta)\mathbf{m}) + \\ & \epsilon(d_1\cos(\theta)\mathbf{t} + d_2\cos(\theta)\mathbf{u} + d_2\sin(\theta)\mathbf{n}), \\ \mathbf{u} = {} & \mathbf{t}\times\mathbf{n}, \quad \|\mathbf{t}\| = 1, \quad (\mathbf{m}\cdot\mathbf{t}) = 0. \end{aligned} \tag{4.6}$$

Here 2θ is the angle of rotation, $2d_1$ the displacement in the **t** direction, and d_2 the displacement in the perpendicular $\mathbf{u} = \mathbf{t}\times\mathbf{n}$ direction.

5. Combining Constraints

5.1. *Parametric Approach*

Consider now moving a part A relative to another part B where there are multiple incidence constraints between features of the two parts. The parametric representations of the relative motion can be used when combining several incidence constraints as follows. Let $F_1, ..., F_r$ be the parametric forms of the residual motion taken separately for each incidence constraints.

By equating the rigid body motions of the F_i, we obtain a system $E_1, ..., E_s$ of implicit equations in the parameters. We solve this system for a set of independent parameters. This is an elimination computation and therefore potentially expensive. Then we can evaluate relative motion by evaluating the dependent parameters as necessary and substituting into F_1, thus obtaining an admissible relative motion.

Example 5.1. Consider a fixed part A with two plane features, $\mathbf{E}_1 = j$ and $\mathbf{E}_2 = i$, namely, the planes $x = 0$ and $y = 0$. On a moving part B we fix the points $\mathbf{P}_1 = 1 + i\epsilon$ and $\mathbf{P}_2 = 1 + j\epsilon$, that is, the points $(1,0,0)$ and $(0,1,0)$, respectively. Evidently $\mathbf{P}_1$ is on $\mathbf{E}_1$ and $\mathbf{P}_2$ is on $\mathbf{E}_2$. We use for the translation in $\mathbf{E}_1$, the vector $\mathbf{s} = (s, 0, t)$ and for the translation in $\mathbf{E}_2$, the vector $\mathbf{s}' = (0, s', t')$. Then the parametric forms for the relative motion, considering the incidence constraints separately, are, in detail,

$$\begin{aligned} F_1 : \mathbf{M}_1 &= \mathbf{q} + \epsilon\big(-sq_1 - tq_3 + (sq_0 - tq_2)i \\ &\qquad + (tq_1 - sq_3 - q_3)j + (sq_2 + tq_0 + q_2)k\big), \\ F_2 : \mathbf{M}_2 &= \mathbf{q}' + \epsilon\big(-s'q_2' - t'q_3' + (s'q_3' - t'q_2' + q_3')i \\ &\qquad + (s'q_0' + t'q_1')j + (t'q_0' - s'q_1' - q_1')k\big). \end{aligned}$$

We equate the parameters $\mathbf{q}$ and $\mathbf{q}'$, and determine the relationships between the other parameters by equating the components of the ϵ quaternion. Accounting for $\|\mathbf{q}\| = 1$, we obtain

$$\begin{aligned} s &= q_0q_3 - q_1q_2 - q_2^2 - q_3^2, \\ s' &= -q_0q_3 - q_1^2 - q_1q_2 - q_3^2, \\ t' &= t + q_0q_1 + q_0q_2 - q_1q_3 + q_2q_3. \end{aligned}$$

Thus, we have four independent parameters. Three of them specify $\mathbf{q}$ and this determines s and s' as well. The fourth parameter is t which, in conjunction with $\mathbf{q}$, determines t'.

Example 5.2. Consider the joint constructed by fitting a tripod of balls into three slots whose center planes intersect in a common line (see Figure 1. at the end of the example). Here we have three point/plane incidence constraints. The features of the fixed part are the three planes

$$\mathbf{E}_1 = j,$$
$$\mathbf{E}_2 = -\frac{\sqrt{3}}{2}i - \frac{1}{2}j,$$
$$\mathbf{E}_3 = \frac{\sqrt{3}}{2}i - \frac{1}{2}j.$$

The respective parametric tangential motions are

$$\mathbf{s}_1 = (s, 0, t),$$
$$\mathbf{s}_2 = (-s', \sqrt{3}s', t'),$$
$$\mathbf{s}_3 = (-s'', -\sqrt{3}s'', t'').$$

The features of the moving part are

$$\mathbf{P}_1 = 1 + \epsilon i,$$
$$\mathbf{P}_2 = 1 + \epsilon(-\frac{1}{2}i + \frac{\sqrt{3}}{2}j),$$
$$\mathbf{P}_3 = 1 + \epsilon(-\frac{1}{2}i - \frac{\sqrt{3}}{2}j).$$

The three transformations obtained are

$$\begin{aligned}
F_1 : \mathbf{M}_1 = \mathbf{q} + \epsilon(&-sq_1 - tq_3 \\
&+ (sq_0 - tq_2)i + (tq_1 - sq_3 - q_3)j + (sq_2 + tq_0 + q_2)k), \\
F_2 : \mathbf{M}_2 = \mathbf{q} + \epsilon(&s'(q_1 - \sqrt{3}q_2) - t'q_3 + \\
&+ (s'(-q_0 + \sqrt{3}q_3) - t'q_2 + \sqrt{3}q_3/2)i \\
&+ (s'(\sqrt{3}q_0 + q_3/2) + t'q_1 + q_3/2)j \\
&+ (s'(-\sqrt{3}q_1 - q_2) + t'q_0 - q_2/2 - \sqrt{3}q_1/2)k), \\
F_3 : \mathbf{M}_3 = \mathbf{q} + \epsilon(&s''(q_1 + \sqrt{3}q_2) - t''q_3 + \\
&+ (s''(-q_0 - \sqrt{3}q_3) - t''q_2 - \sqrt{3}q_3/2)i \\
&+ (s''(-\sqrt{3}q_0 + q_3/2) + t''q_1 + q_3/2)j \\
&+ (s''(\sqrt{3}q_1 - q_2) + t''q_0 + \sqrt{3}q_1/2 - q_2/2)k).
\end{aligned}$$

We equate the coordinates of the three dual quaternion expressions. Note that this results in a linear system of equations in the parameters

s, s', s'', t, t', t''. The equations yield the following

$$\begin{aligned}
2s &= q_1^2 - 3q_2^2 - 2q_3^2, \\
2s' &= -q_1^2 - \sqrt{3}q_1q_2 - q_3^2, \\
2s'' &= -q_1^2 + \sqrt{3}q_1q_2 - q_3^2, \\
2(t - t') &= -3q_0q_2 + 3q_1q_3 - \sqrt{3}(q_0q_1 + q_2q_3), \\
2(t - t'') &= -3q_0q_2 + 3q_1q_3 + \sqrt{3}(q_0q_1 + q_2q_3), \\
q_0q_3 &= 0.
\end{aligned}$$

It follows from these six equations that the relative motion has three degrees of freedom.

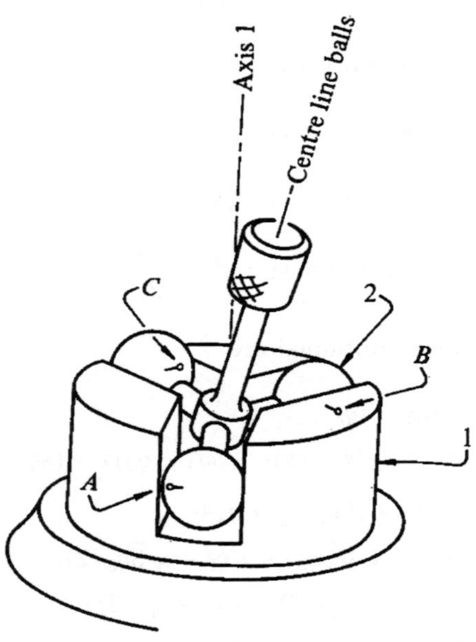

Figure 1. Configuration of Example 5.2 (drawing reproduced from [6])

5.2. *Implicit Approach*

In the implicit approach we express the relative motion as relations on the parameters of the general rigid-body motion. This will allow us to conjoin the parameter relations without having to resort to algebraic elimination.

Implicit Incidence Constraints, Equal Types

We translate the parametric formulations of the incidence constraints into implicit form. We will work with the generic rigid body transformation expression $\mathbf{M}$ of Equation (3.2).

We derived the parametric form of Equation (4.1) for keeping a point $\mathbf{p}$ invariant. Accordingly, the implicit conditions on $\mathbf{M}$ are

$$\mathbf{M_p} = \mathbf{q} + \mathbf{u}\epsilon, \quad \mathbf{p} \times \vec{\mathbf{q}} = \mathbf{u}, \quad \|\mathbf{q}\| = 1. \tag{5.3}$$

These conditions imply in particular that $u_0 = 0$. When $\mathbf{p} = 0$ the point is at the origin and the condition on $\mathbf{u}$ simplifies to $u_0 = u_1 = u_2 = u_3 = 0$.

To fix the plane $\mathbf{E}$, we derived the parametric form of Equation (4.2). It implies the condition $u_0 = 0$. The direction of the rotation axis implied by $\mathbf{q}$ has to be normal to the plane, hence we require $(\mathbf{n} \cdot \vec{\mathbf{q}}) = \sqrt{1 - q_0^2}$. Since both $\mathbf{t}$ and $\mathbf{t} \times \mathbf{n}$ in Equation (4.2) are perpendicular to the plane normal, we obtain the following conditions:

$$\mathbf{M_e} = \mathbf{q} + \mathbf{u}\epsilon, \quad u_0 = 0, \quad (\mathbf{n} \cdot \vec{\mathbf{q}}) = \sqrt{1 - q_0^2}, \quad (\vec{\mathbf{n}} \cdot \mathbf{u}) = 0. \tag{5.4}$$

The second condition degenerates when the motion is a pure translation since, in that case, the right-hand side vanishes. However, in that case the condition $\|\mathbf{q}\| = 1$ forces $q_1 = q_2 = q_3 = 0$, so a pure translation within the plane is implied by the formulation.

Now consider a line $\mathbf{L} = \mathbf{t} + \epsilon\mathbf{m}$ with direction $\mathbf{t}$ and moment vector $\mathbf{m}$. The line is invariant under $\mathbf{M} = \mathbf{q} + \epsilon\mathbf{u}$ if the transformed line $\mathbf{L}' = \mathbf{ML}\overline{\mathbf{M}}$ has the same tangent and moment vectors. This implies the following relations, in which $\mathbf{m}$ and $\mathbf{t}$ are known quaternions:

$$\mathbf{M_l} = \mathbf{q} + \mathbf{u}\epsilon, \quad \mathbf{t} = \mathbf{q}\mathbf{t}\overline{\mathbf{q}}, \quad \mathbf{m} = \mathbf{u}\mathbf{t}\overline{\mathbf{q}} + \mathbf{q}\mathbf{m}\overline{\mathbf{q}} + \mathbf{q}\mathbf{t}\overline{\mathbf{u}}.$$

Implicit Incidence Constraints, Different Types

Consider now keeping a point $\mathbf{P}$ on a plane $\mathbf{E}$, for which we derived the parametric form of Equation (4.4). We derive the implicit condition on $\mathbf{M}$ by requiring that the transformed point $\mathbf{P}'$ is again in the plane $\mathbf{E}$. Let $\mathbf{P} = 1 + \epsilon\mathbf{p}$. We obtain

$$\begin{aligned}
\mathbf{P}' &= (\mathbf{q} + \epsilon\mathbf{u})(1 + \epsilon\mathbf{p})(\overline{\mathbf{q}} - \epsilon\overline{\mathbf{u}}) \\
&= 1 + \epsilon(\mathbf{u}\overline{\mathbf{q}} - \mathbf{q}\overline{\mathbf{u}} + \mathbf{q}\mathbf{p}\overline{\mathbf{q}}) \\
&= 1 + \epsilon(-2(\mathbf{u} \times \vec{\mathbf{q}}) - 2u_0\vec{\mathbf{q}} + 2q_0\vec{\mathbf{u}} + \mathbf{q}\mathbf{p}\overline{\mathbf{q}}) \\
&= 1 + \epsilon\mathbf{p}'.
\end{aligned}$$

An algebraic computation verifies that the real component of $\mathbf{p}'$ is zero, that is, $\mathbf{p}'$ is the position vector of the transformed point. Assuming the original point is in the plane with unit normal vector $\mathbf{n}$, we obtain the condition $(\mathbf{n}\cdot\mathbf{p}) = (\mathbf{n}\cdot\mathbf{p}')$, or equivalently:

$$(\mathbf{n}\cdot\mathbf{p}) = -2(\mathbf{n}\cdot(\mathbf{u}\times\vec{\mathbf{q}}) - 2u_0(\mathbf{n}\cdot\vec{\mathbf{q}}) + 2q_0(\mathbf{n}\cdot\vec{\mathbf{u}}) + (\mathbf{n}\cdot(\mathbf{q}\mathbf{p}\overline{\mathbf{q}})) \tag{5.5}$$

Example 5.6. Consider the plane $\mathbf{E}_1 = j$ and the point $\mathbf{P}_1 = 1 + \epsilon i$ in the plane. Any motion $\mathbf{M}$ that keeps this point in the plane must satisfy according to Equation (5.5)

$$\begin{aligned} q_0q_3 + q_1q_2 - u_0q_2 + u_1q_3 + u_2q_0 - u_3q_1 &= 0 \\ \|\mathbf{q}\| &= 1 \\ (\mathbf{q}\cdot\mathbf{u}) &= 0 \end{aligned}$$

Example 5.7. We consider the points and planes of Example 1. The conditions from the point incidences are then

$$\begin{aligned} q_0q_3 + q_1q_2 - u_0q_2 + u_1q_3 + u_2q_0 - u_3q_1 &= 0 \\ -q_0q_3 + q_1q_2 - u_0q_1 + u_1q_0 - u_2q_3 + u_3q_2 &= 0 \\ \|\mathbf{q}\| &= 1 \\ (\mathbf{q}\cdot\mathbf{u}) &= 0 \end{aligned}$$

Note that a translation in the z-direction, $1+\epsilon dk$, satisfies these conditions.

Example 5.8. We consider the planes and points of Example 2. The incidence conditions, after some simplification, define the equations

$$\begin{aligned} \|\mathbf{q}\| &= 1 \\ (\mathbf{q}\cdot\mathbf{u}) &= 0 \\ 2q_0q_3 - q_1q_2 &= 0 \\ q_0q_3 + q_1q_2 + u_2q_0 - u_0q_2 + u_1q_3 - u_3q_1 &= 0 \\ 2\sqrt{3}(u_0q_1 - u_1q_0 + u_2q_3 - u_3q_2) + \sqrt{3}(q_1^2 - q_2^2) + 2(q_0q_3 + q_1q_2) &= 0 \end{aligned}$$

6. Discussion

The uniformity of the representation and the algebraic nature of the representation are the main attractions when using dual quaternions. Points, lines and planes are simple to represent as dual quaternions, and so are rigid-body motions. Moreover, as we have seen, there is considerable geometric intuition in this representation schema.

Another advantage of dual quaternions, from a computational perspective, is that they describe a general rigid-body motion with only eight parameters, whereas a 4×4 matrix representation would require twelve. Thus, the system of equations describing a particular contact configuration is smaller. A screw representation would lower this to six parameters, but the resulting equations may fail in particular instances and do not differ, in essence, from the dual quaternion representation.

There are some drawbacks to using dual quaternions as well. In the implicit form of the constraint encoding, for instance, the conditions can become fairly complex. An example is the implicit representation of motions that keep a line invariant. Here, the parametric form does better. Moreover, the implicit form we derived has some redundancies. Consider again all conditions on $\mathbf{M}_l$:

$$\mathbf{t} = \mathbf{q}\mathbf{t}\overline{\mathbf{q}} \tag{6.1}$$

$$\mathbf{m} = \mathbf{u}\mathbf{t}\overline{\mathbf{q}} + \mathbf{q}\mathbf{m}\overline{\mathbf{q}} + \mathbf{q}\mathbf{t}\overline{\mathbf{u}} \tag{6.2}$$

$$\|\mathbf{q}\| = 1 \tag{6.3}$$

$$(\mathbf{q} \cdot \mathbf{u}) = 0 \tag{6.4}$$

Conditions (6.1) and (6.2) each yields three scalar equations, giving eight equations in eight variables total. Therefore, there must be two redundant equations. With a pure translation ($|q_0| = 1$), Condition (6.1) is trivial. With a pure rotation ($u_0 = 0$ and $\mathbf{u} = \sin(\theta)\mathbf{m}$), on the other hand, Condition (6.1) is not trivial. Thus, redundancy depends on the parameter values.

Parametric motion representations often lead to motion descriptions in which the system of parameter relations is linear, a computational plus. However, nonlinear relations may ensue, for example on the $\mathbf{q}$ coordinates of the transformation. Here, symbolic algebraic computations may require reasoning that is not entirely automated in, for instance, *Maple*.

Another drawback is that the representation of the relative motion may not include certain special cases. For instance, given a plane $\mathbf{E}$, we may choose a line $(\mathbf{t}, \mathbf{m})$ that lies in the plane and use it as the axis for a rigid-body motion that has a rotation angle of 180°. Those motions also preserve the plane, albeit with a reversal of the plane orientation. Thus, the geometry and the algebra diverge in this case.

In contrast to the parametric expression of relative motion, the implicit formulation does not require intermediate positions to satisfy the constraints. For example, the point $\mathbf{P}$ is required to be on the plane $\mathbf{E}$ only at the start and at the end of the motion. As the motion progresses, it may very well leave the plane **E** at the other times. This is true in particular of the special case of a rotation by 180° about an axis in the plane **E**.

From a computational perspective, dual quaternions do not reduce the number of arithmetic operations that must be done to compute the image of a feature under a given transformation. Using a 4×4 matrix representation, transforming a point requires 12 multiplications and 9 additions. The dual quaternion representation, on the other hand, requires more.

References

1. S. G. Ahlers, J. M. McCarthy. The Clifford algebra and the optimization of robot design. In E. Bayro, G. Sobczyk eds., *Geometric Algebra with Appications in Science and Engineering* (*ACACSE'99, Ixtapa-Zihuatanejo*), 235–251. Birkhäuser, Boston, MA., 2001.
2. W. Blaschke. *Kinematik und Quaternionen.* VEB Deutscher Verlag der Wissenschaften, Berlin, Germany, 1960.
3. M. Chasles. Note sur les propriétés générales du système de deux corps ... *Bull. des Sciences Mathématiques, Astronomiques, Physiques et Chimiques*, **14** (1830), 321–326.
4. X.-S. Gao, C. Hoffmann, W. Yang. Solving spatial basic geometric constraint configurations with locus intersection. In *Proc. 7th ACM Symp. Solid Modeling and Applic.*, 95–104. ACM Press, 2002.
5. X. S. Gao, C. C. Zhu, Y. Huang. Building dynamic mathematical models with Geometry Expert. In W. C. Yang, ed., *3rd Asian Techn. Conf. in Mathematics*, 216–224. Springer, New York, 1998.
6. J. Phillips. *Freedom in Machinery, Vol I: Introducing Screw Theory.* Cambridge University Press, Cambridge, UK, 1984.
7. J. Phillips. *Freedom in Machinery, Vol II: Screw Theory Exemplified.* Cambridge University Press, Cambridge, UK, 1990.
8. H. Pottmann, J. Wallner. *Computational Line Geometry.* Springer, Heidelberg, Germany, 2001.
9. E. Study. Die Geometrie der Dynamen. *Jahresbericht der Deutschen Mathematiker-Vereinigung* **8** (1899), 204–216.
10. J. Wittenburg. *Dynamics of Systems of Rigid Bodies.* B. G. Teubner, Stuttgart, Germany, 1977.

PARAMETRIZATION OF RATIONAL SURFACES

JOSEF SCHICHO*

RISC, Univ. Linz, A-4020 Linz, Austria
email: jschicho@risc.uni-linz.ac.at

We give a survey on the theory of rational parametrizations of algebraic surfaces, available algorithms and open problems. In the following extended abstract, we describe the main themes.

1. History and State of the Art

Rational surfaces are used in CAD/CAM for the modeling of surfaces. The parametric representation is used for many operations like plotting, motion display (computing transformations), computing curvatures or offset surfaces. But there are other operations for which an implicit representation is more convenient, for example, ray tracing. It is therefore convenient to have algorithms that can convert between these two representations.

The parametrization problem is not always solvable as there are non-rational algebraic surfaces. Algorithms for cubic surfaces and for canal surfaces have been given in [AB88,SS87,Pot96]. A general criterion for rationality was given by Castelnuovo [Cas39]. The first algorithm that parametrizes any rational surface was given in [Sch98b]. The main computational tool is adjunction, which has played a fundamental role in the surface theory of the Italian school (see [Enr49]), similar to the role of canonical divisors in more modern treatments (for example, see [Sha65,Kur82]).

In general, the parametrization algorithm [Sch98b] computes a parametrization with complex coefficients. For many applications, a parametrization with real coefficients is required. There are some classes of surfaces for which this is possible (see [Pet97,Sch98a,Sch00a]). In general, it is not known if every real algebraic surfaces with a complex parametrization also has a real parametrization. The decision problem for *proper*

*The author was supported by the Austrian Science fund (FWF) in the frame of the special research area SFB 013.

parametrizations, however, is solved by the theorem of Comesatti [Com12]. For proper parametrizations with coefficients in the field of rational numbers or number fields in general, we refer readers to [MT86,Col87,Sch00b].

Quantitative results are available for the complex case: the paper [Sch99] gives an upper bound for the smallest possible parametrization of a surface of known degree.

2. Stereographic Projection

Let S be a quadric surface in $\mathbb{P}^3$. Let $p \in S$ be a nonsingular point on S. Let $E \subset \mathbb{P}^3$ be a projective plane not containing p. Then the projection $\pi : \mathbb{P}^3 \to E$ is a rational map defined everywhere outside p.

Each line L through p intersects S at one more point q_L. Therefore, the restricted projection $\pi : S \to E$ is birational. We can construct a parametrization by inverting π. Note that the so constructed parametrization is proper. In fact, the inversion formula is the stepping stone for computing the parametrization.

The same method can be applied when we have a surface S of degree d in $\mathbb{P}^3$, and a point p with multiplicity $d - 1$.

3. Pencils of Rational Curves

A *pencil* of curves on a surface S is a one-parameter family of curves on S. Assume that we have a pencil C_t of rational curves parametrized by another rational curve T. Then we can try to parametrize S in two steps. The first parameter, t, fixes a curve C_t in the pencil. A second parameter, s, is used to parametrize C_t in terms of rational functions. The problem is that although these functions depend rationally on s, it cannot be ensured so easily that they also depend rationally on t.

Algebraically, we may treat the pencil C_t as a single curve defined over the function field $\mathbb{R}(t)$. We know that this curve is rational, but it is not known whether it has a parametrization with coefficients in $\mathbb{R}(t)$. Here is a very old theorem telling what can be done without leaving the field of definition.

Theorem 3.1. *Let K be an arbitrary field. Let C be a rational curve which is defined over K. If the degree of C is odd, then C has a parametrization defined over K. If the degree of C is even, then there exists a birational map, defined over K, transforming C to a conic.*

In [Noe70], the theorem was stated and applied to $K = \mathbb{C}(t)$, in order to construct complex surface parameterizations. A modern treatment and an algorithm constructing a parametrization or birational map without leaving K can be found in [SW97].

Corollary 3.2. *Let S be a surface with a pencil of rational curves C_t, the pencil being parametrized by a rational curve T. If the degree of C_t is odd, then S is rational. If the degree of C_t is even, then S is birationally equivalent to a surface with an equation $G(u, v, t) = 0$ that is quadratic in the variables (u, v).*

4. General Theory

A function f from the class of algebraic surfaces to the integers is called a *birational invariant* iff we have $f(S_1) = f(S_2)$ when S_1 and S_2 are birationally equivalent (i.e. there exists a birational map $p : S_1 \to S_2$). For curves, the most important birational invariant is the *genus* (see [Wal78]). For surfaces, the most important invariants are

- p_a, the arithmetic genus;
- p_g, the geometric genus;
- $P_1 = p_g, P_2, P_3, \ldots$, the plurigeni;
- $q := p_g - p_a$, the irregularity;
- $\kappa := \limsup_{n\to\infty} \log P_n$, the Kodaira dimension.

Since a proper parametrization is a birational map to the plane, the existence of a proper parametrization of S implies that S shares all birational invariants with the plane: $p_a = p_g = P_n = q = 0$, $\kappa = -\infty$. Castelnuovo's criterion [Cas39] is a converse that also allows to decide the existence of an improper parametrization.

Theorem 4.1. (*Castelnuovo*) *Let S be an algebraic surface. Then the following are equivalent.*

(1) *S has a complex parametrization.*
(2) *S has a complex proper parametrization.*
(3) *$p_a(S) = P_2(S) = 0$.*

The historically even older investigation [Enr95] studied the possibilities of the construction of a parametrization under the assumption of rationality. The result has been established with the necessary mathematical rigidity in [Man66,Man67].

Theorem 4.2. (*Enriques/Manin*) *Let K be a field of characteristic zero. Let S be a rational surface defined over K. Then one of the following is true.*

(1) *S has a proper parametrization defined over K.*
(2) *S has a pencil of rational curves, defined over K.*
(3) *There is a birational map, defined over K, from S to a so called* Del Pezzo surface *(see* $[dP87, Man74]$*).*

By the above theorem, we can reduce the parametrization of an arbitrary rational surface to the parametrization of a surface with a pencil of rational curves, or to the parametrization of a Del Pezzo surface (and for this class, we refer to [Con39,Man74] for parametrization methods). It remains to make the Enriques/Manin reduction algorithmic. This can be done with the method of adjoints.

Let S be a surface in 3-space with equation $F = 0$. A polynomial is an *m-adjoint* if it vanishes with order at least $m(r-1)$ at each r-fold singular curve of S, and it vanishes with order at least $m(r-2)$ at each r-fold singular point of S. The so-called "infinitely near singularities" have to be taken into account (see [Sch98b]). We consider only polynomials that are reduced with respect to F and we compute orders modulo F.

For any two natural numbers n, m, we define the vector space $V_{n,m}$ as the space of all m-adjoints of degree at most $n + m(d-4)$, where d is the degree of the given surface S. By *adjoint computation*, we mean the computation of a basis of the vector space $V_{n,m}$ for given F, n, m.

Adjoint computation is difficult because it requires a so-called "resolution of the singularities" of S. For algorithms computing adjoints and resolving singularities, we refer to [Sch98b] and [Vil96,BM91,BS00].

Once we can compute adjoints, it is easy to compute the birational invariants mentioned in the previous section. The plurigenus P_m is nothing but $\dim(V_{0,m})$. The parametric genus can also be computed in terms of the dimensions of $V_{n,m}$ (see [Sch98b]).

If $V_{n,m}$ is not the zero space, and if $\ell := \dim(V_{n,m}) - 1$, then any basis $P_0, \ldots, P_\ell$ defines naturally a rational map $f_{n,m} : S \to \mathbb{P}^\ell$ by the evaluation $p \mapsto (P_0(p) : \ldots : P_\ell(p))$. If S is rational, then [Sch98b] shows that there are integers n, m, such that $f_{n,m}$ is either a birational map to a plane, or to a surface with a pencil of lines, or to a surface with a pencil of conics, or to a Del Pezzo surface.

References

AB88. S. S. Abhyankar and B. Bajaj, *Automatic parametrization of curves and surfaces III*, Computer Aided Geometric Design **5** (1988), 309–323.

BM91. E. Bierstone and P. Milman, *A simple constructive proof of canonical resolution of singularities*, Effective methods in algebraic geometry (T. Mora and C. Traverso, eds.), Birkhäuser, 1991, pp. 11–30.

BM97. E. Bierstone and P. Milman, *Canonical desingularization in characteristic zero by blowing up the maximum strata of a local invariant*, Invent. math. **128** (1997), 207–302.

BS00. G. Bodnár and J. Schicho, *Automated resolution of singularities for hypersurfaces*, J. Symb. Comp. **30** (2000), 401–428.

Cas39. G. Castelnuovo, *Sulle superficie di genere zero*, Memorie scelte, Zanichelli, 1939, pp. 307–334.

Col87. J.-L. Colliot-Thélène, *Arithmetic of rational varieties and birational problems*, Proc. ICM 1986, AMS Prov. RI, 1987, pp. 641–653.

Com12. A. Comesatti, *Fondamenti per la geometria sopra le superficie rationali del punto di vista reale*, Math. Ann. **73** (1912), 1–72

Con39. F. Conforto, *Le superfici razionali*, Zanichelli, 1939.

dP87. P. del Pezzo, *On the surfaces of order n in embedded in n-dimensional space*, Rend. mat. Palermo **1** (1887), 241–271.

Enr95. F. Enriques, *Sulle irrazionalita da cui puo farsi dipendere la risoluzione d'un equazione f(xyz)=0 con funzioni razionali di due parametri*, Math. Ann. (1895), 1–23.

Enr49. F. Enriques, *Le superfici algebriche*, Zanichelli, 1949

EV98. S. Encinas and O. Villamayor, *Good points and constructive resolution of singularities*, Acta Math. **181** (1998), 109–158.

EV00. S. Encinas and O. Villamayor, *A course on constructive desingularization and equivariance*, Resolution of Singularities (Obergurgl, 1997) (H. Hauser, ed.), Birkhäuser, 2000, pp. 147–227.

Kur82. H. Kurke, *Vorlesungen über algebraische Flächen*, Teubner, 1982.

Man66. Y. Manin, *Rational surfaces over perfect fields I*, Inst. Hautes Et. Sci. Publ. Math. **30** (1966), 137–186.

Man67. Y. Manin, *Rational surfaces over perfect fields II*, Math. USSR Sb. **1** (1967), 141–168.

Man74. Y. Manin, *Cubic forms*, North-Holland, 1974.

MT86. Y. Manin and M. A. Tsfasman, *Rational varieties: algebra, geometry, arithmetic*, Uspekhi Mat. Nauk. **41** (1986), 43–94.

Noe70. M. Noether, *Über Flächen, welche Scharen rationaler Kurven besitzen*, Math.Ann. **3** (1870), 161–227.

Pet97. M. Peternell, *Rational parametrizations for envelopes of quadric families*, Ph.D. thesis, Techn. Univ. Vienna, 1997.

Pot96. H. Pottmann, *Applications of laguerre geometry in cagd*, Tech. Report 30,31, Techn. Univ. Wien, 1996.

Sch98a. J. Schicho, *Rational parameterization of real algebraic surfaces*, Proc. ISSAC'98, ACM Press, 1998, pp. 302–308.

Sch98b. J. Schicho, *Rational parametrization of surfaces*, J. Symb. Comp. **26** (1998), no. 1, 1–30.

Sch99. J. Schicho, *A degree bound for the parameterization of a rational surface*, J. Pure Appl. Alg. **145** (1999), 91–105.

Sch00a. J. Schicho, *Proper parametrization of real tubular surfaces*, J. Symb. Comp. **30** (2000), 583–593.

Sch00b. J. Schicho, *Proper parametrization of surfaces with a rational pencil*, Proc. ISSAC'2000, ACM Press, 2000, pp. 292–299.

Sha65. I. R. Shafarevich (ed.), *Algebraic surfaces*, Proc. Steklov Inst. Math., 1965, transl. by AMS 1967.

SS87. T. W. Sederberg and J. P. Snively, *Parametrization of cubic algebraic surfaces*, The mathematics of surfaces II (Cardiff 1986), Oxford Univ. Press, 1987, pp. 299–319.

SW97. J.R. Sendra and F. Winkler, *Parametrization of algebraic curves over optimal field extensions*, J. Symb. Comp. **23** (1997), no. 2/3, 191–208.

Vil96. O. Villamayor, *Introduction to the algorithm of resolution*, Algebraic geometry and singularities (La Rábida 1991), Birkhäuser, 1996, pp. 123–154.

Wal78. R. J. Walker, *Algebraic curves*, Springer, 1978.

ALGEBRAIC ALGORITHMS FOR D-MODULES AND NUMERICAL ANALYSIS

TOSHINORI OAKU

Department of Mathematics
Tokyo Women's Christian University

YOSHINAO SHIRAKI

Speech and Motor Control Res. Group
NTT Communication Science Labs

NOBUKI TAKAYAMA

Department of Mathematics
Kobe University

Algorithmic methods in D-modules have been used in the mathematical study of hypergeometric functions and in computational algebraic geometry. In this paper, we show that these algorithms give correct algorithms to perform several operations for holonomic functions and also generate substantial information for numerical evaluation of holonomic functions.

1. Introduction

As was observed by Castro and Galligo [4,6], the Buchberger algorithm for computing Gröbner bases of ideals of the polynomial ring applies also to the Weyl algebra, which is a ring of differential operators with polynomial coefficients. This generalization of the Buchberger algorithm has turned out to be very fruitful in the computational approach to the theory of D-modules. Its goal is an algebraic treatment for systems of linear partial (or ordinary) differential equations and its theoretical foundation was laid by Bernstein, Kashiwara, M. Sato, and many others.

The aim of this paper is to show that such an algorithmic approach to the D-module theory, which essentially depends on the Buchberger algorithm, enables us to solve some fundamental problems in symbolic computation. These problems are related to computations with so-called holonomic functions and our motivation comes from signal processing and numerical analysis. We will sketch some applications of computation of holonomic functions to these areas.

A system of linear differential equations $P_1 u = \cdots = P_r u = 0$, where $P_1, \ldots, P_r$ are elements of the Weyl algebra $D_n = \mathbf{C}\langle x_1, \ldots, x_n, \partial_1, \ldots, \partial_n \rangle$ over the field of the complex numbers with $\partial_i = \partial_{x_i} = \frac{\partial}{\partial x_i}$, and whose solutions, including higher order solutions (see Remark below), form a finite dimensional vector space, is called *holonomic*. Holonomic systems play a key role in the theory of D-modules. A function u is called holonomic, roughly speaking, if u satisfies a holonomic system.

Since a linear ordinary differential equation is always holonomic, special functions of one variable, such as the Gauss hypergeometric function and the Bessel function are holonomic by definition. Moreover, rational functions in an arbitrary number of variables and their exponentials are simple examples of holonomic functions. As nontrivial examples, the expression f^λ for an arbitrary polynomial f and an arbitrary complex number λ, and GKZ-hypergeometric systems (see, for example, [19]) are holonomic.

We can expect to obtain substantial information on a holonomic function by studying the differential equations which it satisfies, rather than dealing with the function itself. This holonomic approach to special function identities was initiated by Zeilberger *et al.* [1,17,23,24].

We are concerned with the following computational issues on holonomic functions: (1) Given two holonomic functions f, g and two differential operators P, Q, find a holonomic system which the function $Pf + Qg$ satisfies; (2) Given two holonomic functions f, g, find a holonomic system which the function fg satisfies; (3) Given a holonomic function $f(t, x)$, find a holonomic system which the integral $\int_C f(t, x)\, dt$ satisfies.

We give answers to the three problems (under a technical condition for the third one) by using the Buchberger algorithm applied to the Weyl algebra. The class of holonomic functions are stable under these three operations (addition, multiplication, integration) and two more operations of restriction and localization [14]. We give explicit algorithms for these constructions. Partial answers to the above three problems were given in [1,20,23,24].

Remark 1.1. Throughout this paper, the words "system" and "holonomic system" will be used for several related objects. A left ideal I of D_n is called holonomic if the module D_n/I is holonomic, and a system of differential equations $P_1 u = 0, \ldots, P_r u = 0$ is said to be holonomic if the left ideal generated by $P_1, \ldots, P_r$ is holonomic. This definition is equivalent to the definition of holonomic system in terms of higher order solution spaces. Note that the finite dimensionality of the classical solution space does not imply holonomy. A counter example is

$$(x^3 - y^2)\partial_x + 3x^2, (x^3 - y^2)\partial_y - 2y,$$

for which the classical solution space is finite dimensional, but it is not holonomic. When all $\mathcal{E}\mathrm{xt}_D^k(D/I, \mathcal{O})$ (higher order solutions) are finite dimensional, then the system is holonomic. See, for example, [2,10,19].

2. Holonomic Functions

Definition 2.1. A multi-valued analytic function f defined on (the universal covering of) $\mathbf{C}^n \setminus S$, where S is an algebraic subset of $\mathbf{C}^n$, is called a *holonomic function* if there exists a left ideal I of D_n so that $M = D_n/I$ is a holonomic system and $Pf = 0$ holds on $\mathbf{C}^n \setminus S$ for any $P \in I$.

We set $\mathrm{Ann}(f) := \{P \in D_n \mid Pf = 0 \text{ on } \mathbf{C}^n \setminus S\}$. Then f is holonomic if and only if $D_n/\mathrm{Ann}(f)$ is holonomic.

Proposition 2.2. [2] *Let $f \in \mathbf{C}[x]$ be a nonzero polynomial and let λ be an arbitrary complex number. Then f^λ is holonomic.*

Algorithms to compute a holonomic system which f^λ satisfies are given in [11] and [9].

Proposition 2.3. *Let f and g be holonomic functions and $P, Q \in D_n$. Then $Pf + Qg$ and fg are holonomic.*

We shall give an algorithmic proof to this proposition in Section 3.3. The class of the holonomic functions is not closed under the division [24].

Proposition 2.4. *Let $f \in \mathbf{C}(x)$ be a rational function. Then f, $\exp(f)$, and $\log f$ are holonomic.*

Proof. Suppose $f = p/q$ with $p, q \in \mathbf{C}[x]$. Then by Proposition 2.2, p and q^{-1} are holonomic. Hence f is holonomic by Proposition 2.3. The holonomicity of $\exp(f)$ is a special case of the proposition below. To prove that $u := \log f$ is holonomic, we may assume that f is a polynomial. Then u satisfies $f\partial_i u = f_i$ with $f_i := \partial_i(f)$. Let f_i be of degree $n_i - 1$ with respect to x_i. Then we have $\partial_i^{n_i}(f\partial_i u) = 0 \quad (i = 1, \ldots, n)$. Then, this system is identified with the left D_n-module $M = D_n/(D_nP_1 + \cdots + D_nP_n)$ with

$P_i := \partial_i^{n_i} f \partial_i$ and M is a holonomic system on $\{x \in \mathbf{C}^n \mid f(x) \neq 0\}$ since the characteristic variety $\mathrm{Char}(M) \subset \{(x, \xi) \mid \xi_i f(x) = 0 \ (i = 1, \ldots, n)\}$. In view of Theorem 3.1 of Kashiwara [8], the localization $M[1/f]$ is holonomic. Since $M[1/f]$ is isomorphic to M outside $f = 0$, we are done. □

We note that an algorithmic method for the localization is given in [14].

Proposition 2.5. [1] *Let f be a multi-valued analytic function and assume that $(\partial f/\partial x_i)/f$ is a rational function for every $i = 1, \ldots, n$. Then f is holonomic.*

Proof. Put $a_i := (\partial f/\partial x_i)/f = p_i/q_i$ with $p_i, q_i \in \mathbf{C}[x]$. Then f satisfies $M\colon (q_i \partial_i - p_i) f = 0 \quad (i = 1, \ldots, n)$. Let q be the least common multiple of $q_1, \ldots, q_n$. Then M is holonomic outside the hypersurface defined by $q = 0$. This implies that f is holonomic in the same way as in the proof of the preceding proposition. □

Example 2.6. For two polynomials $f_1(x)$, $f_2(x)$ in $\mathbf{C}[x_1, \ldots, x_n]$, put $f(x) = \exp(f_1(x)/f_2(x))$. The system of differential equations M above is not holonomic in general (consider, for example, $\exp(1/(x_1^3 - x_2^2 x_3^2))$). A holonomic system for $f(x)$ can be found by the method in [14].

Let f be a holonomic function. By definition, it is a multi-valued analytic function defined on $\mathbf{C}^n \setminus S$. The algebraic set S is contained in the singular locus of the annihilating ideal I of f. The singular locus is the zero set of $\left(\mathrm{in}_{(\mathbf{0},\mathbf{1})}(I) : (\xi_1, \ldots, \xi_n)^\infty\right) \cap \mathbf{C}[x_1, \ldots, x_n]$, generators of which are computable by the Buchberger algorithm in D_n from generators of I. See [10] and [19, §1.4] for notations above and algorithms.

3. Four Operations on Holonomic Functions

3.1. *Restriction to $x_{m+1} = \cdots = x_n = 0$*

Let $u(x)$ be a holonomic function and suppose that a left ideal I of D_n is explicitly given so that $M := D_n/I$ is a holonomic system. Then $M_Y = M/x_n M$ is a holonomic system. This holonomic system is called the *restriction* of M to $x_n = 0$. As a left D_{n-1}-module, M_Y is generated by the residue classes of $1, \partial_n, \ldots, \partial_n^{k_0}$. Hence, there exists a submodule J such that $D_{n-1}^{k_0+1}/J \simeq M_Y$; J is a system of equations for $u(x', 0), (\partial_n u)(x', 0), \ldots, (\partial_n^{k_0} u)(x', 0)$, where $x' = (x_1, \ldots, x_{n-1})$. An algorithm of finding generators of J from those of I is given in [12]. By an elimination algorithm [19, §5.2], we can find a system of equations for $u(x', 0)$ from J.

Take an integer m such that $0 \leq m < n$. Let Z be the algebraic set $\{(x_1, \ldots, x_n) \mid x_{m+1} = \cdots = x_n = 0\}$ and M a left D_n-module D_n^r/I where I is a left submodule of D_n^r. The *restriction* of M to Z is defined by $M/(x_{m+1}M + \cdots + x_n M)$ and is denoted by M_Z as in the case of the restriction to a hypersurface. It follows from the definition we have

$$M/(x_{n-1}M + x_n M) \simeq (M/x_n M)/x_{n-1}(M/x_n M),$$
$$M/(x_{n-2}M + x_{n-1}M + x_n M) \simeq$$
$$\big((M/x_n M)/x_{n-1}(M/x_n M)\big)/x_{n-2}\big((M/x_n M)/x_{n-1}(M/x_n M)\big),$$

and so on. Therefore, the iterative application of the restriction algorithm for the hypersurface case provides an algorithm to get the restriction M_Z. Yet another algorithm that uses weight vectors to compute the restriction M_Z without the iteration is given in [19, §5.5]. It is an interesting question to compare the two methods from the efficiency point of view. We finally note that the book [19] and our discussion consider only the case of a left D_n-module D_n/I where I is a left ideal in D_n, but it is straightforward to generalize to the case of D_n^r/J where J is a left submodule of D_n^r.

3.2. *Integrals of Holonomic Functions with Parameters*

Let $f(x)$ be a holonomic function and let I be a left ideal of D_n such that $M := D_n/I$ is a holonomic system and $I \subset \mathrm{Ann}(f)$. For the sake of simplicity, let us assume that $f(x)$ is infinitely differentiable on $\mathbf{R}^n$ and rapidly decreasing with respect to x_n, that is, $\lim_{x_n \to \infty} x_n^j \partial_n^k f(x) = 0$ holds for any $x' := (x_1, \ldots, x_{n-1}) \in \mathbf{R}^{n-1}$ and $j, k \in \mathbf{N}$. Put

$$g_k(x') := \int_{-\infty}^{\infty} x_n^k f(x', t)\, dt, \qquad (k \in \mathbf{N}).$$

Then $g_0(x'), g_1(x'), \ldots, g_{k_0}(x')$ are solutions of the holonomic system $M/\partial_n M$ where k_0 is the maximal non-negative integral root of the associated b-function (see also [1,20] although only g_0 is considered there). Computation of $M/\partial_n M$ can be reduced to that of $M/x_n M$ by an isomorphism of D_n induced by the Fourier transform. See, for example, [19, §5.5] for details.

3.3. *Sum and Product of Holonomic Functions*

Let u be a holonomic function and suppose that a left ideal I of D_n is given so that $I \subset \mathrm{Ann}(u)$ and $M := D_n/I$ is holonomic. First, for a given

$Q \in D_n$, we show that we can compute a holonomic system for Qu. The fact that Qu is holonomic follows from $D_n Qu \subset D_n u$. Let $P_1, \ldots, P_r$ be generators of I. Then for $P \in D_n$, $PQ \in I$ holds if and only if there exist $Q_1, \ldots, Q_r \in D_n$ such that $PQ + Q_1 P_1 + \cdots + Q_r P_r = 0$. By computing a Gröbner basis of the ideal generated by $Q, P_1, \ldots, P_r$, we can obtain generators of their syzygy module

$$S := \{(P, Q_1, \ldots, Q_r) \in D_n^{r+1} \mid PQ + Q_1 P_1 + \cdots + Q_r P_r = 0\}.$$

Then the projections of generators of S to the first component generate the left ideal $I : Q = \{P \in D_n \mid PQ \in I\}$. Thus we have $I : Q \subset \mathrm{Ann}(Qu)$. The left D_n-homomorphism $D_n \to D_n$ defined by $P \mapsto PQ$ induces a homomorphism $D_n/(I : Q) \to D_n/I$, which is injective by the definition of $I : Q$. Hence $D_n/(I : Q)$ is holonomic.

Now let v be another holonomic function with an explicitly given left ideal $J \subset \mathrm{Ann}(v)$ so that D_n/J is a holonomic system. Our first aim is to compute a holonomic system for $Pu + Qv$ for given $P, Q \in D_n$. Since the holonomic systems for $I : P$ and $J : Q$ are computed in the way described above, we may assume that $P = Q = 1$. Then we have $I \cap J \subset \mathrm{Ann}(u + v)$. This ideal intersection can be computed by the Buchberger algorithm in the same way as in the polynomial ring (see, e.g., [5]). $D_n/(I \cap J)$ is a holonomic system since the homomorphism $D_n \to D_n^2$ defined by $P \mapsto (P, P)$ induces an injective homomorphism

$$D_n/(I \cap J) \longrightarrow (D_n/I) \bigoplus (D_n/J).$$

Next let us consider an algorithm to find a holonomic system for the product uv. Let G_u and G_v be finite sets of generators of I and J respectively. Put $D_{2n} = \mathbf{C}[x, y]\langle \partial_x, \partial_y \rangle$ with $y = (y_1, \ldots, y_n)$ and $\partial_x := (\partial_{x_1}, \ldots, \partial_{x_n})$, $\partial_y := (\partial_{y_1}, \ldots, \partial_{y_n})$. Let $I_{u \otimes v}$ be a left ideal of D_{2n} generated by both

$$G_u(x) := \{P(x, \partial_x) \mid P \in G_u\}, \quad \text{and} \quad G_v(y) := \{P(y, \partial_y) \mid P \in G_v\}.$$

Then it is easy to see that $I_{u \otimes v} \subset \mathrm{Ann}(u(x)v(y))$ and that $M_{u \otimes v} := D_{2n}/I_{u \otimes v}$ is holonomic. Put $\Delta := \{(x, y) \in \mathbf{C}^{2n} \mid x = y\}$. Then the restriction of M to Δ:

$$M_\Delta := D_{2n}/((x_1 - y_1)D_{2n} + \cdots + (x_n - y_n)D_{2n} + I_{u \otimes v})$$

can be computed by performing coordinate transformation $x_i - y_i \to y_i$ and $x_i \to x_i$ and then applying the restriction algorithm with respect to the variables $y_1, \ldots, y_n$. Note that M_Δ is holonomic since holonomicity is preserved under restriction. In fact, M_Δ is nothing but the tensor product of D_n/I and D_n/J over $\mathbf{C}[x]$, and the above algorithm was introduced in [13]. From M_Δ, we can compute a left ideal I_{uv} of D_n so that D_n/I_{uv} is a holonomic system for $u(x)v(x)$ by elimination.

The above algorithm for I_{uv} is for general purpose but is not efficient since it involves restriction to the n-dimensional linear space in the $2n$-dimensional space. Hence possible short cuts for some particular cases would be worth mentioning. For one such case, consider $v := e^f u$ for a holonomic function u and a polynomial f. Let $I \subset \mathrm{Ann}(u)$ be a left ideal such that D_n/I is holonomic. Put $f_i := \partial f/\partial x_i$. Then the left ideal J of D_n generated by

$$\{P(x_1,\ldots,x_n;\partial_1 - f_1,\ldots,\partial_n - f_n) \mid P(x_1,\ldots,x_n;\partial_1,\ldots,\partial_n) \in I\}$$

is contained in $\mathrm{Ann}(v)$ since $(\partial_i - f_i)\bullet(e^f u) = e^f(\partial_i \bullet u)$. The characteristic variety of D_n/J is

$$\{(x,\xi_1 - f_1(x),\ldots,\xi_n - f_n(x)) \in \mathbf{C}^{2n} \mid (x,\xi) \in \mathrm{Char}(D_n/I)\}.$$

Hence D_n/J is holonomic. For more cases, the product of a holonomic function and the Heaviside function will be discussed later.

4. Holonomic Distributions and Their Integrals

Since some important analytic holonomic functions are expressed as definite integrals of distributions, the notion of holonomic function should be generalized; we will introduce holonomic distributions. They are closed under four operations if the result of an operation is well-defined. Computation of these operations can be done by the same algorithms as in the case of holonomic functions.

Definition 4.1. Let u be a distribution (in the sense of Schwartz) defined on $\mathbf{R}^n$. Then u is said to be a *holonomic distribution* if there is a left ideal I of D_n so that D_n/I is holonomic and $Pu = 0$ holds as distribution for any $P \in I$.

For example, the Dirac's delta function $\delta(x) = \delta(x_1)\cdots\delta(x_n)$ is a holonomic distribution since $x_1\delta(x) = \cdots = x_n\delta(x) = 0$. Let us introduce the Heaviside function $Y(x_1)$ defined by $Y(x_1) = 0$ for $x_1 < 0$ and $Y(x_1) = 1$ for $x_1 \geq 0$. Then we have $\partial_1 Y(x_1) = \delta(x_1)$ as distribution derivative. The Heaviside function is a holonomic distribution since it satisfies the holonomic system $x_1\partial_1 Y(x_1) = \partial_2 Y(x_1) = \cdots = \partial_n Y(x_1) = 0$. As another example of a holonomic distribution, let $f(x)$ be a polynomial with real coefficients and let λ be a complex number. Then we introduce the symbol

$$f(x)_+^\lambda := \begin{cases} f(x)^\lambda & \text{if } f(x) \geq 0, \\ 0 & \text{if } f(x) < 0. \end{cases}$$

It is easy to see that $f(x)_+^\lambda$ is well-defined as a tempered distribution if the real part of λ is positive by the pairing

$$\langle f(x)_+^\lambda, \psi(x) \rangle = \int_{\mathbf{R}^n} f(x)_+^\lambda \psi(x)\, dx$$

for rapidly decreasing smooth functions $\psi(x)$. By virtue of the identity of the Bernstein-Sato polynomial

$$P(\lambda) f(x)_+^{\lambda+1} = b_f(\lambda) f(x)_+^\lambda$$

with the Bernstein-Sato polynomial $b_f(s) \in \mathbf{C}[s]$ of $f(x)$ and some $P(s) \in D_n[s]$, the tempered distribution $f(x)_+^\lambda$ can be analytically continued to the whole complex plane as a meromorphic function with respect to the parameter λ. The possible poles are contained in the set

$$\{r - \nu \mid r \in \mathbf{C},\ b_f(r) = 0,\ \nu = 0, 1, 2, \ldots\}, \tag{4.2}$$

which is in fact a subset of the negative rational numbers according to the celebrated theorem of Kashiwara [7]. Let

$$\mathrm{Ann}(f^s) := \{P(s) \in D_n[s] \mid P(s) f^s = 0\}.$$

Then the algorithm in [11] produces a set G of generators of $\mathrm{Ann}(f^s)$. If λ does not belong to the exceptional set (4.2), then we have $P(\lambda) f(x)_+^\lambda = 0$ for any $P(s) \in G$. This follows easily from the definition of the action of $P(s)$ on f^s viewed as a multi-valued analytic function together with analytic continuation. However, even if $P \in D_n$ annihilates f^λ as an analytic function, it does not necessarily annihilates $f(x)_+^\lambda$ as a distribution. For example, we have $\partial_x(1) = 0$ but $\partial_x 1_+ = \partial_x Y(x) = \delta(x) \neq 0$ with $n = 1$. Anyway, it is known that the ideal generated by $\{P(\lambda) \mid P(s) \in \mathrm{Ann}(f^s)\}$ is holonomic [7, Prop 6.1]. Hence the distribution $f(x)_+^\lambda$ is holonomic if λ does not belong to (4.2).

The integral of a holonomic distribution with respect to some variables is again holonomic and can be computed by the integration algorithm. In general, let $u = u(x_1, \ldots, x_m)$ be a holonomic distribution on $\mathbf{R}^n$ such that the projection $\pi_m : \mathbf{R}^n \to \mathbf{R}^m$ defined by $x \mapsto (x_1, \ldots, x_m)$ restricted to the support of u is proper. Then the integral

$$v(x_1, \ldots, x_m) := \int_{\mathbf{R}^{n-m}} u(x_1, \ldots, x_m, x_{m+1}, \ldots, x_n)\, dx_{m+1} \cdots dx_n$$

is well-defined as a distribution on $\mathbf{R}^m$. In fact, it is defined by the pairing

$$\langle v, \psi \rangle := \langle u, 1 \otimes \psi \rangle$$

for a smooth function $\psi(x_1, \dots, x_m)$ with compact support, where $1 \otimes \psi$ means regarding $\psi(x_1, \dots, x_m)$ as a function on $\mathbf{R}^n$. We have

$$\langle \partial_i P u, 1 \otimes \psi \rangle = \langle u, -P^* \partial_i (1 \otimes \psi) \rangle = 0$$

for any $P \in D_n$ and $i = m+1, \dots, n$, where P^* denotes the formal adjoint of P. It follows that v satisfies the integral of the D-module for u. In particular, if u is a holonomic distribution, then so is its integral v.

Example 4.3. Put $u = \delta(t - x_1^4 - x_2^4)$ and

$$v(t) := \int_{\mathbf{R}^2} \delta(t - x_1^4 - x_2^4)\, dx_1 dx_2.$$

By the integration algorithm, we know that the distribution $v(t)$ satisfies $(2t\partial_t + 1)v(t) = 0$ on $\mathbf{R}$. From definition, it follows that $u(t) = 0$ on $t < 0$. Hence $v(t)$ is written in the form $v(t) = Ct_+^{-1/2}$ with some constant C.

5. Definite Integral by Using the Heaviside Function

We can compute the definite integral of the form

$$\int_a^b u(x)\, dx_1 = \int_{-\infty}^{\infty} Y(x_1 - a) Y(b - x_1) u(x)\, dx_1,$$

where $u(x)$ is a smooth function defined on an open neighborhood of $[a, b] \times U$ with an open set U of $\mathbf{R}^{n-1}$. The integrand $Y(x_1 - a)Y(b - x_1)u(x)$ is well-defined as a distribution on $\mathbf{R} \times U$ with a proper support with respect to the projection to U. In the extreme case $b = \infty$, we can define

$$v(x_2, \dots, x_n) := \int_a^{\infty} u(x)\, dx_1 = \int_{-\infty}^{\infty} Y(x_1 - a) u(x)\, dx_1,$$

which is a smooth function on U if $u(x)$ is a smooth function on a neighborhood of $[a, \infty) \times U$ which is rapidly decreasing as x_1 tends to infinity. More precisely, we assume that $\lim_{x_1 \to \infty} Pu(x) = 0$ for any $P \in D_n$ and $(x_2, \dots, x_n) \in U$. The distribution $Y(x_1 - a)u(x)$ satisfies a holonomic system $M = D_n / \mathrm{Ann}(Y(x_1 - a)u(x))$. Then we can see that $v(x_2, \dots, x_n)$ satisfies the integral $M/\partial_1 M$ in the same way as for a distribution with proper support discussed in the previous section. A possible bottleneck in this computation is that of the product of $Y(x_1 - a)u(x)$. So let us present a short cut for this computation. Let I be a left ideal of D_n which annihilates $u(x)$ such that D_n/I is holonomic. We assume $a = 0$ for the sake of simplicity. First recall the formulae

$$x_1^j \delta^{(k)}(x_1) = \begin{cases} (-1)^j k(k-1)\cdots(k-j+1)\delta^{(k-j)}(x_1) & \text{if } j \le k, \\ 0 & \text{if } j > k; \end{cases}$$

$$\partial_1^m (Y(x_1)u(x)) = Y(x_1)\partial_1^m u(x) + \sum_{k=1}^{m} \binom{m}{k} \delta^{(k-1)}(x_1)\partial_1^{m-k} u(x).$$

Let P be an element of I whose order with respect to the weight vector $(-1, 0, \ldots, 0; 1, 0, \ldots, 0)$ is m. Using the above formulae, we get

$$\begin{aligned} P(Y(x_1)u(x)) &= Y(x_1)Pu(x) + \sum_{k=1}^{\max\{m,0\}} \delta^{(k-1)}(x_1)Q_k u(x) \\ &= \sum_{k=1}^{\max\{m,0\}} \delta^{(k-1)}(x_1)Q_k u(x) \end{aligned}$$

with some $Q_1, \ldots, Q_m \in D_n$. It follows that

$$\begin{aligned} \widetilde{I} &:= \{\mathrm{sat}(P) := x_1^{\max\{m,0\}} P \mid P \in I,\, m = \mathrm{ord}_{(-1,0,\ldots,0;1,0\ldots,0)} P\} \\ &\subset \mathrm{Ann}(Y(x_1)u(x)). \end{aligned}$$

We conjecture that $D_n/\widetilde{I}$ is holonomic. In practice, we can take a generating set G of I and compute $\widetilde{G} := \{\mathrm{sat}(P) \mid P \in G\}$, which generates an ideal contained in $\mathrm{Ann}(Y(x_1)u(x))$. We can easily extend the arguments so far to integrals of the form

$$\int_{a_1}^{b_1} \cdots \int_{a_m}^{b_m} u(x)\, dx_1 \cdots dx_m.$$

Example 5.1. Let t, x be real variables and put

$$v(x) := \int_0^\infty e^{(-t^3+t)x}\, dt,$$

which is a smooth function on $x > 0$. Then $u := e^{(-t^3+t)x}$ satisfies a holonomic system

$$(\partial_t + (3t^2 - 1)x)u = (\partial_x + t^3 - t)u = 0.$$

By using the argument above, we know that $Y(t)u$ satisfies

$$(t\partial_t + (3t^3 - t)x)u = (\partial_x + t^3 - t)u = 0.$$

By the integration algorithm, we can conclude that $v(x)$ satisfies

$$(27x^3\partial_x^3 - 4x^3\partial_x + 54x^2\partial_x^2 - 4x^2 - 3x\partial_x + 3)v(x) = 0.$$

6. Mellin Transform and z-Transform

Let C be a path in the complex plane. The *C-Mellin transform* of a function $f(x)$ is defined as

$$f(x) \longmapsto g[k] = \int_C f(x)x^{k-1}dx. \tag{6.1}$$

When the path C can be regarded as a twisted cycle with respect to $f(x)x^{k-1}$, we have the following identities:

$$\begin{aligned}(k-1)E_k^{-1} \bullet g[k] &= -\int_C (\partial_x f(x))x^{k-1}dx, \\ E_k \bullet g[k] &= \int_C xf(x)x^{k-1}dx,\end{aligned}$$

where $E_k \bullet g[k] = g[k+1]$. The identities induce the correspondences

$$(k-1)E_k^{-1} \longleftrightarrow -\partial_x \quad \text{and} \quad E_k \longleftrightarrow x.$$

In other words, if the function $f(x)$ is a solution of a differential equation

$$\sum_{i=0}^{m} a_i(x)\partial_x^i f = 0,$$

then the function $g(k)$ satisfies the difference equation

$$\sum_{i=0}^{m} a_i(E_k)(-(k-1)E_k^{-1})^i f = 0.$$

Conversely, if the function $g(k)$ satisfies a difference equation

$$\sum_{i=0}^{m} b_i[k-1]E_k^i g = 0,$$

then the function $f(x)$ satisfies the differential equation

$$\sum_{i=0}^{m} b_i(-\partial_x x)x^i g = 0.$$

Following these observations, we can prove, by a purely algebraic discussion, that $\mathbf{C}\langle k, E_k\rangle \simeq \mathbf{C}\langle -\theta_x, x\rangle$ and

$$\mathbf{C}\langle k, E_k, E_k^{-1}\rangle \simeq \mathbf{C}\langle x, \partial_x\rangle. \tag{6.3}$$

Let us consider a function $f[k,n]$ which satisfies a system of difference operators J. We apply the Mellin transform

$$k \leftrightarrow -\theta_x,\ E_k \leftrightarrow x,\ -E_k^{-1}k \leftrightarrow \partial_x,\ n \leftrightarrow -\theta_y,\ E_n \leftrightarrow n,\ -E_n^{-1}n \leftrightarrow \partial_y$$

to J and obtain the ideal $\widehat{J}$ in the ring of differential operators.

Theorem 6.4. *We assume $f[k,n]=0$ for a sufficiently large $|k|$. Put*

$$I = \left(\widehat{J} + (x-1)D_2\right) \cap \mathbf{C}\langle y, \partial_y \rangle.$$

By applying the inverse Mellin transform to I, we obtain a difference equation for $F[n] = \sum_k f[k,n]$.

Example 6.5. Put $f[k,n] = \binom{n}{k}$. Then, we have

$$(E_n - 2)\sum_k f[k,n] = 0.$$

The function $f[k,n]$ satisfies the system of difference equations

$$\{(n-k+1)E_n - (n+1)\}f = 0, \text{ and}$$
$$\{(k+1)E_k - (n-k)\}f = 0.$$

Let J be the ideal generated by the two difference operators above. Consider the inverse Mellin transform of J. Apply the algorithm of restriction to obtain the restriction $\widehat{J} + (x-1)D_2$. From the output of the algorithm, we can see that the ideal I is generated by $-y^2\partial_y + 2y\partial_y - 2 = -y\theta_y + 2\theta_y - 2$. Hence, the sum is annihilated by $E_n n - 2n - 2 = (n+1)(E_n - 2)$.

The inverse Mellin transform is called the z-transform in the theory of signal processing. Let $\{s[k]\}$ be a sequence of complex numbers indexed by $k = (k_1, k_2, \ldots, k_n) \in \mathbf{Z}^n$, which we call a (multidimensional) *discrete signal.* The z-transform of $\{s[k]\}$ is the formal series

$$Z(s)(z) = \sum_{k \in \mathbf{Z}^n} s[(k_1, \ldots, k_n)] z_1^{k_1} \cdots z_n^{k_n}.$$

If the z-transform $S(z) = Z(s)(z)$ is convergent around $z = 0$, then we have

$$s[k] = \frac{1}{(2\pi\sqrt{-1})^n} \int_C z_1^{-k_1-1} \cdots z_n^{-k_n-1} S(z) dz_1 \cdots dz_n$$

by the residue theorem where C is the product of n circles centered at 0.

The inverse z-transform is nothing but a multi-variable generalization of the C-Mellin transform. A signal $s[k]$ is called *bounded* when $s[k] = 0$ for $k_1, \ldots, k_n \ll 0$.

A bounded discrete signal is called *holonomic* if the annihilating set of the difference operators of the signal is holonomic under the n-variable generalization of the isomorphism (6.3).

Let $x[k]$ and $y[k]$ be one dimensional holonomic signals and $X(z)$ and $Y(z)$ be the z-transforms of $x[k]$ and $y[k]$ respectively. Since we have

$$Z(x[k] * y[k]) = X(z)Y(z)$$

and

$$Z(x[k]y[k]) = \frac{1}{2\pi\sqrt{-1}} \int_C X(w)Y(z/w)w^{-1}dw,$$

the product and the convolution of holonomic signals are again holonomic signal. It follows from discussions of the previous and this section that we have the following theorem.

Theorem 6.6. *Holonomic signals are closed under the operations listed below if they are well-defined, and holonomic systems for new signals under these operations are computable.*

$S(z)$	$Z^{-1}(S)[k]$
sum and subtraction	*sum and subtraction*
product	*convolution*
multiplicative convolution	*(element-wise) product*
restriction to $z_n = 1$	*Sum with respect to* k_n
integration w.r.t. z_n (*the inverse of z-transform*)	

We consider a one dimensional discrete signal system with the impulse response $h[k]$. Then, for an input signal $x[k]$, the system outputs the signal $y[k] = h[k] * x[k]$. Let $H(z)$, $X(z)$, and $Y(z)$ be the z-transforms of $h[k]$, $x[k]$, and $y[k]$ respectively. Then, we have $Y(z) = H(z)X(z)$. The function $H(z)$ is called the *transfer function* of the system. In the theory of discrete signals, rational functions usually appear as transfer functions and a beautiful theory is established for this class of transfer functions. We may try to replace rational functions by holonomic functions. This idea is not only mathematically natural, but has also been used in signal processing. An example is the Kaiser window, which is expressed in terms of the zeroth-order modified Bessel function of the first kind (see [16]); the Bessel function is no longer rational, but it is a holonomic function. Our holonomic approach will give a systematic framework to design filters out of rational functions. As the first step, numerical evaluation of holonomic functions is necessary to design and evaluate a new filter. In the next section, we will see that our holonomic approach gives an effective method of numerical evaluations of holonomic functions.

7. Numerical Evaluation of Holonomic Functions

Let us compare several computational techniques to evaluate a definite integral. We consider the problem of getting an approximate value of the expression on the left hand side of the identity [3].

$$F\left(\frac{1}{12}, \frac{5}{12}, \frac{1}{2}; \frac{1323}{1331}\right) = \frac{3}{4}\sqrt[4]{11}$$

where

$$F(\alpha, \beta, \gamma; z) = \frac{\Gamma(\gamma)}{\Gamma(\beta)\Gamma(\gamma-\beta)} \int_0^1 t^{\beta-1}(1-t)^{\gamma-\beta-1}(1-tz)^{-\alpha}dt.$$

The function F is a holonomic function with respect to z; F satisfies the Gauss differential equation

$$z(1-z)f'' + (\gamma - (\alpha+\beta+1)z)f' - \alpha\beta f = 0, \quad f(0) = 1. \tag{7.1}$$

Let us try a numerical integration over $[0, 1]$ by the adaptive Gauss method; we do not utilize the differential equation. Since the integrand is singular at the boundary, we use the following contiguity relation and evaluate the two hypergeometric functions below [22]

$$\begin{aligned} &F\left(\frac{1}{12}, \frac{5}{12}, \frac{1}{2}; \frac{1323}{1331}\right) \\ &= -\frac{555146934690291893170809321}{77265229938688} F\left(-\frac{31}{12}, \frac{37}{12}, \frac{13}{2}; \frac{1323}{1331}\right) \\ &+ \frac{2300849705553019085468253191 9}{4017791956811776} F\left(-\frac{31}{12}, \frac{37}{12}, \frac{15}{2}; \frac{1323}{1331}\right) \end{aligned}$$

It takes about 9 seconds to get the value accurate to 10^{-4}.

Let us evaluate the value by solving (7.1). The fourth order adaptive Runge Kutta method [18] takes about 2 seconds to get the value accurate to 10^{-4}.

We can find the series solution of (7.1) in an algorithmic way. The evaluation of the series expansion at $z = \frac{1323}{1331}$ gives the value accurate to 10^{-4} in less than 1 second [22].

This example shows that differential equations give substantial information for effective numerical evaluations and leads us to the following method to evaluate a holonomic function f at $x = b$ numerically.

(1) Find a system of differential equations for the holonomic function f. Let r be the rank of the system of differential equations.

(2) Choose a point $x = a$. Evaluate $f(a)$, $f^{(1)}(a)$, ..., $f^{(r-1)}(a)$. *This step is not algorithmic.*

(3) Find the value $f(b)$ by an adaptive Runge-Kutta method by the system of differential equations and the initial values at $x = a$.

If we can find series solution at $x = a$ and it converges at $x = b$ rapidly, we may replace the last step by a computation of a series solution and its evaluation. As to methods to find series solutions, see the Chapter 2 of [19] and references of it. However, there remains some fundamental unsolved problems. As a demonstration of our method, we close this paper with an example showing the graph of a solution of a Bessel differential equation in two variables [15], which is drawn by using our method.

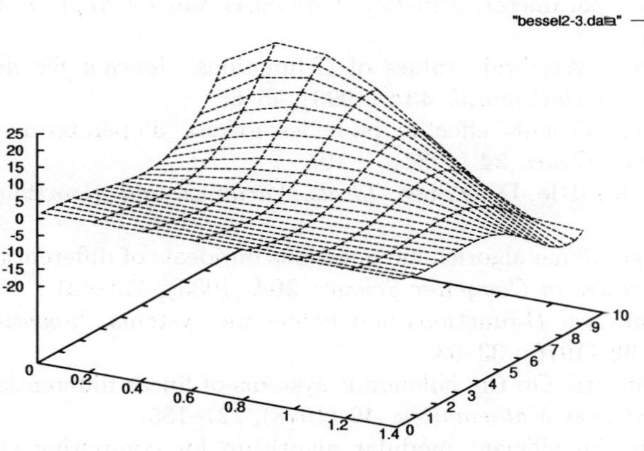

Figure 1. Example 7.2: Bessel function in two variables

Example 7.2. (Bessel function in two variables.) We consider the integral

$$f(a; x, y) = \int_C \exp(-\frac{1}{4}t^2 - xt - y/t)t^{-a-1}dt,$$

where

$$C = \overrightarrow{01} + \{e^{2\pi\sqrt{-1}\theta} \mid \theta \in [0, 2\pi]\} + \overrightarrow{10}\,.$$

The function $f(a; x, y)$ satisfies the holonomic system defined by the operators

$$\partial_x\partial_y + 1,$$
$$\partial_x^2 - 2x\partial_x + 2y\partial_y + 2a,$$
$$2y\partial_y^2 + 2(a+1)\partial_y - \partial_x + 2x.$$

The rank of the system is 3. Take $a = 1/2$. It admits a unique solution of the form $y^{-a}g(x, y)$ such that g is holomorphic at the origin and $g(0, 0) = 1$. This is the graph of g for $(x, y) \in [0, 1.4] \times [0, 9]$ (Figure 1). The function $f(1/2; x, y)$ is a constant multiple of $y^{-a}g(x, y)$. The normal form computation in D is used to derive the ODE's to which we apply the adaptive Runge-Kutta method.

References

1. G. Almkvist, D. Zeilberger. The method of differentiating under the integral sign. *Journal of Symbolic Computation* **10** (1990), 571–591.
2. J. Bernstein. The analytic continuation of generalized functions with respect to a parameter. *Functional Analysis and its Application* **6** (1972), 273–285.
3. F. Beukers. Algebraic values of G-functions. *Journal für die reine und angewandte Mathematik* **434** (1993), 45–65.
4. F. Castro. Calculs effectifs pour les idéaux d'opérateurs différentiels. *Travaux en Cours* **24** (1987), 1–19.
5. D. Cox, J. Little, D. O'Shea. *Ideals, Varieties, and Algorithms.* Springer, 1992.
6. A. Galligo. Some algorithmic questions on ideals of differential operators. *Lecture Notes in Computer Science* **204** (1985), 413–421.
7. M. Kashiwara. B-functions and holonomic systems, *Inventiones Mathematicæ* **38** (1976), 33–53.
8. M. Kashiwara. On the holonomic systems of linear differential equations, II. *Inventiones Mathematicæ* **49** (1978), 121–135.
9. M. Noro. An efficient modular algorithm for computing the global b-function, *Mathematical Software* (*ICMS 2002*) (2002), 147–157.
10. T. Oaku. Computation of the characteristic variety and the singular locus of a system of differential equations with polynomial coefficients. *Japan Journal of Industrial and Applied Mathematics* **11** (1994), 485–497.
11. T. Oaku. Algorithms for the b-function and D-modules associated with a polynomial. *Journal of Pure and Applied Algebra* **117** (1997), 495–518.
12. T. Oaku. Algorithms for b-functions, restrictions, and algebraic local cohomology groups of D-modules. *Advances in Applied Mathematics* **19** (1997), 61–105.
13. T. Oaku, N. Takayama. Algorithms for D-modules—restriction, tensor product, localization, and algebraic local cohomology groups. *Journal of Pure and Applied Algebra* **156** (2001), 267–308.
14. T. Oaku, N. Takayama, U. Walther. A localization algorithm for D-modules, *Journal of Symbolic Computation* **29** (2000), 721–728.
15. K. Okamoto, H. Kimura. On particular solutions of the Garnier systems and the hypergeometric functions of several variables. *Quarterly Journal of Mathematics* **37** (1986), 61–80.

16. A. V. Oppenheim, R. W. Schafer. *Discrete-Time Signal Processing.* Prentice Hall, 1989.
17. M. Petkovsek, H. S. Wilf, D. Zeilberger. $A = B$. A K Peters, Ltd., 1996.
18. W. Press, S. Teukolsky, W. Vetterling, B. Flannery. *Numerical Recipes in C++.* Cambridge University Press, 1988.
19. M. Saito, B. Sturmfels, N. Takayama. *Gröbner Deformations of Hypergeometric Differential Equations.* Springer, 2000.
20. N. Takayama. An algorithm of constructing the integral of a module, *Proceedings of International Symposium on Symbolic and Algebraic Computation* (1990), 206–211.
21. N. Takayama. Kan: a system for computation in algebraic analysis. Source code available at **http://www.openxm.org**, Version 1 (1991), Version 2 (1994). The latest version is 3.021108 (2002).
22. Y. Tamura. A design and implementation of a digital formula book for generalized hypergeometric functions. Thesis, Kobe University, 2003.
23. H. S. Wilf, D. Zeilberger. An algorithmic proof theory for hypergeometric (ordinary and q-) multisum/integral identities, *Inventiones Mathematicæ* **108** (1992), 575–633.
24. D. Zeilberger. A holonomic systems approach to special function identities. *Journal of Computational and Applied Mathematics* **32** (1990), 321–368.

ON THE DIVISION OF GENERALIZED POLYNOMIALS

NOR'AINI ARIS, ALI ABD RAHMAN

Department of Mathematics
Faculty of Science
Universiti Teknologi Malaysia
E-mail: noraini,@mel.fs.utm.my, aar1619@yahoo.com

The objective of this work is to design an efficient algorithm for the division of two polynomials represented with respect to a given orthogonal basis. The theoretical developments toward achieving this goal are investigated, extended and presented. From the theories developed, the algorithm for determining whether a polynomial divides another polynomial is devised. The theoretical aspects of its computing time is also investigated. The ultimate aim is to incorporate this divisor test into the modular generalized polynomial GCD algorithm, which serves as a termination criterion whether the primes chosen are lucky primes and whether the final correct solution has been achieved; thus contributing to the effectiveness of the *MOPBGCD* (Modular Orthogonal Polynomial Basis GCD) algorithm.

1. Introduction

The theory on polynomials in forms other than the power form is recovered from several works of Barnett and Maroulas, such as in [3,5,6]. The application of polynomials in the generalized form arises in the computation of the Smith form of a polynomial matrix and problems associated with the minimal realization of a transfer function matrix. Barnett [4] describes the theory and applications of polynomials in linear control systems. In [14] a modular algorithm is presented for computing the exact greatest common divisor of polynomials relative to some orthogonal basis, given the respective defining terms of the defining relations of the basis. The design of the algorithm is based on the results pertaining to an application of the companion matrix which is proposed by Barnett and Maroulas [3].

2. Problem Description

Consider a set of orthogonal polynomials over the field of rational numbers given by

$$p_i(x) = \sum_{j=0}^{i} p_{ij}x^j. \qquad (2.1)$$

where the set $\{p_i(x)\}$, known as an orthogonal basis, is generated by the three-term recurrence relations [2]:

$$\begin{aligned}
&p_0(x) = 1, \quad p_1(x) = \alpha_0 x + \beta_0, \\
&p_{i+1}(x) = (\alpha_i x + \beta_i)\, p_i(x) - \gamma_i\, p_{i-1}(x) \qquad (i = 1, \dots\ n-1), \qquad (2.2) \\
&\alpha_i > 0, \quad \gamma_i, \beta_i \geq 0.
\end{aligned}$$

Any given n-th degree integral polynomial $a(x)$ can be uniquely expressed in the form

$$a(x) = a_0 p_0(x) + a_1 p_1(x) + \dots + a_n p_n(x) \tag{2.3}$$

Assume without loss of generality that $a_n = 1$ and for subsequent convenience, define the monic polynomial (as in [2]):

$$\widetilde{a}(x) = a(x)/\alpha_0\alpha_1 \cdots \alpha_{n-1} = \widetilde{a}_0 + \widetilde{a}_1 x + \cdots + \widetilde{a}_{n-1} x^{n-1} + x^n. \tag{2.4}$$

Let $a(x)$ and $b(x)$ be in the form (2.3) such that $n = \deg(a) \geq \deg(b) = m$. We can assume, again without loss of generality (see [3]), that if $m = n$, then $b(x)$ can be replaced by :

$$\Big(b_0 - \frac{a_0 b_n}{a_n}\Big) p_0(x) + \cdots + \Big(b_{n-1} - \frac{a_{n-1} b_n}{a_n}\Big) p_{n-1}(x).$$

A general algorithm applying the companion matrix approach in calculating the GCD of two polynomials with respect to an orthogonal basis is presented in [1]. If the degrees of the input polynomials are sufficiently large (say for degrees ≥ 10), these entries may involve multiprecision rational numbers or integers after multiplying through by a suitable integer to obtain an equivalent integral systems. As an example, let

$$\widetilde{a}(x) = x^{10} + x^9 - 2x^8 - x^7 - 16x^6 - 11x^5 + 7x^4 - 4x^3 + 31x^2 - 21x + 3$$

be a dense polynomial over Z in power series form. In the generalized form, the coefficients of a in a sequence of ascending order of the Legendre basis is given by

$$a = \Big(\frac{44799131}{20160}, \frac{-68321929}{13440}, \frac{12068249}{4032}, \frac{-6777509}{5760}, \frac{-752267}{1120}, \\ \frac{-1232891}{5040}, \frac{-543647}{2520}, \frac{361}{48}, \frac{119}{90}, \frac{19}{10}, 1\Big).$$

The coefficients of a has at most 9 digits in the numerator even though the coefficients of the power series form of the polynomial are relatively small. In such cases, the components of the companion matrix [3] and the corresponding coefficient matrix may include rational numbers with large numerators and denominators. Therefore, multiprecision operations are inherent in the exact methods applied. Since multiprecision operations are rather slow and may lead to the problem of intermediate expression swell, our intention is to minimize the use of such operations by the application of the modular technique proposed in [14].

In applying the modular technique, prior knowledge of a bound to the solutions in the original domain is required to ensure that a unique solution for the Chinese Remainder Algorithm (CRA) has been achieved. Alternatively, one may use some direct test to decide whether the solution obtained so far should be the final solution and then proceed to check on the correctness of the solution. We refer the readers to the examples of applications of the modular approach in [8] and [12], where, in the computation of the polynomial GCD, the solution modulo $P = p_0 \cdots p_{k-1}$ (each p_i being a beta digit or machine prime) from successive applications of the CRA can be checked for its correctness when the results of two successive applications of the CRA are equal, or when the Rational Number Reconstruction Algorithm (RNRA) gives a nonempty solution. In the GCD computation of two integral polynomials, it suffices to check if the candidate solution divides both a and b. In this paper, we elaborate on the technique used in the divisor test, that is, if c is a candidate for the GCD of a and b, we would like to test if c actually divides both a and b. The division can be done without converting to the power series representation.

3. The Multiplication $x\,b(x)$

Let $a(x)$ and $b(x)$ be polynomials of the form (2.3) such that $\deg(a) = n$ and $\deg(b) = m$. If $m = n$, the division of $a(x)$ by $b(x)$ can be done directly in the generalized form. Otherwise, if $m < n$, we can multiply $b(x)$ by x^{n-m}. The remainder of $a(x)$ divided by $x^{n-m}b(x)$ can then be calculated using successive computation of the determinant of a two by two matrix involving the coefficients of these polynomials of equal degree [4].

Let $\{p_i(x)\}$ be a set of orthogonal basis described in (2.1) satisfying (2.2). Let $b(x) = b_m p_m(x) + b_{m-1}p_{m-1}(x) + \cdots + b_0 p_0(x)$. The coefficients of $x\,b(x)$ can then be calculated from the $(m+1) \times (m+2)$ matrix

$$C = \begin{pmatrix} \frac{1}{\alpha_m} & \frac{-\beta_m}{\alpha_m} & \frac{\gamma_m}{\alpha_m} & 0 & \cdots & 0 & 0 & 0 \\ 0 & \frac{1}{\alpha_{m-1}} & \frac{-\beta_{m-1}}{\alpha_{m-1}} & \frac{\gamma_{m-1}}{\alpha_{m-1}} & \cdots & 0 & 0 & 0 \\ \vdots & \vdots & \vdots & \vdots & \vdots & \vdots & \vdots & \vdots \\ 0 & 0 & 0 & 0 & \cdots & \frac{1}{\alpha_1} & \frac{-\beta_1}{\alpha_1} & \frac{\gamma_1}{\alpha_1} \\ 0 & 0 & 0 & 0 & \cdots & 0 & \frac{1}{\alpha_0} & \frac{-\beta_0}{\alpha_0} \end{pmatrix} \quad (3.1)$$

and the matrix multiplication (see [4, p. 396])

$$\left(b_{1,m+1}, b_{1,m}, \ldots, b_{1,0} \right) = \left(b_m, b_{m-1}, \ldots, b_0 \right) C, \tag{3.2}$$

giving $x\, b(x) = b_{1,m+1}p_{m+1} + b_{1,m}p_m(x) + \cdots + b_{1,0}p_0(x)$.

4. Division of $a(x)$ by $x^{n-m}b(x)$

Following the assumption that $\deg(a) = n \geq m = \deg(b)$, let

$$x^{n-m}b(x) = b_{n-m,n}p_n(x) + b_{n-m,n-1}p_{n-1}(x) + \cdots + b_{n-m,0}p_0(x).$$

If $b_{n-m,n} \neq 0$, let the remainder when $a(x)$ is divided by $x^{n-m}b(x)$ be

$$r(x) = r_{n-1}p_{n-1}(x) + r_{n-2}p_{n-2}(x) + \cdots + r_0p_0(x).$$

The coefficients are tabulated and can be calculated as:

	$p_n(x)$	$p_{n-1}(x)$	$\ldots$	$p_0(x)$
$a(x)$	1	a_{n-1}	$\ldots$	a_0
$x^{n-m}b(x)$	$b_{n-m,n}$	$b_{n-m,n-1}$	$\ldots$	$b_{n-m,0}$

$$r_i = \frac{-1}{b_{n-m,n}} \begin{vmatrix} 1 & a_i \\ b_{n-m,n} & b_{n-m,i} \end{vmatrix} \tag{4.1}$$

for $i = n-1, \ldots, 0$.

Theorem 4.2. *Let $a(x)$ and $b(x)$ be integral polynomials relative to an orthogonal basis $\{p_i(x)\}_{i=0}^n$. Let degree of $a(x) = n$ and degree of $b(x) = m$ such that $m \leq n$. Let $r(x)$ be the remainder when $a(x)$ is divided by $x^{n-m}b(x)$. Then $r(x)$ is a multiple of $b(x)$ if and only if $b(x)$ divides $a(x)$.*

Proof. By the remainder theorem, $r(x) = a(x) - q(x)x^{n-m}b(x)$ for some $q(x)$ in $\mathbb{Z}[x]$.
($\Rightarrow$) Let $r(x)$ be a multiple of $b(x)$. From the preceding equation, we can factor out $b(x)$, so that

$$r(x) = b(x)\Big(\frac{a(x)}{b(x)} - q(x)x^{n-m}\Big).$$

From the assumption, this implies that $\frac{a(x)}{b(x)} - q(x)x^{n-m}$ is in $\mathbb{Z}[x]$. Therefore, $\frac{a(x)}{b(x)}$ is in $\mathbb{Z}[x]$, which implies that $a(x)$ is a multiple of $b(x)$.
($\Leftarrow$) Let $b(x)$ divide $a(x)$. Then there exists $q'(x) \in \mathbb{Z}[x]$ such that $a(x) = q'(x)b(x)$. So, $r(x) = b(x)\,(q'(x) - q(x)x^{n-m})$, implying that $r(x)$ is a multiple of $b(x)$. □

Example 4.3. Let

$$a(x) = p_4(x) - \frac{7}{4}p_3(x) - \frac{55}{6}p_2(x) + \frac{119}{8}p_1(x) - \frac{119}{24}p_0(x),$$
$$b(x) = p_2(x) - \frac{9}{2}p_1(x) + \frac{7}{2}p_0(x),$$

be represented relative to the Legendre basis $\{L_i(x)\}$ defined by

$$\alpha_0 = 1, \quad \beta_0 = 0; \quad \alpha_i = \frac{2i+1}{i+1}, \quad \beta_i = 0, \quad \gamma_i = \frac{i}{i+1} \qquad (i \geq 1).$$

Show that $b(x)$ divides $a(x)$.

Solution: Since the degree of $b(x)$ is less than the degree of $a(x)$, we multiply x to $b(x)$ until we obtain a polynomial $x^t b(x)$ which has degree equal to degree of $a(x)$. In this case, $t = 2$. The remainder when $a(x)$ is divided by $x^2 b(x)$ can be calculated as $r(x) \equiv p_3(x) - 5p_2(x) + \frac{13}{2}p_1(x) - \frac{5}{2}p_0(x)$. To apply Theorem 4.2, we need to show that $r(x)$ is a multiple of $b(x)$. We divide $r(x)$ by $x^e b(x)$ for some $e \geq 0$ in which case $e = 1$. The results are as follows:

	$p_3(x)$	$p_2(x)$	$p_1(x)$	$p_0(x)$
$r(x)$	1	-5	$\frac{13}{2}$	$-\frac{5}{2}$
$b(x)$		1	$-\frac{9}{2}$	1
$xb(x)$	1	-5	$\frac{13}{2}$	$-\frac{5}{2}$
$r'(x)$		0	0	0

Since $r'(x) = 0$, the division of $r(x)$ by $xb(x)$ gives no remainder. This implies that $r(x)$ is a multiple of $b(x)$. Applying Theorem 4.2, we conclude that $b(x)$ divides $a(x)$.

Example 4.4. Let

$$a(x) = -17p_0(x) - 26p_1(x) - 8p_2(x) + 2p_3(x) + p_4(x),$$
$$b(x) = p_0(x) + p_1(x),$$

where $\{p_i(x)\}_{i=0}^4$ are the Chebyshev basis of the first kind. Show that $b(x)$ divides $a(x)$.

Solution: To divide $a(x)$ by $b(x)$, we have to first multiply $b(x)$ by x^e, for some $e \geq 0$ to obtain a polynomial of degree 4, which is the degree of $a(x)$. In this case $e = 3$. The results are tabulated below:

	$p_4(x)$	$p_3(x)$	$p_2(x)$	$p_1(x)$	$p_0(x)$
$a(x)$	1	2	-8	-26	-17
$b(x)$				1	1
$xb(x)$			1	2	1
$x^2b(x)$		1	2	3	2
$x^3b(x)$	1	2	4	6	3
$r_1(x)$			1	$\frac{8}{3}$	$\frac{5}{3}$

We need to show that $r_1(x)$ is a multiple of $b(x)$ by dividing $r_1(x)$ by $b(x)$ as follows:

	$p_2(x)$	$p_1(x)$	$p_0(x)$
$r_1(x)$	1	$\frac{8}{3}$	$\frac{5}{3}$
$b(x)$		1	1
$xb(x)$	1	2	1
$r_2(x)$		1	1

Again, we need to check if $r_2(x)$ is a multiple of $b(x)$. Since $r_2(x) = b(x)$, Theorem 4.2 implies that $r_1(x)$ is a multiple of $b(x)$. By Theorem 4.2, this again implies that $a(x)$ is a multiple of $b(x)$.

5. Algorithm OPDIV

Given a polynomial b and a represented in the generalized form with respect to its orthogonal basis, Algorithm OPDIV will check if b divides a. We first describe the input set for the algorithm and some properties of the length of the integers and rational numbers that will be applied in the time complexity analysis for the algorithm.

If S denotes the set of valid inputs x to an algorithm $\mathcal{A}$, let $t_{\mathcal{A}} = t_{\mathcal{A}}(x)$ denote the computation time for the algorithm $\mathcal{A}$ to compute the output when applied to the input x. Let S be the set of generalized polynomials of the form (2.3) such that $\{p_i(x)\}_{i=0}^{n}$ corresponds to the set of Legendre basis as described in Example 4.3. We consider the decomposition of S into the sets,

$$\begin{aligned} S(u,v,n) = \{a = \textstyle\sum_{i=0}^{k} a_i p_i(x) \in S : &|\mathrm{num}(a_i)| \leqslant u, 0 < \mathrm{den}(a_i) \leqslant v, \\ &\ni \gcd(u,v) = \gcd(\mathrm{num}(a_i), \mathrm{den}(a_i)) = 1 \text{ and } k \leqslant n\}, \end{aligned} \tag{5.1}$$

where num (resp. den) is the numerator (resp. denominator) of a_i. For any beta digit degree $n \geq 2$, by the properties of the set of the integers $\mathbb{Z}$, there exist integers u and v such that $\mathrm{num}(a_i) \leq u$ and $\mathrm{den}(a_i) \leq v$. In the empirical input values, we observe that $u > v$. In order that $\gcd(u, v) = 1$, we must have $u \geq \max(\text{num}(a_i))$ and $v \geq \max(\text{den}(a_i))$, for $0 \leq i \leq n$.

For an integer $a \neq 0$, represented as $(a_0, \ldots, a_{n-1})$, the length n of a is denoted by $\mathrm{L}(a)$. This definition can be extended to the rational numbers and polynomials. If $a > 0$ and $b > 0$ are in $\mathbb{Z}$ and $\gcd(a, b) = 1$, then we let $\mathrm{L}(\frac{a}{b}) = \max(\mathrm{L}(a), \mathrm{L}(b))$. From this definition, let $x > 0$ and $y > 0$ belong to $\mathbb{Z}$ be such that $\gcd(x, y) = k > 1$. This implies that $\max(\mathrm{L}(x), \mathrm{L}(y)) = \max(\mathrm{L}(ka), \mathrm{L}(kb))$ for some $a, b \in \mathbb{Z}$. From the properties of the length on the integers, $\mathrm{L}(ka) \sim \mathrm{L}(k) + \mathrm{L}(a) \sim \mathrm{L}(a)$ since $2 \leq k < a$. This gives $\max(\mathrm{L}(x), \mathrm{L}(y)) \sim \max(\mathrm{L}(a), \mathrm{L}(b)) = \mathrm{L}(\frac{a}{b})$. But $\frac{a}{b} = \frac{x}{y}$. Therefore, $\max(\mathrm{L}(x), \mathrm{L}(y)) \sim \mathrm{L}(\frac{x}{y})$. Now, for $0 \leq i \leq n$ let a_i be the coefficients of $a \in S(u, v, n)$. Then $\mathrm{L}(\text{num}(a_i)) \leq \mathrm{L}(u)$ and $\mathrm{L}(\text{den}(a_i)) \leq \mathrm{L}(v)$. This implies that $\mathrm{L}(a_i) = \max(\mathrm{L}(\text{num}(a_i)), \mathrm{L}(\text{den}(a_i))) \leq \max(\mathrm{L}(u), \mathrm{L}(v)) = \mathrm{L}(u)$. Assuming classical matrix addition and multiplication, we apply the results on the computing time for the rational number product, sum, quotient and negative denoted by the algorithms RNPROD, RNSUM, RNQ and RNNEG respectively, which are given in [9], [10] and [15].

Theorem 5.2. $\mathrm{L}(\prod_{i=1}^{n} a_i) \leq \sum_{i=1}^{n} \mathrm{L}(a_i)$, *for all* $a_i \in \mathbb{Z}$

Proof. Refer to [9]. □

Theorem 5.3. *Let* $Y \neq 0, X \in \mathbb{Z}$ *such that* $X \geq Y$. *For* $1 \leq i \leq n$, *let* x_i *and* $y_i \neq 0$ *be single precision integers. Then* $\mathrm{L}(\sum_{i=1}^{n} \frac{Xx_i}{Yy_i}) \preceq n\mathrm{L}(X) + n^2$.

Proof. Let $\sum_{i=1}^{n} \dfrac{Xx_i}{Yy_i} = \dfrac{\sum_{i=1}^{n} Xx_i \prod_{j\neq i} y_j}{Y \prod_{i=1}^{n} y_i}$. We obtain

$$\begin{aligned} \mathrm{L}(\sum_{i=1}^{n} Xx_i \prod_{j\neq i} y_j) &\preceq \sum_{i=1}^{n} \mathrm{L}(Xx_i \prod_{j\neq i} y_j) \\ &= \mathrm{L}(Xx_1 \prod_{j\neq 1} y_j) + \cdots + \mathrm{L}(Xx_n \prod_{j\neq n} y_j) \\ &\preceq n\mathrm{L}(X) + \sum_{i=1}^{n} \mathrm{L}(x_i) + n \sum_{i=1}^{n} \mathrm{L}(y_i) \sim n\mathrm{L}(X) + n^2 \end{aligned}$$

and $\mathrm{L}(Y \prod_{i=1}^{n} y_i) \leq \mathrm{L}(Y) + n$. Therefore, $\mathrm{L}(\sum_{i=1}^{n} \frac{Xx_i}{Yy_i}) \preceq n\mathrm{L}(X) + n^2$. □

In the following, we present the algorithm OPDIV pertaining to the purpose of incorporating it into the modular algorithm MOPBGCD [14].

Algorithm CMULTX($alph, bet, gamm, m$)

[The inputs are $alph$, bet and $gamm$ which correspond to the lists of coefficients $\{\alpha_i\}_{i=0}^{n-1}, \{\beta_i\}_{i=0}^{n-1}, \{\gamma_i\}_{i=0}^{n-1}$ of Eqn. (2.2) defining the polynomial basis $\{p_0, \ldots, p_n\}$; m is the degree of the polynomial $b = (b_m, \ldots, b_0)$ to be multiplied to x. The output is the $(m+1) \times (m-2)$ matrix C in Eqn. (3.1) to be post multiplied to b in order to obtain the desired polynomial $x\,b(x)$ according to Eqn. (3.2).]

1. generate the diagonal entries $C_{i,i}$ for $i = 0$ to m
2. generate the diagonal entries $C_{i,i+1}$ for $i = 0$ to m
3. generate the diagonal entries $C_{i,i+2}$ for $i = 0$ to $m - 1$
4. convert matrix C to list representation and output(C)

Confining the analysis to the Legendre polynomial basis, the costs of computing the values of $C_{i,i} = \frac{(m-i)+2}{2(m-i)+3}$ and $C_{i,i+1} = 0$ for $1 \leqslant i \leqslant m+1$, and $C_{i,i+2} = \frac{(m-i)+1}{2(m-i)+3}$ for $1 \leqslant i \leqslant m$, are constant. For $C_{i,j} \neq 0$, $\mathrm{L}(C_{i,j}) \sim 1$ from which we see that $t_{\mathrm{CMULTX}}(alph, bet, gamm, m) \sim m$.

Algorithm PMULTX($alph, bet, gamm, a$)

[Given $alph$, bet and $gamm$ as in CMULTX and $a = (a_0, \ldots, a_m)$, the polynomial to be multiplied to x, this function outputs a monic polynomial $x * a = (xmulta_0, \ldots, xmulta_{m+1})$.]

1. $m \leftarrow$ LENGTH(a)
2. obtain the coefficient matrix C from algorithm CMULTX
3. construct INV$(a) = (a_m, \ldots, a_0)$ (reversing the order)
4. compute $xmultp =$ INV$(a) * C$ (in decreasing order)
5. find the inverse of $xmultp$
6. output(INV($xmultp$)) (a polynomial in increasing order)

Let the input polynomial be $a \in S(u, v, n)$ defined by (5.1). Let t_i be the computing time for Step i. $t_1 \sim 1$ and $t_2 \preceq m$. For each $0 \leq i \leq m$, $\mathrm{L}(a_i) \leq \mathrm{L}(u)$. So, $t_3 \preceq m\mathrm{L}(u)$. Let C be the matrix obtained from CMULTX. For $1 \leq i \leq m+1$, and $1 \leq j \leq m+2$, $\mathrm{L}(C_{i,j}) \sim 1$. This implies that for $0 \leq i \leq m$ and $0 \leq j \leq m+1$, $t_{\mathrm{RNPROD}}(a_i, C_{i,j}) \preceq \mathrm{L}(u)\mathrm{L}(\mathrm{L}(u))$ whereby $\mathrm{L}(a_i C_{i,j}) \preceq \mathrm{L}(u)$. The computing time for aC is the sum of the computing times of the rational number multiplications in the products $a_i C_{i,j}$ for $0 \leq i \leq m$ and $0 \leq j \leq m+1$ and the rational number additions in the sum $\sum_{i=0}^{m} a_i C_{i,j}$ for $0 \leq j \leq m+1$. Therefore, $t_4 \preceq m^2\mathrm{L}^2(u)\mathrm{L}(\mathrm{L}(u))$. For $0 \leq j \leq m+2$, $xmultp(j) = \sum_{i=0}^{m} a_i C_{i,j}$. Applying Theorem 5.3, we have, for a fix j, $\mathrm{L}(xmultp(j)) \preceq \sum_{i=0}^{m} \mathrm{L}(a_i C_{i,j}) \preceq m\mathrm{L}(u) + m^2$. Therefore, adding the cost for all $0 \leq j \leq m+1$, $t_5 \preceq m^2\mathrm{L}(u) + m^3$. For any valid input $x = (alph, bet, gamm, a)$, $t_{\mathrm{PMULTX}}(x) \preceq m^2\mathrm{L}^2(u)\mathrm{L}(\mathrm{L}(u)) + m^3$.

Algorithm ORTPREM(a, b, deg)

[Given a and b in the form of generalized polynomials with respect to an orthogonal basis such that $\deg(a) = \deg(b) = deg$. The function outputs the remainder when a is divided by b. Here $a(i) = a_{i-1},\ \ b(i) = b_{i-1}$ for $1 \le i \le deg + 1$.]

1. $c \leftarrow \dfrac{-1}{b(deg+1)}$
2. for $i = deg$ to 1, $elt(i) \leftarrow c * \{b(i) - b(deg+1) * a(i)\}$
3. if $elt(i) = 0$ then for $l < i \le deg$, assign $oprem = \frac{1}{elt(l)} * elt$
4. output(oprem)

Suppose $a \in S(u, v, n)$ defined by (5.1) and $\|b\|_\infty = B$. Then $deg = n$ and $\mathrm{L}(c) \le \mathrm{L}(B)$. From this and $\mathrm{L}(b(n+1)a(i)) \preceq \mathrm{L}(B) + \mathrm{L}(u)$, we obtain $t_{\mathrm{RNSUM}}(b(i), b(n+1)a_i) \preceq (\mathrm{L}(B) + \mathrm{L}(u))\mathrm{L}(B)\mathrm{L}(\mathrm{L}(B) + \mathrm{L}(u))$. For $1 \le i \le n$, let $y = b(i) + b(n+1)a(i)$. Then $y \le B + B\frac{u}{v}$ which implies that $\mathrm{L}(y) \preceq \mathrm{L}(u) + \mathrm{L}(v) + \mathrm{L}(B) \sim \mathrm{L}(u) + \mathrm{L}(B)$. Therefore $t_{\mathrm{RNPROD}}(c, y) \preceq (\mathrm{L}(u) + \mathrm{L}(B))\mathrm{L}(B)\mathrm{L}(\mathrm{L}(u) + \mathrm{L}(B))$. $t_2 \preceq n(\mathrm{L}(u) + \mathrm{L}(B))\mathrm{L}(B)\mathrm{L}(\mathrm{L}(u) + \mathrm{L}(B))$. $\mathrm{L}(elt(i)) \le \mathrm{L}(cy)$ such that $cy \le \frac{(u+v)B}{vb_{n+1}}$. This implies that $\mathrm{L}(elt(i)) \preceq \mathrm{L}(u) + \mathrm{L}(B)$ which gives $t_{\mathrm{RNPROD}}(\frac{1}{elt(l)}, elt(i)) \preceq (\mathrm{L}(u) + \mathrm{L}(v))^2\mathrm{L}(\mathrm{L}(u) + \mathrm{L}(v))$. Therefore, $t_3 \preceq l(\mathrm{L}^2(u) + \mathrm{L}^2(B) + \mathrm{L}(u)\mathrm{L}(B))\mathrm{L}(\mathrm{L}(u) + \mathrm{L}(B))$. $t_{\mathrm{ORTPREM}}(a, b, deg) \preceq n(\mathrm{L}^2(u) + \mathrm{L}^2(B) + \mathrm{L}(u)\mathrm{L}(B))\mathrm{L}(\mathrm{L}(u) + \mathrm{L}(B))$.

Algorithm OPDIV($alph, bet, gamm, a, c$)

[Let a and c be in the orthogonal basis such that $deg(a) = n - 1 \ge deg(c) = m - 1$ and $alph, bet, gamm$ represents the respective list of defining terms. The function returns 1 if c divides a and 0 otherwise.]

1. $n \leftarrow$ LENGTH(a), $m \leftarrow$ LENGTH(c), $xmsoln \leftarrow c$
2. if $n > m$ then {
   $DIV \leftarrow NIL$
   for $i = 1$ to $n - m${
      $xmsoln \leftarrow$ PMULTX($alph, bet, gamm, xmsoln$)
      $DIV \leftarrow$ COMP($xmsoln, DIV$)}
   $DIV \leftarrow$ INV(DIV)}
3. DIVIDE: $remainder \leftarrow$ ORTPREM($a, xmsoln, n-1$)
4. set k=LENGTH($remainder$)
5. while($remainder(k) == 0$ && $k > 0$) $k--$
6. if $k = 0$ then output(1)
7. else if $k > 0$ then {
   if LENGTH($remainder$)$\ge$ LENGTH(c) then{
      $a \leftarrow remainder$ and $n \leftarrow$ LENGTH(a)
      if $n - m \ne 0$ then $xmsoln = DIV(n-m)$
      else $xmsoln \leftarrow c$
      goto DIVIDE}
   else output(0)}

The algorithm OPDIV generates the generalized polynomial remainder sequence $(a = r_0, r_1, \ldots, r_k)$ in which r_{i+1} is the remainder when r_i is divided by $x^{\deg(r_i)-m} * c$ and $n = \deg(r_0) > \deg(r_1) > \cdots > \deg(r_{k+1})$. The algorithm terminates upon encountering that the polynomial $r_{k+1} = 0$ or whenever $\deg r_k < m$. In step 2, the coefficients of the polynomials $x^i * c$ are calculated for $1 \leq i \leq n - m$. The coefficients of $x^i * c$ are computed by performing a vector-matrix multiplication with an $i+1$ and $i+2$ matrix with a cost proportional to i^2 operations. Computing $x^i * c$ for the $n-m$ iterations in step 2 requires a total cost proportional to n^3 operations. Since $\deg(r_i) < n$ for each $i > 1$, $x^{\deg(r_i)-m}$ can be obtained from $DIV(\deg(r_i) - m)$. It is possible to generate the remainder sequences r_i until $\deg(r_{k+1}) = m$. Since each application of ORTPREM is dominated by n, $n - m$ executions of OPDIV gives a total cost in the order of n^2 operations. This implies that, regardless of the length of the rational numbers, the number of operations required by OPDIV is $O(n^3)$. If the polynomials have to be converted to power series form, the coefficients of the basis $\{p_i(x)\}$ for $2 \leq i \leq n$ have to be generated from the equations (2.2) at a cost of $O(n)$ in the computation of $p_n(x)$. The cost of converting to the respective basis requires $O(n^3)$ from solving an $n \times n$ linear system. Division of a by c is $Om(n - m)$ if $\deg(c) = m$. This also gives a total cost proportional to n^3 operations.

However, a detailed analysis of the computing time for OPDIV looks into the bound for the length of the inputs of ORTPREM each time it is executed. The first argument of ORTPREM are the remainder sequences r_i and the input parameter b are the divisors $x^{\deg(r_i)-m} * c$. The bound on the length of $x^{\deg(r_{i+1})-m} * c$ depends on the bound of the length of the result of the previous iteration, that is $x^{\deg(r_i)-m} * c$.

Theorem 5.4.
Assuming the correctness of algorithms ORTPREM, PMULTX *and* CMULTX *respectively, algorithm* OPDIV *works correctly, returning 1 if c divides a and returning 0 if c does not divide a.*

Proof. The correctness of ORTPREM and PMULTX and CMULTX is assumed from (3.1) and (4.1). Let LENGTH$(a) = n \geq m =$ LENGTH(c). In step 2 of OPDIV, if $n > m$, PMULTX computes $x^i * c$ for $1 \leq i \leq n - m$ and in step 3 ORTPREM computes the coefficients of the remainder when a is divided by $x^{n-m} * c$. If $n = m$, the remainder when a is divided by c is calculated, skipping step 2. Steps 4 to 6 checks if this remainder is 0 which means that a is a multiple of c and so the algorithm returns 1.

Else if this remainder is not equal to 0, then $k > 0$ and step 7 checks if degree(*remainder*) $\geq$ degree(c). If degree(*remainder*) $\geq$ degree(c), the process of dividing the remainder with $x^{\deg(remainder)-m}$ is repeated by going back to step 3. Again, when repeating ORTPREM, the result is either $k = 0$ or $k > 0$ since k is never less than 0. If $k > 0$, OPDIV will be repeated if degree(*remainder*) $\geq$ degree(c), each time reducing the degree of the remainder when the remainder of the previous step is divided by c. If $k = 0$, OPDIV returns 1. If $k > 0$ and degree(*remainder*) is not any more greater than degree(c), then degree(*remainder*) $<$ degree(c) which implies that the remainder is not a multiple of c and the algorithm returns 0. □

6. Conclusion

The ability to check on the correctness of a potential solution is a very essential step in ensuring the effectiveness of the modular algorithm [14]. The incorporation of OPDIV into MOPBGCD algorithm enables the detection of unlucky primes. Thus, the polynomials need not ever be converted to power series form while much of the properties of polynomials in the generalized form can be made applicable in performing basic polynomial operations such as addition, multiplication, division or GCD computations.

Acknowledgments

An appreciation is due Professor George Collins and RISC, Johannes Kepler University, Linz, Austria for granting permission to a copy of SACLIB. The research is funded by the short term research grant RMC vot 71565, Universiti Teknologi Malaysia.

References

1. N. Aris, A. A. Rahman. On the computation of the GCD of polynomials relative to an orthogonal basis. Technical Report LT/M Bil.1/2001, April 2001, Department of Mathematics, Faculty of Science, UTM, Malaysia.
2. S. Barnett. A companion matrix analogue for orthogonal polynomials. *Linear Algebra and its Applications* **12** (1975), 197–208.
3. S. Barnett, J. Maroulas. Greatest common divisor of generalized polynomial and polynomial matrices. *Linear Algebra and its Applications* **22** (1978), 195–210.
4. S. Barnett. *Polynomials and linear control systems.* Marcel Dekker Inc. New York and Basel, 1983.
5. S. Barnett, J. Maroulas. Polynomials with respect to a general basis, I Theory. *J. of Math. Analysis and Applications* **72** (1979), 177–194.

6. S. Barnett, J. Maroulas. Polynomials with respect to a general basis, II Applications. *J. of Math. Analysis and Applications* **72** (1979), 599–614.
7. S. Barnett, J. Maroulas Further results on the qualitative theory of generalized polynomial. *J. Inst. Maths Applics.* **23** (1979), 33–42.
8. I. Borosh, A. S. Fraenkel. On modular approaches to solving systems of linear equations with rational coefficients. *Math. Comp.* **20** (1966), 107–112.
9. G. E. Collins. The computing of the Euclidean algorithm. *SIAM J. Comput.* **3**(1) (1974), 1–10.
10. G. E., Collins, M. Mignotte, F. Winkler. Arithmetic in basic algebraic domains. *Computing, Suppl.* **4** (1982), Springer Verlag, 189–220.
11. G. E. Collins, *et al. SACLIB 1.1 User's Guide.* RICS-Linz Report Series Technical Report Number 93-19, Research Institute for symbolic Computation, Johannes Kepler University, A-4040 Linz, Austria, 1993.
12. G. E. Collins, M. J. Encarnacion. Efficient rational number reconstruction. *J. Symbolic Computation* **20** (1995), 287–297.
13. M. Lauer. Computing by homomorphic images. *Computing* Suppl. **4** (1982). In B. Buchberger, G. E. Collins, R. Loos in cooperation with R. Albrecht eds., *Computer Algebra, Symbolic and Algebraic Computation*, 139–168. Springer Verlag, Wien-New York, 1982.
14. A. A. Rahman, N. Aris. Alkhwarizmi Modulo Pembahagi Sepunya Terbesar Dua Polinomial Relatif terhadap Asas Berortogon. Simposium Kebangsaan Matematik ke–9, UKM, Bangi, Malaysia, 18–20 Julai 2001.
15. C. M. Rubald. *Algorithms for Polynomials over a Real Algebraic Number Field.* Ph.D. Thesis, University of Wisconsin, 1973.

ON ONE PROPERTY OF HURWITZ DETERMINANTS

LARISA A. BURLAKOVA

Institute of Systems Dynamics and Control Theory,
Siberian Branch, Russian Academy of Sciences
134, Lermontov str., Irkutsk, 664033, Russia
E-mail: irteg@icc.ru

The paper discusses a conjecture about a special form of representation of Hurwitz determinants. The conjecture has been verified on the $n = 5, \ldots, 11$ degree polynomials. The property of the determinants obtained has allowed us to soften the well-known sufficient condition of the Hurwitzean property for the polynomial of an arbitrary degree n.

1. Introduction

Lyapunov's theorems on stability in the first approximation are widely applied in investigations of stability and stabilization of motions of interconnected mechanical systems. In such cases a linear differential equation with the matrix, whose elements are dependent on the parameters of the initial system, represents the object of investigation. The parametric analysis is complicated by virtue of the necessity to compute and analyze high-order Routh-Hurwitz determinants for the characteristic polynomials of differential equations of motion.

It is expedient to recall that for the polynomial

$$\varphi(\lambda) = a_0\lambda^n + a_1\lambda^{n-1} + \cdots + a_{n-1}\lambda + a_n, \tag{1.1}$$

the Routh-Hurwitz criterion below (Eqns. (1.2, 1.3)), under the assumption that $a_0 > 0$, gives both necessary and sufficient conditions of the fact that all the roots of the polynomial (1.1) have negative real parts [5].

$$\Delta_{n,1} = a_1 > 0; \quad \Delta_{n,2} = \begin{vmatrix} a_1 & a_3 \\ a_0 & a_2 \end{vmatrix} > 0; \quad \Delta_{n,3} = \begin{vmatrix} a_1 & a_3 & a_5 \\ a_0 & a_2 & a_4 \\ 0 & a_1 & a_3 \end{vmatrix} > 0; \quad \ldots \tag{1.2}$$

$$\Delta_{n,n} = \Delta_n = \begin{vmatrix} a_1 & a_3 & a_5 & a_7 & \cdots & 0 \\ a_0 & a_2 & a_4 & a_6 & \cdots & 0 \\ 0 & a_1 & a_3 & a_5 & \cdots & 0 \\ 0 & a_0 & a_2 & a_4 & \cdots & 0 \\ \vdots & \vdots & \vdots & \vdots & & \vdots \\ 0 & 0 & 0 & 0 & \cdots & a_n \end{vmatrix} = a_n \Delta_{n,n-1} > 0 \qquad (1.3)$$

A polynomial satisfying this property is called Hurwitzean, and the linear differential system, for which this polynomial is the characteristic one, possesses the property of asymptotic stability.

The Routh-Hurwitz criterion is one of several alternative formulations of the criterion, and is frequently used in problems of stability, when the characteristic polynomial coefficients a_i are functions of the parameters of the system under scrutiny. If the polynomial coefficients are given numerically then it is preferable to use other schemes to verify the Hurwitzean property [1,2]. There are many publications concerning the classical problem of localization of polynomial roots. This problem is quite important for applications, and it has been an object of investigations for many authors (see, for example, [7]).

The paper [6] suggests a conjecture, and in [3] a theorem was proved showing that the Routh-Hurwitz conditions for the polynomial (1.1) hold when

$$a_i > 0 \ , \ (i = 0, \ldots, n), \qquad (1.4)$$

$$1 > \frac{a_0 a_3}{a_1 a_2} + \frac{a_1 a_4}{a_2 a_3} + \cdots + \frac{a_{n-3} a_n}{a_{n-2} a_{n-1}}. \qquad (1.5)$$

In accordance with this theorem, (1.4)–(1.5) are sufficient conditions for the polynomial (1.1) to be Hurwitzean. Furthermore, for the 3rd and 4th degree polynomials these conditions are also necessary. The simple condition (1.5) allows one to avoid the need to investigate the positiveness of the high-order determinants. But it is important to know the degree of proximity of the sufficient condition to the necessary one. This made us reconsider the conditions in [6]. That was especially reasonable due to the fact that presently one has efficient aids of computer algebra systems available. We managed to obtain a recurrent relation with the aid of which analytical verification of rather high-degree polynomials (1.1) has been conducted. As a result, for the polynomials of degrees $n = 5, \ldots, 11$, analytical expressions, which enables the estimation of the proximity of the sufficient condition (1.5) to the necessary one, have been obtained.

2. Conjecture on the Representation of Hurwitz Determinants

Let us from now on assume that the condition (1.4) holds, that is, in the polynomial (1.1) all the coefficients are positive. Introduce the notation:

$$b_n = 1 - \left(\frac{a_0 a_3}{a_1 a_2} + \frac{a_1 a_4}{a_2 a_3} + \cdots + \frac{a_{n-3} a_n}{a_{n-2} a_{n-1}}\right),$$
$$c_n = (a_1 a_2 \cdots a_{n-1}) b_n.$$

Note that

$$c_n = a_{n-1} c_{n-1} - \nu_n, \tag{2.1}$$

where $\nu_n = a_1 \cdots a_{n-4} a_{n-3}^2 a_n$ for $n > 4$. Thus c_{n-1} is a formal denotation of c_* for the polynomial, whose maximum degree is one less than the degree n of the polynomial under scrutiny.

Results from computational experiments on high-degree polynomials ($n > 4$) suggest the following conjecture:

Conjecture 2.2. *The main diagonal minors of the Hurwitz determinant for the $n \geq 5$ degree polynomial may be represented in the form:*

$$\Delta_{n,n-1} = m_{n,n-1} + c_n f_n, \quad \Delta_{n,i} = m_{n,i} + c_n f_{n,i}/(a_{i+1} \cdots a_{n-1}) \tag{2.3}$$

where $f_n > 0$, $m_n > 0$, $f_{n,i} > 0$, $m_{n,i} > 0$ $(i = 2, \ldots, n-2;\ n = 5, \ldots)$.

Let us show how the expressions m_n, f_n, $m_{n,i}$, $f_{n,i}$ can be constructed.

The Recurrent Relation

The structure of the Hurwitz determinants is such that

$$\Delta_{n,i} = \Delta_{n-1,i} + a_n \eta_{n,i} \quad (i = 2, \ldots, n-1), \tag{2.4}$$

where $\Delta_{n-1,i}$ is the main diagonal i-order minor of the Hurwitz determinant for the $n-1$ degree polynomial, and consequently, does not contain any terms with a_n.

Suppose now $n > 5$, and for the $n-1$ degree equation, all the Hurwitz determinants are represented in the form (2.3). Consequently, in this case, the functions f_{n-1}, m_{n-1}, $f_{n-1,i}$, $m_{n-1,i}$, $(i = 2, \ldots, n-2)$, can be determined. For $n = 4$ we have $f_4 = 1$, $m_4 = 0$ (see Example 3.1 for $n = 4$).

For the Hurwitz determinant $\Delta_{n,n-1}$, we define the function

$$f_n = f_{n-1} + g_n - g_{n-1} + h_n \tag{2.5}$$

and for the Hurwitz determinants $\Delta_{n,i}$, we define

$$f_{n,i} = f_{n-1,i} + g_{n,i} - g_{n-1,i} + h_{n,i} \qquad (i = 2, \ldots, n-2). \tag{2.6}$$

Here

$$g_n = \left(\Delta_{n,n-1}/(a_1 \cdots a_{n-1}) - b_n\right)^+, \tag{2.7}$$

$$g_{n,i} = \left(\Delta_{n,i}/(a_1 \cdots a_i) - b_n\right)^+, \quad (i = 2, \ldots, n-2) \tag{2.8}$$

where the superscript + notation in $(\ )^+$ means that all the negative terms in the expression ( ) have been removed. The terms $h_n, h_{n,i}$ will be defined shortly.

In accordance with the assumption, the first term in the relation (2.4) has been reduced to the form (2.3) involving c_{n-1}. Now we only have to perform the transformation of the terms containing a_n in $\Delta_{n,i}$ to the desired form. So, when taking account of the relationships (2.1), (2.4), we determine the values of $h_n > 0$ and $h_{n,i} > 0$ from the conditions of absence of negative terms in the explicit expressions:

$$\mu_n = \nu_n f_{n-1} + a_n \eta_{n,n-1} - c_n(g_n - g_{n-1} + h_{n,n-1}), \tag{2.9}$$

$$\mu_{n,i} = \nu_n f_{n-1,i}/(a_{i+1} \cdots a_{n-1}) + a_n \eta_{n,i} \tag{2.10}$$
$$-c_n(g_{n,i} - g_{n-1,i} + h_{n,i})/(a_{i+1} \cdots a_{n-1}), \ (i = 2, \ldots, n-2).$$

The number of elements, which additively enter in the solution is not less than the number of negative terms in the initial expression for $\mu_{n,i}$. For example, for the 6-th degree polynomial we have managed to construct a solution for h_6, which contains one term; for the 7-th degree polynomial, $h_{7,6}$ contains eight terms, $h_{7,5}$ contains four terms; for the 8-th degree polynomial $h_{8,7}$ contains 12 terms $h_{8,6}$ contains five terms; and for the 9-th degree polynomial, the solution $h_{9,8}$ includes 80 terms (without grouping like terms). It is possible to employ different algorithms for removing negative terms. In the examples below we intend to show one such algorithm.

Having constructed $f_n > 0$ and $f_{n,i} > 0$, we compute $m_{n,i}$ by the formulas:

$$m_{n,n-1} = \Delta_{n,n-1} - c_n f_n,$$
$$m_{n,i} = \Delta_{n,i} - c_n f_{n,i}/(a_{i+1} \cdots a_{n-1}) \qquad (i = 2, \ldots, n-2), \tag{2.11}$$

and make certain that these expressions are positive because there are no terms having the negative sign in the explicated expressions.

That is how analytical computations have been conducted for the $n = 5$ to $n = 11$ (inclusive) degree polynomials. So, the validity of the conjecture for these polynomials has been demonstrated. There are no principal restrictions imposed on the degree of the polynomial. But such restriction may appear by virtue of limitation of computers. For example, for the polynomials, with symbolic coefficients of degree $n > 14$, complete expressions of the Hurwitz determinants $\Delta_{n,i}$, $f_{n,i}$ and $m_{n,i}$ are very bulky.

Remark 2.12. If there is a necessity to perform the decomposition (2.3) for the given degree n^*, it is possible not to use the recurrent formula but to immediately construct the Lagrange functions

$$m_{n^*-1} = \Delta_{n^*,n^*-1} - c_{n^*}(1 + \ell_{n^*}),$$
$$m_{n^*,i} = \Delta_{n^*,i} - c_{n^*}(1 + \ell_{n^*,i})/(a_{i+1} \cdots a_{n^*-1}) \qquad (i = 2, \dots, n^* - 2),$$

where ℓ_{n^*}, $\ell_{n^*,i}$ must be determined from the condition of absence of the terms with negative signs in m_{n^*-1}, $m_{n^*,i}$. But in this case, the computational process will be substantially longer.

3. Results of Computational Experiments

Example 3.1. (**n** $=$ **4**). For the 4-th degree polynomial

$$\varphi_4 = a_0\lambda^4 + a_1\lambda^3 + a_2\lambda^2 + a_3\lambda + a_4$$

we have

$$c_4 = a_1a_2a_3 - a_0a_3^2 - a_1^2a_4; \quad \Delta_4 = \begin{vmatrix} a_1 & a_3 & 0 & 0 \\ a_0 & a_2 & a_4 & 0 \\ 0 & a_1 & a_3 & 0 \\ 0 & a_0 & a_2 & a_4 \end{vmatrix} = a_4\Delta_{4,3};$$

$$\Delta_{4,3} = \begin{vmatrix} a_1 & a_3 & 0 \\ a_0 & a_2 & a_4 \\ 0 & a_1 & a_3 \end{vmatrix} = a_1a_2a_3 - a_0a_3^2 - a_1^2a_4 = c_4;$$

$$\Delta_{4,2} = a_1a_2 - a_0a_3 = c_3 = a_1^2a_4/a_3 + c_4/a_3.$$

Consequently, $m_4 = 0$, $f_4 = f_{4,2} = 1, m_{4,3} = a_1^2a_4/a_3$. These values will be used in the computations for the $n = 5$ degree polynomial. If $c_4 > 0$ then $\Delta_4 > 0$, $\Delta_{4,3} > 0$, $\Delta_{4,2} > 0$. When the inequality $c_4 > 0$ fails to hold, the Routh-Hurwitz criterion is violated, and hence the sufficient condition (1.5) is also the necessary condition for the Hurwitz property of the 4-th degree polynomial with positive coefficients [6].

Example 3.2. (**n** $=$ **5**). For the 5-th degree polynomial

$$\varphi(\lambda) = a_0\lambda^5 + a_1\lambda^4 + a_2\lambda^3 + a_3\lambda^2 + a_4\lambda + a_5$$

we have

$$b_5 = 1 - \left(\frac{a_0a_3}{a_1a_2} + \frac{a_1a_4}{a_2a_3} + \frac{a_2a_5}{a_3a_4}\right);$$
$$c_5 = a_1a_2a_3a_4 - a_0a_3^2a_4 - a_1^2a_4^2 - a_1a_2^2a_5;$$
$$\nu_5 = a_1a_2^2a_5;$$
$$\Delta_5 = a_5\Delta_{5,4} = a_5\big(a_1a_2a_3a_4 - a_0a_3^2a_4 - a_1^2a_4^2 - a_1a_2^2a_5 + a_0a_2a_3a_5 + 2a_0a_1a_4a_5 - a_0^2a_5^2\big);$$
$$\Delta_{5,4} = \Delta_{4,4} + a_5\eta_{5,4},$$

where $\eta_{5,4} = -a_1a_2^2 + a_0a_2a_3 + 2a_0a_1a_4 - a_0^2a_5$.

Now let us compute g_5 (2.7). To this end let us construct the bundle

$$\frac{\Delta_{5,4}}{a_1a_2a_3a_4} - b_5 = \frac{2a_0a_5}{a_2a_3} + \frac{a_0a_5}{a_1a_4} - \frac{a_0^2a_5^2}{a_1a_2a_3a_4}$$

in which the terms having the negative sign will be removed:

$$g_5 = \frac{2a_0a_5}{a_2a_3} + \frac{a_0a_5}{a_1a_4}.$$

Such computations in the previous case have given $g_4 = g_{4,3} = 0$. Hence, in accordance with (2.5) $f_5 = f_4 + g_5 + h_5$, where $h_5 > 0$ is obligatory dependent on a_5 and is determined from the condition of absence of negative terms in (2.9):

$$\begin{aligned}\mu_5 &= \nu_5 f_4 + a_5\eta_{5,4} - c_5(g_5 - g_4 + h_{5,4})\\ &= -h_{5,4}a_1a_2a_3a_4 + h_{5,4}a_0a_3^2a_4 + h_{5,4}a_1^2a_4^2 + h_{5,4}a_1a_2^2a_5 + \frac{a_0^2a_3^2a_5}{a_1}\\ &\quad + a_0a_1a_4a_5 + \frac{2a_0^2a_3a_4a_5}{a_2} + \frac{2a_0a_1^2a_4^2a_5}{a_2a_3} - a_0^2a_5^2 + \frac{2a_0a_1a_2a_5^2}{a_3} + \frac{a_0a_2^2a_5^2}{a_4}.\end{aligned}$$

This bundle is positive when $h_{5,4}$ is chosen to be such that:

$$h_{5,4} = \frac{a_0a_5^2}{a_3^2a_4}.$$

This is the solution of the equation $h_{5,4}a_0a_3^2a_4 - a_0^2a_5^2 = 0$. Having substituted this value $h_{5,4}$ into μ_5, we obtain:

$$\begin{aligned}\mu_5 = {}& \frac{a_0^2a_3^2a_5}{a_1} + a_0a_1a_4a_5 + \frac{2a_0^2a_3a_4a_5}{a_2} + \frac{2a_0a_1^2a_4^2a_5}{a_2a_3}\\ &+ \frac{a_0a_1a_2a_5^2}{a_3} + \frac{a_0a_2^2a_5^2}{a_4} + \frac{a_0a_1^2a_4a_5^2}{a_3^2} + \frac{a_0a_1a_2^2a_5^3}{a_3^2a_4}.\end{aligned}$$

Remark 3.3. When constructing equations, it is advisable to avoid variants that give solutions for $h_{n,i}$ containing a_0 or a_n in the denominator.

Finally, we have

$$f_5 = 1 + \frac{2a_0a_5}{a_2a_3} + \frac{a_0a_5}{a_1a_4} + \frac{a_0a_5^2}{a_3^2a_4} > 0.$$

Using the formula (2.11), for this value of f_5 we obtain $m_{5,4}$:

$$m_{5,4} = \frac{1}{a_1a_2a_3^2a_4}\big(a_0a_5\big(a_0a_2a_3^4a_4 + a_1^2a_2a_3^2a_4^2 + 2a_0a_1a_3^3a_4^2$$
$$2a_1^3a_3a_4^3 + a_1a_2^3a_3^2a_5 + a_1^2a_2^2a_3a_4a_5 + a_1^3a_2a_4^2a_5 + a_1^2a_2^3a_5^2\big)\big) > 0.$$

Consequently, the following representation of (2.3) is obtained:

$$\Delta_{5,4} = m_{5,4} + c_5f_5, \quad m_{5,4} > 0, \; f_5 > 0.$$

Let us find the representation of (2.3) for the 3-rd order determinant.

$$\Delta_{5,3} = a_1a_2a_3 - a_0a_3^2 - a_1^2a_4 + a_0a_1a_5.$$

Construct the function (2.6):

$$f_{5,3} = f_{4,3} + g_{5,3} - g_{4,3} + h_{5,3},$$

where $f_{4,3} = 1$, $g_{4,3} = 0$ (what follows from the example 3.1 with n=4). Due to the formula (2.8)

$$g_{5,3} = \big(\Delta_{5,3}/(a_1a_2a_3) - b_5\big)^+ = \frac{a_0a_5}{a_2a_3} + \frac{a_2a_5}{a_3a_4}.$$

The value of $h_{5,3}$ will be found from the condition that there are no negative terms in (2.10):

$$\mu_{5,3} = \nu_5 f_{4,3}/a_4 + a_5\eta_{5,3} - c_5(g_{5,3} - g_{4,3} + h_{5,3})/a_4.$$

On account of the fact that $\eta_{5,3} = a_0a_1$, we have

$$\mu_{5,3} = -h_{5,3}a_1a_2a_3 + h_{5,3}a_0a_3^2 + h_{5,3}a_1^2a_4 + \frac{a_1^2a_2a_5}{a_3} + \frac{a_0^2a_3a_5}{a_2}$$
$$+\frac{h_{5,3}a_1a_2^2a_5}{a_4} + \frac{a_0a_2a_3a_5}{a_4} + \frac{a_0a_1^2a_4a_5}{a_2a_3} + \frac{a_1a_2^3a_5^2}{a_3a_4^2} + \frac{a_0a_1a_2a_5^2}{a_3a_4}.$$

Hence, obviously, it is sufficient to choose $h_{5,3} = 0$. After all, we obtain

$$f_{5,3} = 1 + \frac{a_0a_5}{a_2a_3} + \frac{a_2a_5}{a_3a_4} > 0.$$

Now compute

$$m_{5,3} = \Delta_{5,3} - c_5f_{5,3}/a_4 = \frac{(a_2^2 + a_0a_4)a_5(a_0a_3^2a_4 + a_1^2a_4^2 + a_1a_2^2a_5)}{a_2a_3a_4^2} > 0.$$

Hence we obtain the representation (2.3):

$$\Delta_{5,3} = m_{5,3} + c_5 f_{5,3}/a_4 \quad m_{5,3} > 0, \; f_{5,3} > 0.$$

Having performed similar computations for the second order determinant, we have, with $f_{5,2} = 1$,

$$\Delta_{5,2} = \frac{a_1(a_1a_4^2 + a_2^2a_5) + c_5}{a_3a_4}.$$

As obvious from the representation (2.3), all the main diagonal minors of the Hurwitz determinant for the 5-th degree polynomial will remain positive when $c_5 = 0$. If we put

$$c_5 > \max\{-a_1(a_1a_4^2 + a_2^2a_5), -(m_{53}a_4/f_{53}), -m_{54}/f_{53}\},$$

then the domain of values for the parameters under which the system is asymptotically stable will be wider.

Example 3.4. (**n = 6**). For the polynomial

$$\varphi_6 = a_0\lambda^6 + a_1\lambda^5 + a_2\lambda^4 a_3\lambda^3 + a_4\lambda^2 + a_5\lambda + c_6$$

we have

$$c_6 = a_1a_2a_3a_4a_5 - a_0a_3^2a_4a_5 - a_1^2a_4^2a_5 - a_1a_2^2a_5^2 - a_1a_2a_3^2a_6;$$

$$f_6 = 1 + \frac{2a_0a_5}{a_2a_3} + \frac{a_0a_5}{a_1a_4} + \frac{a_0a_5^2}{a_3^2a_4} + \frac{2a_1a_6}{a_3a_4} + \frac{a_1a_6}{a_2a_5}$$
$$+ \frac{a_0a_3^2a_6}{a_1a_2a_4a_5} + \frac{a_1a_6^2}{a_4^2a_5} > 0;$$

$$\Delta_{65} = m_{65} + c_6 f_6;$$

$$m_{65} = \frac{1}{a_1a_2a_3^2a_4^2a_5}(a_0^2a_2a_3^4a_4^2a_5^3 + a_0a_1^2a_2a_3^2a_4^3a_5^3 + 2a_0^2a_1a_3^3a_4^3a_5^3$$
$$+2a_0a_1^3a_3a_4^4a_5^3 + a_0a_1a_2^3a_3^2a_4a_5^4 + a_0a_1^2a_2^2a_3a_4^2a_5^4 + a_0a_1^3a_2a_4^3a_5^4$$
$$+a_0a_1^2a_2^3a_4a_5^5 + a_0^2a_3^6a_4^2a_5a_6 + 2a_0a_1^2a_3^4a_4^3a_5a_6 + a_1^4a_3^2a_4^4a_5a_6$$
$$+2a_0a_1a_2^2a_3^4a_4a_5^2a_6 + a_1^3a_2^2a_3^2a_4^2a_5^2a_6 + a_0a_1^2a_2a_3^3a_4^2a_5^2a_6$$
$$+2a_1^4a_2a_3a_4^3a_5^2a_6 + 2a_1^3a_2^3a_3a_4a_5^3a_6 + a_0a_1^2a_2^2a_3^2a_4a_5^3a_6$$
$$+ + a_0a_1a_2a_3^6a_4a_6^2a_1^3a_2a_3^4a_4^2a_6^2 + a_1^3a_2^2a_3^3a_4a_5a_6^2$$
$$+a_0a_1^2a_2a_3^4a_4a_5a_6^2 + a_1^3a_2^3a_3^2a_5^2a_6^2 + a_1^3a_2^2a_3^4a_6^3) > 0;$$

$$f_{6,4} = 1 + \frac{2a_0a_5}{a_2a_3} + \frac{a_0a_5}{a_1a_4} + \frac{a_0a_5^2}{a_3^2a_4} + \frac{a_1a_6}{a_3a_4} + \frac{a_3a_6}{a_4a_5} > 0;$$

$$\begin{aligned} m_{6,4} = {} & \frac{1}{a_1 a_2 a_3^2 a_4 a_5^2}(a_0^2 a_2 a_3^4 a_4 a_5^3 + a_0 a_1^2 a_2 a_3^2 a_4^2 a_5^3 + 2a_0^2 a_1 a_3^3 a_4^2 a_5^3 \\ & + 2a_0 a_1^3 a_3 a_4^3 a_5^3 + a_0 a_1 a_2^3 a_3^2 a_5^4 + a_0 a_1^2 a_2^2 a_3 a_4 a_5^4 + a_0 a_1^3 a_2 a_4^2 a_5^4 \\ & + a_0 a_1^2 a_2^3 a_5^5 + a_0 a_1 a_2 a_3^5 a_4 a_5 a_6 + a_1^3 a_2 a_3^3 a_4^2 a_5 a_6 + a_1^2 a_2^3 a_3^3 a_5^2 a_6 \\ & + a_0 a_1 a_2^2 a_3^4 a_5^2 a_6 + 2a_0 a_1^2 a_2 a_3^3 a_4 a_5^2 a_6 + a_1^4 a_2 a_3 a_4^2 a_5^2 a_6 \\ & + a_1^3 a_2^3 a_3 a_5^3 a_6 + a_0 a_1^2 a_2^2 a_3^2 a_5^3 a_6 + a_1^2 a_2^2 a_3^5 a_6^2 + a_1^3 a_2^2 a_3^3 a_5 a_6^2) > 0; \end{aligned}$$

$$\Delta_{6,4} = m_{6,4} + c_6 f_{6,4}/a_5;$$

$$f_{6,3} = f_{5,3} > 0; \qquad f_{6,2} = f_{5,2} = 1.$$

4. Theorem on the Sufficient Condition

As obvious from the representation (2.3) and from above examples, at least for $n = 5$ ($n = 6$) all the main diagonal minors of the Hurwitz determinant are positive when the sufficient condition (1.5) holds with the boundary: $c_n \geq 0$. This gives the possibility to prove the following theorem:

Theorem 4.1. *If*

$$b_n = 1 - \left(\frac{a_0 a_3}{a_1 a_2} + \frac{a_1 a_4}{a_2 a_3} + \cdots + \frac{a_{n-3} a_n}{a_{n-2} a_{n-1}}\right) \geq 0, \tag{4.2}$$

then the polynomial (1.1) with positive coefficients is Hurwitzean.

The proof of the theorem with the condition $b_n > 0$ by the method of mathematical induction is given in [3]. We are going to repeat this proof, while introducing some necessary additions bound up with the changes in the theorem's conditions.

Proof. The theorem is valid for the $n = 5$ degree polynomial. Suppose that the theorem is valid for all the $n - 1 > 5$ degree polynomials. Let us show that the n-th degree polynomial (1.1), which possesses the property (4.2), is also Hurwitzean.

Likewise in [3], let us construct the following additional polynomial

$$\psi(\lambda) = p_0 \lambda^{n-1} + p_1 \lambda^{n-2} + \cdots + p_{n-2}\lambda + p_{n-1}, \tag{4.3}$$

whose coefficients are determined from the formulas:

$$p_0 = a_1; \quad p_1 = a_2 - \frac{a_0 a_3}{a_1}; \quad p_2 = a_3; \quad p_3 = a_4 - \frac{a_0 a_5}{a_1};$$

$$p_4 = a_5 \quad p_5 = a_6 - \frac{a_0 a_7}{a_1}; \quad p_6 = a_7 \quad \ldots.$$

Due to the condition (4.2), all the coefficients $p_i > 0$. As obvious from the alternative formulation of the Routh-Hurwitz theorem [4], the polynomials (1.1) and (4.3) are simultaneously Hurwitzean (or simultaneously unstable).

Remark 4.4. If at least one of the coefficients $p_i < 0$ then the polynomial (1.1) is unstable. Consequently, the condition

$$\frac{a_0 a_\ell}{a_1 a_{\ell-1}} > 1, \quad \ell = \begin{cases} 3, 5, \ldots, n, \text{ when } n = 2s+1 \\ 3, 5, \ldots, n-1, \text{ when } n = 2s \end{cases}$$

for any ℓ is sufficient for the instability of the polynomial.

Let us show that for the polynomial (4.3)

$$b^*_{n-1} = 1 - \left(\frac{p_0 p_3}{p_1 p_2} + \frac{p_1 p_4}{p_2 p_3} + \cdots + \frac{p_{n-4} p_{n-1}}{p_{n-3} p_{n-2}}\right) \geq 0$$

due to the condition (4.2). Likewise in [3], introduce the notation:

$$q_i = \frac{a_i a_{i+3}}{a_{i+1} a_{i+2}} \quad (i = 0, 1, \ldots, n-3);$$

$$d_1 = \frac{1 - q_0 q_2}{1 - q_0}; \; d_2 = \frac{1 - q_0 q_2 q_4}{1 - q_0 q_2}; \; d_3 = \frac{1 - q_0 q_2 q_4 q_6}{1 - q_0 q_2 q_4}; \ldots$$

Hence

$$b_n = 1 - (q_0 + q_1 + q_2 + q_3 + q_4 + \cdots + q_{n-3});$$

$$b^*_{n-1} = 1 - (q_1 d_1 + \frac{q_2}{d_1} + q_3 d_2 + \frac{q_4}{d_2} + \cdots).$$

Due to the condition (4.2), the inequalities

$$\begin{aligned} q_i &< 1 & (i &= 0, 1, \ldots, n-3); \\ d_j &> 1 & (j &= 1, \ldots, [n/2 - 1]); \\ (1 - q_0 q_2)^{-1} &> d_j & (j &= 2, \ldots, [n/2 - 1]) \end{aligned}$$

hold (where $[k]$ is the integer part of k). So, it is easy to trace the following chain of relations:

$$\begin{aligned} b^*_{n-1} &= d_1 b_n + 1 + (q_2 + q_4 + \cdots)(d_1 - 1) - d_1 + q_0 d_1 \\ &\quad - q_3(d_2 - d_1) - q_5(d_3 - d_1) + \cdots \\ &\geq 1 - d_1 + q_0 d_1 - q_3(d_2 - d_1) - q_5(d_3 - d_1) + \cdots \\ &\geq 1 - d_1 + q_0 d_1 - (q_3 + q_5 + \cdots)\left(\frac{1}{(1 - q_0 q_2)} - d_1\right) \\ &\geq 1 - d_1 + q_0 d_1 - \left(\frac{1}{(1 - q_0 q_2)} - d_1\right) \end{aligned}$$

$$
\begin{aligned}
&= \frac{q_0}{(1-q_0)(1-q_0q_2)}\big((q_2b_n + \quad (1-q_2)^2 + q_0^2q_2^2 \\
&\qquad +q_2(q_1+q_3+q_4+q_5+q_6+\cdots))\big) \\
&> 0.
\end{aligned}
$$

Consequently, in accordance with the assumption, the polynomial (4.3) (as an $n-1$ degree polynomial) is Hurwitzean. And hence the polynomial (1.1), which has positive coefficients and possesses the property (4.2), is also Hurwitzean. □

5. Conclusion

A conjecture of special representation (2.3) of the Hurwitz determinant and of its main diagonal minors has been formulated. This conjecture has been verified for the polynomials of degrees of $n = 5, \ldots, 11$. These values correspond to those values of the degrees for which practical operating with the polynomials having symbolic coefficients is possible. For the values of degrees indicated above we have obtained analytical expressions for the components f_n, $f_{n,i}$, m_n, $m_{n,i}$ of the representation (the expressions are given for $n = 5$, $n = 6$ in Examples 3.2 and 3.4). The result obtained may be verified with the aid of any computer algebra system.

As follows from the above examples for $n = 5$, $n = 6$, the sufficient condition (4.2) $b_n \geq 0$ $(n \geq 5)$, for which all the Hurwitz determinants for the polynomial having the coefficients $a_i > 0$, are positive, are "rough" (this condition is far from the necessary one). But application of this inequality in problems of parametric analysis of asymptotic stability is substantiated by its relative simplicity. "Softer" stability conditions can be obtained from the representation (2.3). The representation (2.3) suggests a direct dependence of Hurwitz determinants on c_n. This allows one to estimate their values for various c_n, and consequently, to estimate their proximity to zero.

References

1. D. K. Anand, R. B. Zmood. *Introduction to Control Systems.* 3rd ed. Butterworth-Heinemann, 1995.
2. E. Kaltofen, G. Villard. Computing the sign or the value of the determinant of an integer matrix. A complexity survey. *J. Computational Applied Math.* (2002). To appear, 17 pages. *Special issue on Congrés International Algébre Linéaire et Arithmétique: Calcul Numérique, Symbolique et Paralléle* (*Rabat, Morocco, May 2001*). Available at `http://www4.ncsu.edu:8030/~kaltofen/bibliography/index.html`

3. A. F. Kleptsyn, A sufficient condition for stability of a polynomial. *Avtomatika i Telemehanika* **10** (1984), 175–176. English transl. in *J. of Automation and Remote Control* **10** (1984).
4. G. A. Korn, T. A. Korn. *Mathematical Handbook for Scientists and Engineers.* McGraw-Hill, New York-Toronto-London, 1961.
5. P. Lankaster. *Theory of Matrices.* Academic Press, New York-London, 1969.
6. V. V. Maslennikov. A hypothesis on existence of a simple analytical sufficient condition for stability. *Avtomatika i Telemehanika* **2** (1984), 160–161. English transl. in *J. of Automation and Remote Control* **2** (1984).
7. A. I. Perov. New conditions for the stability of linear systems with constant coefficients. (Russian) *Avtomatika i Telemehanika* **2** (2002), 22–33. English transl. in *J. of Automation and Remote Control* **63**(2) (2002), 189–199.

INTERVAL PARAMETRIZATION OF PLANAR ALGEBRAIC CURVES

FALAI CHEN,[†] LIN DENG

Department of Mathematics
University of Science and Technology of China
Hefei, Anhui 230026, People's Republic of China
[†]*Email: chenfl@ustc.edu.cn*

In this paper, we propose a new concept called *interval parametrization* of algebraic curves, that is, we find an interval Bézier curve which bounds a given algebraic curve such that the bound is as tight as possible. An algorithm is presented to compute the interval parametrization of algebraic curves. The algorithm starts by finding the algebraic conditions that the interval Bézier curve bounds the algebraic curve, and then converts the problem into solving a non-linear programming problem with four variables. The non-linear programming is then approximately simplified to solving two non-linear programming problems with two variables. Some examples are provided to demonstrate the algorithm.

Key words: algebraic curve, parametrization, interval arithmetic, interval Bézier curve

1. Introduction

Parametric curves/surfaces and algebraic curves/surfaces are two common types of representations of geometric objects in Computer Aided Geometric Design. Both of these representations have their own advantages and disadvantages.

For example, it is relatively easy to generate points on parametric curves/surfaces, thus efficient algorithms exist to render parametric curves/surfaces; on the other hand, with algebraic curves/surfaces, it is convenient to determine if a point is on, inside or outside a solid using an implicit representation. Thus it is valuable to have both representations at the same time.

It is well known from classic algebraic geometry that every parametric curve has an implicit representation, but not all algebraic curves admit parametric representations, and only algebraic curves with genus zero have rational parametrizations. Thus, for an algebraic curve whose genus is not zero, it is of practical interest to find an approximate parametrization. So far, few papers have focused on this topic ([9,13,7]). One problem with approximate parametrization is that it doesn't deal with the numerical gaps between the approximate parametric representation and the algebraic curve. Because of the gaps, the approximate solutions may be unreliable in geometrical computation and interrogation, and/or may make the geometry and topology of geometric objects inconsistent. To solve this problem, in this paper, we adopt the interval representation of geometric objects and put forward a new concept called *interval parametrization* of algebraic curves, that is, we find an interval parametric curve which bounds the given algebraic curve such that the bound is as tight as possible.

Interval representation of geometric objects that embodies a complete description of coefficient errors were proposed by S.P. Mudur *et al.* [5] and Sederberg *et al.* [12]. A recent paper [6] in this area suggests that such a representation greatly helps to increase the numerical stability in geometric computations, and thus to enhance the robustness in current CAD/CAM systems.

Although there are many works on interval polynomials in power form, few literatures have discussed the problem of bounding one type of curves with another type of interval polynomial curves. Sederberg *et al.* [10] and Lin *et al.* [3] presented methods to bound a parametric curve with a fat arc (a pair of arcs). One of the present authors [1] proposed algorithms to bound an interval Bézier curve with a lower degree interval Bézier curve. However, as far as the authors are aware, there is no similar work in the literature which discusses the problem of bounding an algebraic curve with an interval Bézier curve (or equivalently, a pair of parametric curves). In this paper, we present an efficient algorithm to solve the problem by solving two non-linear programming problems, each with two variables.

We organize the paper as follows. In the next section, some preliminary knowledge about interval arithmetic and interval Bézier curves is introduced. Then in Section 3, we present an efficient algorithm to find the interval Bézier representation of a given algebraic curve. The algorithm starts by finding an approximate Bézier curve which is served as the center of the interval Bézier curve, and then computes the width of the interval Bézier curve by solving two simple non-linear programming problems. Finally, we provide some examples to demonstrate the algorithm in Section 4 and conclude the paper in Section 5.

2. Interval Arithmetic and Interval Bézier Curves

In this section, we first briefly review the definitions of interval arithmetic and interval Bézier curves.

2.1. *Interval Arithmetic*

An *interval* $[a, b]$ is a set of real numbers defined by

$$[a, b] = \{x \mid a \leq x \leq b\}.$$

The *width* of $[a, b]$ is $w([a, b]) = b - a$. If $\mathbf{A} = [a, b]$ and $\mathbf{B} = [c, d]$ are two intervals, and $\circ \in \{+, -, \times, /\}$ is an operator, then $\mathbf{A} \circ \mathbf{B}$ is defined by

$$\mathbf{A} \circ \mathbf{B} = \{x \circ y \mid x \in \mathbf{A}, y \in \mathbf{B}\}.$$

More specifically [4],

$$\begin{aligned} [a, b] + [c, d] &= [a + c, b + d], \\ [a, b] - [c, d] &= [a - d, b - c], \\ [a, b] \times [c, d] &= [\min(ac, ad, bc, bd), \max(ac, ad, bc, bd)], \\ [a, b]/[c, d] &= [a, b] \times [1/d, 1/c], \qquad \text{if } 0 \notin [c, d]. \end{aligned} \tag{2.1}$$

It is easy to verify that addition and multiplication are commutative and associative, but that multiplication does not, in general, distribute over addition. For details, the reader is referred to [4].

2.2. *Interval Bézier Curves*

An *interval polynomial* is a polynomial with interval coefficients:

$$[p](t) = \sum_{k=0}^{m} [a_k, b_k] B_k^m(t), \qquad 0 \leq t \leq 1, \tag{2.2}$$

where $B_k^m(t) = \binom{m}{k} t^k (1 - t)^{m-k}$, $k = 0, 1, ..., m$ are the *Bernstein basis functions.* An interval polynomial can also be rewritten as

$$[p](t) := [p_{min}(t), p_{max}(t)], \tag{2.3}$$

where

$$p_{min}(t) = \sum_{k=0}^{m} a_k B_k^m(t), \qquad p_{max}(t) = \sum_{k=0}^{m} b_k B_k^m(t). \tag{2.4}$$

We refer to $p_{min}(t)$ and $p_{max}(t)$ as the *lower bound polynomial* and the *upper bound polynomial* respectively.

The *width* of an interval polynomial is defined as

$$w([p](t)) = ||p_{max}(t) - p_{min}(t)||, \tag{2.5}$$

where $||\cdot||$ is some standard norm such as $||\cdot||_2$.

An *interval Bézier curve* is a Bézier curve with interval control points:

$$[\mathbf{P}](t) = \sum_{k=0}^{m} [\mathbf{P}_k] B_k^m(t), \qquad 0 \le t \le 1, \tag{2.6}$$

where $[\mathbf{P}_k] = ([a_k, b_k], [c_k, d_k])$ are vector-valued intervals. A sample interval Bézier curve is illustrated in the following figure.

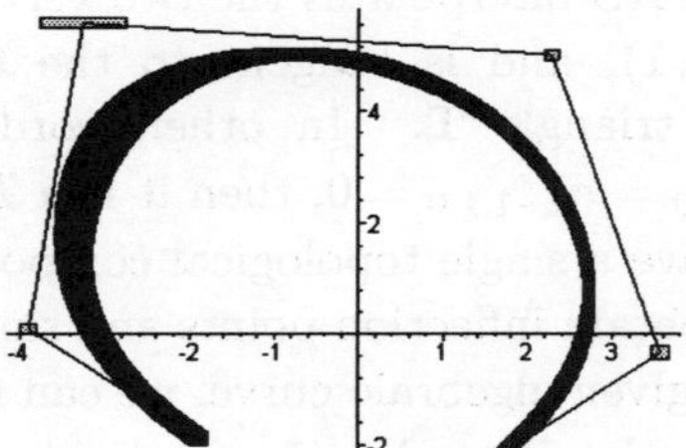

Figure 1. A sample interval Bézier curve

Interval Bézier curve (2.6) can be extended to a more general case where the control points $[\mathbf{P}_k]$, $k = 0, 1, \ldots, m$ are any closed areas in the plane, for example, triangles, circles and line segments. In the following, we will use interval Bézier curves whose control points are line segments.

3. Interval Parametrization of Algebraic Curves

Before we come to the interval parametrization problem of an algebraic curve, we begin by introducing the concept of piecewise algebraic curves.

3.1. *Piecewise Algebraic Curves*

A planar algebraic curve of degree n can be expressed as

$$f(x, y) := \sum_{i+j \le n} a_{ij} x^i y^j = 0. \tag{3.1}$$

In the above expression, the coefficients a_{ij} do not have any geometric significance. For example, if we change the value of one of the coefficients, we have no idea where the algebraic curve will go. To better control the algebraic curve, Sederberg [8] trims the algebraic curve using a triangle $\mathbf{T} = \triangle \mathbf{T}_1\mathbf{T}_2\mathbf{T}_3$ and defines the algebraic curve segment based on Bernstein-Bézier (B-B for short) representation:

$$f(\mathbf{P}) := \sum_{i+j+k=n} c_{ijk} \binom{n}{ijk} u^i v^j w^k = 0, \tag{3.2}$$

where (u, v, w) are barycentric coordinates of a point $\mathbf{P} = (x, y)$ with respect to the given triangle $\mathbf{T}$, and the c_{ijk} are called the Bézier ordinates of the algebraic curve. The algebraic curve segment lying inside the triangle $\mathbf{T}$ is called a *piecewise algebraic curve* or shortly *PAC*. A *PAC* is called a *TPAC* if the *PAC* interpolates the two vertices $(u, v, w) = (1, 0, 0)$ and $(u, v, w) = (0, 0, 1)$, and is tangent to the two sides $u = 0$ and $w = 0$ of the base triangle $\mathbf{T}$. In other words, if a *PAC* satisfies $c_{00n} = c_{0,1,n-1} = c_{n,0,0} = c_{n-1,1,0} = 0$, then it is a *TPAC*. In practice, we require a *TPAC* to have a single topological component in the triangle $\mathbf{T}$, and that it does not contain inflection points and singular points (except at the endpoints). For a given algebraic curve, we can first subdivide it into a series of PACs. For the details on how to subdivide an algebraic curve into PACs, the reader is referred to [9,11,2].

Fig. 2 illustrates an example of a *PAC* and a *TPAC*.

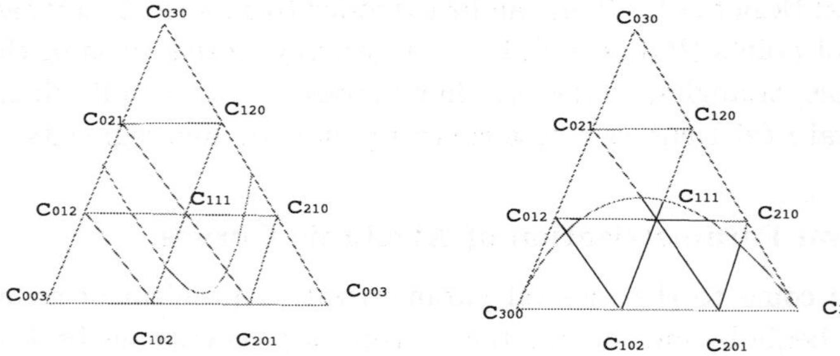

Figure 2. Left: a *PAC*; Right: a TPAC

3.2. *Interval Parametrization of TPCs*

The problem of *interval parametrization* of an algebraic curve can be stated as follows:

Problem 3.3. Given an algebraic curve segment (TPAC), find an interval Bézier curve $[\mathbf{Q}](t)$ such that $[\mathbf{Q}](t)$ bounds the given TPAC, and the width of $[\mathbf{Q}](t)$ is as small as possible.

To be precise, we have to find the conditions for which the interval Bézier curve bounds the given algebraic curve, and the criteria for which the bound is optimal.

3.2.1. *Bounding conditions*

Suppose the TPAC is defined as in (3.2) which interpolates $\mathbf{T}_1$ and $\mathbf{T}_3$, and is tangent to $\mathbf{T}_1\mathbf{T}_2$ and $\mathbf{T}_3\mathbf{T}_2$, respectively, where $\mathbf{T}_i$, $i = 1, 2, 3$ are the vertices of the base triangle $\mathbf{T}$. We will find a cubic interval Bézier curve

$$[\mathbf{Q}](t) = \sum_{i=0}^{3} [\mathbf{Q}_i] B_i^3(t), \quad 0 \le t \le 1 \tag{3.4}$$

to bound the TPAC (see Fig.3 for an illustration).

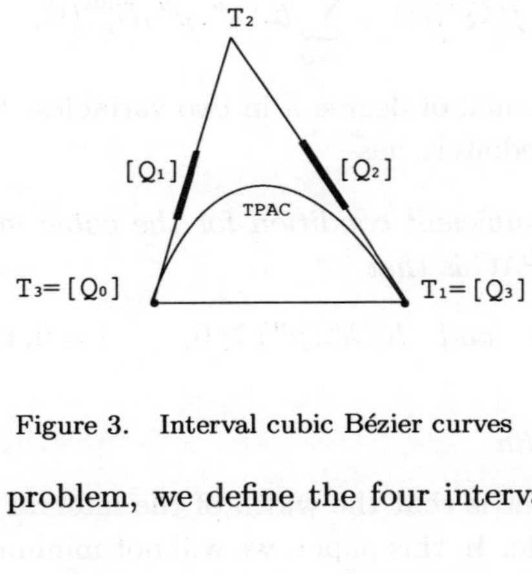

Figure 3. Interval cubic Bézier curves

For our specific problem, we define the four interval control points of $[\mathbf{Q}](t)$ as:

$$[\mathbf{Q}_0] = \mathbf{T}_3, \qquad [\mathbf{Q}_1] = \mathbf{T}_3 + [\lambda](\mathbf{T}_2 - \mathbf{T}_3),$$
$$[\mathbf{Q}_2] = \mathbf{T}_1 + [\mu](\mathbf{T}_2 - \mathbf{T}_1), \qquad [\mathbf{Q}_3] = \mathbf{T}_1,$$

where $[\lambda]$ and $[\mu]$ are two interval parameters:

$$[\lambda] = [\lambda^\ell, \lambda^u], \quad [\mu] = [\mu^\ell, \mu^u], \quad 0 < \lambda^\ell \le \lambda^u < 1,\ 0 < \mu^\ell \le \mu^u < 1.$$

Let

$$\mathbf{Q}_0^\ell = \mathbf{T}_3, \quad \mathbf{Q}_1^\ell = \mathbf{T}_3 + \lambda^\ell(\mathbf{T}_2 - \mathbf{T}_3),$$
$$\mathbf{Q}_2^\ell = \mathbf{T}_1 + \mu^\ell(\mathbf{T}_2 - \mathbf{T}_1), \quad \mathbf{Q}_3^\ell = \mathbf{T}_1,$$
$$\mathbf{Q}_0^u = \mathbf{T}_3, \quad \mathbf{Q}_1^u = \mathbf{T}_3 + \lambda^u(\mathbf{T}_2 - \mathbf{T}_3),$$
$$\mathbf{Q}_2^u = \mathbf{T}_1 + \mu^u(\mathbf{T}_2 - \mathbf{T}_1), \quad \mathbf{Q}_3^u = \mathbf{T}_1.$$

Then $[\mathbf{Q}_i]$ is the line segment connecting $\mathbf{Q}_i^\ell$ and $\mathbf{Q}_i^u$, $i = 0, 1, 2, 3$. We write $[\mathbf{Q}_i] = [\mathbf{Q}_i^\ell, \mathbf{Q}_i^u]$. The boundary curves of the interval Bézier curve $[\mathbf{Q}](t)$ are the two cubic Bézier curves $\mathbf{Q}^\ell(t) = \sum_{i=0}^3 \mathbf{Q}_i^\ell B_i^3(t)$ and $\mathbf{Q}^u(t) = \sum_{i=0}^3 \mathbf{Q}_i^u B_i^3(t)$. To require $[\mathbf{Q}](t)$ to bound the TPAC, it is equivalent to ensure that the TPAC lies between $\mathbf{Q}^\ell(t)$ and $\mathbf{Q}^u(t)$. Thus a bounding condition is

$$f(\mathbf{Q}^\ell(t)) \leq 0 \quad \text{and} \quad f(\mathbf{Q}^u(t)) \geq 0, \quad 0 \leq t \leq 1, \tag{3.5}$$

where $f(\mathbf{Q}^\ell(t))$ and $f(\mathbf{Q}^u(t))$ are degree $3n$ polynomials. These can be written in Bernstein form

$$f(\mathbf{Q}^\ell(t)) = \sum_{i=0}^{3n} h_i(\lambda^\ell, \mu^\ell) B_i^{3n}(t),$$

$$f(\mathbf{Q}^u(t)) = \sum_{i=0}^{3n} h_i(\lambda^u, \mu^u) B_i^{3n}(t),$$

where h_i is a polynomial of degree n in two variables. From the above two equations, one immediately has

Theorem 3.6. *A sufficient condition for the cubic interval Bézier curve (3.4) to bound a TPAC is that*

$$h_i(\lambda^\ell, \mu^\ell) \leq 0 \quad \text{and} \quad h_i(\lambda^u, \mu^u) \geq 0, \qquad i = 0, 1, \ldots, 3n. \tag{3.7}$$

3.2.2. *Bounding width*

Our next requirement is that the width of the interval Bézier curve $[\mathbf{Q}](t)$ is as small as possible. In this paper, we will not minimize the actual width of $[\mathbf{Q}](t)$, but instead we will define

$$w([\mathbf{Q}](t)) = (w([\lambda]) + w([\mu]))/2. \tag{3.8}$$

It's obvious that the approximation is better if the bounding width is smaller. Especially, if the width satisfies $w([\mathbf{Q}](t)) = 0$, the interval Bézier curve $[\mathbf{Q}](t)$ degenerates to a normal Bézier curve. In this case, $[\mathbf{Q}](t)$ is an accurate parametrization of the algebraic curve.

3.2.3. *Optimal problem*

Now the problem of interval parametrization of algebraic curves can be precisely stated as follows:

Problem 3.9. Given a TPAC defined in (3.2), find an interval Bézier curve (3.4) such that

(1) (3.7) holds.
(2) The width (3.8) of the interval Bézier curve is minimized.

That is, the interval parametrization problem can be converted to the following optimization problem:

$$\begin{aligned} &\text{Min}\,(\lambda^u + \mu^u - \lambda^\ell - \mu^\ell) \\ &s.t.\quad h_i(\lambda^\ell, \mu^\ell) \le 0, \quad i = 0, 1, \ldots, 3n \\ &\qquad\ h_i(\lambda^u, \mu^u) \ge 0, \quad i = 0, 1, \ldots, 3n \\ &\qquad\ 0 < \lambda^\ell \le \lambda^u < 1 \\ &\qquad\ 0 < \mu^\ell \le \mu^u < 1. \end{aligned} \tag{3.10}$$

The above optimization problem is a non-linear programming problem with four variables. To more efficiently solve the problem in the following, we modify it to two non-linear programming problems, each with only two variables. The idea is as follows. We first find a cubic Bézier curve $\mathbf{Q}^c(t)$ to approximate the given TPAC. Suppose the control points of $\mathbf{Q}^c(t)$ are

$$\mathbf{Q}_0^c = \mathbf{T}_3,\ \mathbf{Q}_1^c = \mathbf{T}_3 + \lambda^c(\mathbf{T}_2 - \mathbf{T}_3),\ \mathbf{Q}_2^c = \mathbf{T}_1 + \mu^c(\mathbf{T}_2 - \mathbf{T}_1),\ \mathbf{Q}_3^c = \mathbf{T}_1.$$

We perturb $\mathbf{Q}^c(t)$ to get the two boundary curves $\mathbf{Q}^\ell(t)$ and $\mathbf{Q}^u(t)$ of the interval Bézier curve $[\mathbf{Q}](t)$ by

$$\begin{aligned} \mathbf{Q}_0^\ell &= \mathbf{Q}_0^u = \mathbf{T}_3, \\ \mathbf{Q}_1^\ell &= \mathbf{T}_3 + (\lambda^c - \varepsilon^\ell)(\mathbf{T}_2 - \mathbf{T}_3), \\ \mathbf{Q}_1^u &= \mathbf{T}_3 + (\lambda^c + \varepsilon^u)(\mathbf{T}_2 - \mathbf{T}_3), \\ \mathbf{Q}_2^\ell &= \mathbf{T}_1 + (\mu^c - \delta^\ell)(\mathbf{T}_2 - \mathbf{T}_1), \\ \mathbf{Q}_2^u &= \mathbf{T}_1 + (\mu^c + \delta^u)(\mathbf{T}_2 - \mathbf{T}_1), \\ \mathbf{Q}_3^\ell &= \mathbf{Q}_3^u = \mathbf{T}_1, \end{aligned}$$

or equivalently, by

$$\begin{aligned} \lambda^\ell &= \lambda^c - \varepsilon^\ell, \quad \lambda^u = \lambda^c + \varepsilon^u, \\ \mu^\ell &= \mu^c - \delta^\ell, \quad \mu^u = \mu^c + \delta^u, \end{aligned}$$

where $\varepsilon^\ell, \varepsilon^u, \delta^\ell, \delta^u$ are all nonnegative real numbers. The bounding condition (3.7) now is changed to

$$\begin{aligned} g_i(\varepsilon^\ell, \delta^\ell) &:= h_i(\lambda^\ell, \mu^\ell) \le 0, \qquad i = 0, 1, \ldots, 3n; \\ g_i(\varepsilon^u, \delta^u) &:= h_i(\lambda^u, \mu^u) \ge 0, \qquad i = 0, 1, \ldots, 3n. \end{aligned} \tag{3.11}$$

Instead of solving the optimization problem (3.10), we will solve the following two non-linear programming problems:

$$\begin{aligned} \text{Min} \quad & \varepsilon^\ell + \delta^\ell \\ s.t. \quad & g_i(\varepsilon^\ell, \delta^\ell) \leq 0, \qquad i = 0, 1, \ldots, 3n \\ & \varepsilon^\ell \geq 0, \quad \delta^\ell \geq 0. \end{aligned} \tag{3.12}$$

and

$$\begin{aligned} \text{Min} \quad & \varepsilon^u + \delta^u \\ s.t. \quad & g_i(\varepsilon^u, \delta^u) \geq 0, \qquad i = 0, 1, \ldots, 3n \\ & \varepsilon^u \geq 0, \quad \delta^u \geq 0. \end{aligned} \tag{3.13}$$

The above non-linear programming problems can be easily solved by some software package such as *MATLAB*.

3.2.4. *Approximate parametric representations*

Before solving the optimization problems (3.12) and (3.13), we will first have to find the approximate parametric representation of the algebraic curve (3.2). We will use the idea from [7] to solve the problem.

For a given TPAC as defined in (3.2), we want to find a cubic Bézier curve $\mathbf{Q}^c(t)$ such that $\mathbf{Q}^c(t)$ gives a good approximation to the TPAC. We define the approximation error by

$$\Delta(\lambda^c, \mu^c) = \sqrt{\int_0^1 f(\mathbf{Q}^c(t))^2 dt}. \tag{3.14}$$

Our aim is to find λ^c, μ^c such that $\Delta(\lambda^c, \mu^c)$ is as small as possible. If $\Delta(\lambda^c, \mu^c) = 0$, then $\mathbf{Q}^c(t)$ gives an exact parametrization of the TPAC.

One can easily compute $\Delta(\lambda^c, \mu^c)$, which is a polynomial of degree $2n$. Standard solvers such as Newton iteration can be used to compute the parameter values (λ^c, μ^c) at which Δ is minimized.

4. Examples

In this section, we will provide some examples to illustrate the algorithm for interval parametrization of algebraic curves.

Example 4.1. Given a cubic TPAC whose Bézier ordinates are as follows

$$c_{003} = 0, \quad c_{012} = 0, \quad c_{021} = -1, \quad c_{030} = -1, \quad c_{102} = 1,$$
$$c_{111} = 0, \quad c_{120} = -1, \quad c_{201} = 1, \quad c_{210} = 0, \quad c_{300} = 0.$$

The vertices of the base triangle $\mathbf{T}$ are $\mathbf{T}_1 = (6,1), \mathbf{T}_2 = (3,3), \mathbf{T}_3 = (0,1)$. We will find a cubic interval Bézier curve to bound the TPAC.

The first step is to find an approximate parametrization $\mathbf{Q}^c(t)$ to the TPAC. By applying the algorithm in Section 3.2.4, one gets

$$\lambda^c = 0.4251, \qquad \mu^c = 0.4245.$$

The next step is to solve the optimization problems (3.12) and (3.13), resulting in:

$$\varepsilon^\ell = 0, \quad \delta^\ell = 0.0134, \quad \varepsilon^u = 0.0078, \quad \delta^u = 0.0133.$$

The control points (in barycentric coordinates) of the final interval Bézier curve are thus obtained as

$$[\mathbf{Q}_0] = (0,0,1), \quad [\mathbf{Q}_1] = [(0, 0.4251, 0.5749), (0, 0.4329, 0.5671)],$$
$$[\mathbf{Q}_2] = [(0.5889, 0.4111, 0), (0.5623, 0.4377, 0)], \quad [\mathbf{Q}_3] = (1,0,0).$$

The width of $[\mathbf{Q}](t)$ is 0.0172. Fig.4 depicts the TPAC and the interval Bézier curve $[\mathbf{Q}](t)$.

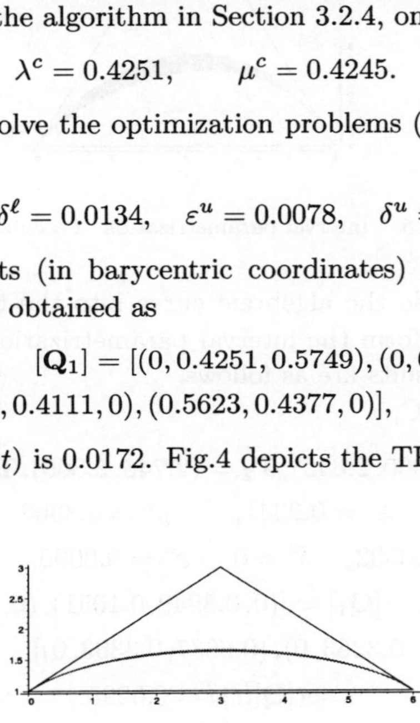

Figure 4. Interval parametrization of a cubic algebraic curve

Example 4.2. Let the Bézier ordinates of the given TPAC be

$$c_{003} = 0, \quad c_{012} = 0, \quad c_{021} = -2, \quad c_{030} = -4, \quad c_{102} = 3,$$
$$c_{111} = 1, \quad c_{120} = -4, \quad c_{201} = 2, \quad c_{210} = 0, \quad c_{300} = 0.$$

The base triangle $\mathbf{T}$ is given by: $\mathbf{T}_1 = (6,1), \mathbf{T}_2 = (4,6), \mathbf{T}_3 = (2,1)$. By the algorithm presented in the last section, one can directly compute

$$\lambda^c = 0.6463, \qquad \mu^c = 0.1844$$
$$\varepsilon^\ell = 0.0546, \quad \delta^\ell = 0.0076, \quad \varepsilon^u = 0.0425, \quad \delta^u = 0.0584;$$
$$[\mathbf{Q}_0] = (0,0,1), \quad [\mathbf{Q}_1] = [(0, 0.5917, 0.4083), (0, 0.6888, 0.3112)],$$
$$[\mathbf{Q}_2] = [(0.8232, 0.1768, 0), (0.7572, 0.2428, 0)], \quad [\mathbf{Q}_3] = (1,0,0)$$
$$w([\mathbf{Q}](t)) = 0.0816.$$

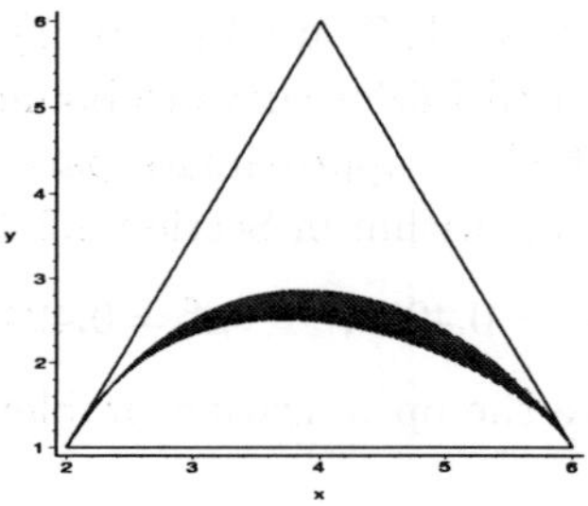

Figure 5. Interval parametrization of a cubic TPAC

Now we subdivide the algebraic curve into the two TPACs along the line $u = w$, and perform the interval parametrization for the two TPACs respectively. The results are as follows.

For the left TPAC,

$$\mathbf{T}_1 = (4.000, 2.640), \mathbf{T}_2 = (2.746, 2.866), \mathbf{T}_3 = (2, 1);$$
$$\lambda^c = 0.9341, \qquad \mu^c = 0.3363,$$
$$\varepsilon^\ell = 0.0392, \quad \delta^\ell = 0, \quad \varepsilon^u = 0.0099, \quad \delta^u = 0;$$
$$[\mathbf{Q}_0] = (0, 0, 1), \quad [\mathbf{Q}_1] = [(0, 0.8949, 0.1051), (0, 0.9440, 0.0560)],$$
$$[\mathbf{Q}_2] = [(0.6637, 0.3363, 0), (0.6637, 0.3363, 0)], \quad [\mathbf{Q}_3] = (1, 0, 0);$$
$$w([\mathbf{Q}](t)) = 0.0295.$$

For the right part,

$$\mathbf{T}_1 = (6, 1), \mathbf{T}_2 = (5.449, 2.378), \mathbf{T}_3 = (4.000, 2.640);$$
$$\lambda^c = 0.5112, \qquad \mu^c = 0.5902,$$
$$\varepsilon^\ell = 0.0028, \quad \delta^\ell = 0.0021, \quad \varepsilon^u = 0.0093, \quad \delta^u = 0.0043;$$
$$[\mathbf{Q}_0] = (0, 0, 1), \quad [\mathbf{Q}_1] = [(0, 0.5917, 0.4083), (0, 0.6038, 0.3962)],$$
$$[\mathbf{Q}_2] = [(0.8232, 0.1768, 0), (0.7572, 0.2428, 0)], \quad [\mathbf{Q}_3] = (1, 0, 0);$$
$$w([\mathbf{Q}](t)) = 0.0093.$$

From the examples we have tested, the width of the interval parametric curve decreases very quickly after each subdivision. Hence if the width of the interval parametric curve is too large, we can recursively subdivide the original TPAC and find interval parametrization for each part of the TPAC until the width of each interval parametric curve is less than some given tolerance.

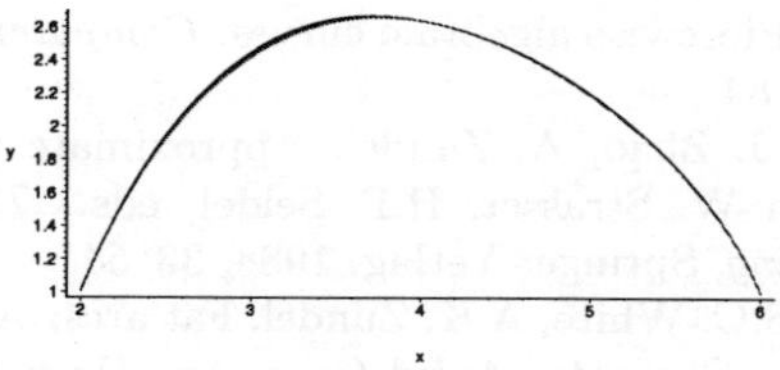

Figure 6. Interval parametrization after subdivision

5. Conclusion

In this paper, we propose a new concept called *interval parametrization* of algebraic curves, and develop an algorithm to compute an interval parametrization of a given algebraic curve. The interval parametrization overcomes some shortcomings in approximate parametrization where unreliability exists in subsequent geometric computation and integrations. Experimental results suggest the algorithm generally produces tight interval parametrization. In practice, interval parametrization of algebraic surfaces is more useful, we will discuss this problem in another paper.

Acknowledgments

This work is supported by NKBRSF on Mathematical Mechanics (no. G1998030600), National Science Foundation of China (no. 60225002, 19971087), the TRAPOYT in Higher Education Institute of MOE of China and the Doctoral Program of MOE of China (no. 20010358003). The authors are thankful for the referees' helpful comments.

References

1. F. Chen, W. Lou. Degree reduction of interval Bézier curves. *Computer Aided Design* **32** (2000), 571–582.
2. R.T. Farouki. The characterization of parametric suface sections. *Computer Vision, Graphics and Image Processing* **33** (1986), 72–84.
3. Q. Lin, J. Rokne. Approximation by fat arcs and fat biarcs. *Computer Aided Design* **34** (2002), 969–979.
4. R.E. Moore. *Interval analysis.* Englewood Cliffs, NJ, Prentice-Hall, 1966.
5. S. P. Mudur, P. A. Koparkar. Interval methods for processing geometric objects. *IEEE Comput. Graphics and Appl.* **4** (1984), 7–17.
6. N.M. Patrikalakis. Robustness issues in geometric and solid modeling. *Computer Aided Design* **32** (2000), 629–689.
7. Y. Qu, J. Sun, F. Chen. Approximate parametrization of algebraic curves. *J. of University of Science and Technology of China* 1997.

8. T.W. Sederberg. Piecewise algebraic curves. *Computer Aided Geometric Design* **1** (1984), 72–84.
9. T.W. Sederberg, J. Zhao, A. Zundel. Approximate parametrization of algebraic curves. In W. Strasser, H.P. Seidel, eds., *Theory and Practice of Geometric Modeling*, Springer Verlag, 1988, 33–54.
10. T.W. Sederberg, S.C. White, A.K. Zundel. Fat arcs: A bounding region with cubic convergence. *Computer Aided Geometric Design* **6** (1989), 205–218.
11. T.W. Sederberg. Algorithm for algebraic curve intersection. *Computer Aided Design* **21** (1989), 547–554.
12. T.W. Sederberg, R.T. Farouki. Approximation by interval Bézier curves. *IEEE Comput. Graph. Appl.* **12** (1992), 87–95.
13. W.N. Waggenspack, D.C. Anderson. Piecewise approximation to algebraic curves. *Computer Aided Geometric Design* **6** (1989), 33–53.

BLENDING QUADRIC SURFACES VIA A BASE CURVE METHOD

JINSAN CHENG
Institute of Mathematics, Jilin University, Changchun,130012, P. R. China
Institute of Systems Science, AMSS, CAS, Beijing, 100080, P. R. China
E-mail: jcheng@mmrc.iss.ac.cn

A method for blending surfaces (implicit or parametric) is introduced. The blending surface is defined by a collection of curves generated through the same base curve and has a parametric representation. Here the given surfaces are not restricted to any particular type of surface representation as long as they have a well-defined and continuous normal vector at each point of their blending boundaries. In this paper, we mainly discussed the blending problems of quadratic surfaces. In particular, we derive the uniform parametric blending surface for six quadratic surfaces with closed blending boundaries at the first time. We also use the method to solve n-way quadratic closed surfaces blending. The method is extensible to blend general surfaces, although we concentrate on quadratic surfaces.

1. Introduction

One of the fundamental tasks of CAGD is surface blending. There are several methods to solve the problem. For example, Hoffmann and Hopcroft [6] proposed the potential method in 1986; Warren [8] proposed the ideal theory method in 1989; Bloor and Wilson [2] proposed PDE method in 1989; Bajaj and Ihm [1] proposed Hermite interpolation method in 1992; Wu and Wang [10] proposed Wu's method in 1994; Zhu and Jin [11] proposed the generatrix method in 1998; Wu and Zhou [9] in 1995; Hartmann in 1995 and 2001 [4,5]; Rossignac and Requicha [7] proposed rolling ball method in 1984; Chen *et al.* [3] use piecewise algebraic surfaces to blend pipe surfaces, and so on.

Hartmman [5] introduced a method for constructing G^n-continuous transition surfaces between two given normal ringed surfaces based on a recent G^n-blending method for parametric curves. Here a ringed surface is a surface generated by sweeping a circle with a non constant radius along a curve. The ringed surface is called *normal* if the circle is contained in the normal planes of the curve. But the method is only fit for a special kind of surfaces. Chen *et al.* [3] presented a scheme to blend three cylinders with piecewise cubic algebraic surfaces. They used six algebraic surfaces to form the whole blending algebraic surface with degree three. But to get one part of the blending surface, one needs complicated computations. And it is not easy to get the range of the parameters of the blending surface we need. Zhu and Jin [11] presented a method which was based on generatrix for blending round or elliptical tubes. The basic idea of it is to design a basic generatrix and then change the parameter of the generatrix to form the blending surface. Wu and Wang [10] studied the blending problem of several quadrics by using Wu's method and gave some examples in transition of pipelines. In these examples, the method can be used to find all possible blending surfaces with given degree. However, one has to do complicated symbolic computations with this way. One is not sure which surface is a "good" surface that can be used in practice. When drawing it on computer, one has to seek a parametric representation of the implicit algebraic surface. Also, the blending surfaces may be difficult (or not) to be adjusted. The problem still exists in Wu and Zhou's method [9]. However they reduced the problem of finding blending algebraic surfaces to one of solving a linear system. The major advantage of our method is that we can give the explicit equation for the blending surface, while most other methods only give the blending surface under certain conditions.

In this paper, we mainly discuss the smoothly joint problem of two quadric surfaces and derive the corresponding explicit formula. We note that the method can be expanded to n-way blending problems. The method is called the *base curve method.* It works as follows. We first construct a base curve connecting the two axes of the surfaces to be blended. Based on the curve, we construct a collection of curves. The blending surface is defined by these curves. Examples given in this paper show that this method gives nice solution to the problem. To get the blending surfaces, we only need the normal vector of the given surfaces at each point on their blending boundaries. Here the boundary curves are regular and continuous, and the normal vector at each point on the boundaries are well-defined and continuous. That means the blending surfaces are only defined by the boundary conditions. This is an extrusive advantage of the method. Further more, we can adjust the shape of the blending surface by adjusting the base curve. The method can be easily extended to solve other blending problems. Moreover, the blending surfaces have parametric representations that make them easy to be realized on computers or industrial applications. But the blending surfaces are non-rational.

2. The Blending Surface of Two Quadric Surfaces

Definition 2.1. Two C^1 continuous surfaces meet along a common boundary. The two surfaces are said to have *G^1-continuity* (or *tangent plane continuity*) if they have the same tangent plane at each point of the boundary and the unit normal vector is continuous along the common boundary.

A space curve (resp. surface)[a] is called *regular* if the tangent vector (resp. normal vector) at every point on the curve (resp. surface) exists and is unique and nonzero. For example, if a space curve

$$P(t) = (x(t), y(t), z(t)), \quad t \in [0,1]$$

has a tangent vector $Q(t) = \left(\frac{\partial x(t)}{\partial t}, \frac{\partial y(t)}{\partial t}, \frac{\partial z(t)}{\partial t}\right)$ and $Q(t) \neq (0,0,0)$ for all t in $[0,1]$, then $P(t)$ is a regular curve. A regular space curve is called a *base curve of a curve* (resp. *the base curve of a surface*), or *base curve* for short, if the curve (resp. surface) is constructed through the space curve based on some given rules. For example, we can regard the X-axis as the base curve of the surface of revolution $(t, t^2\cos\theta, t^2\sin\theta)$, $(t \in [1,4], \theta \in [0, 2\pi])$ and the rule is revolving $y = t^2$ along the X-axis. It is similar to the spine curve of a canal surface.

Theorem 2.2. *Let S_1, S_2 be regular surfaces, let $C = C(t) = S_1 \bigcap S_2$ be a regular space curve, and let $N = N(t)$ be the normal vector of S_1 at the point $C(t)$ on C. Suppose for each point $P = C(t_0) \in C$, there exists a regular space curve $C_2 = C_2(s) \subset S_2$ and $C_2(s_0) = P$ and furthermore, suppose*

$$\frac{C'(t_0) \times C_2'(s_0)}{\|C'(t_0) \times C_2'(s_0)\|} = \pm \frac{N(t_0)}{\|N(t_0)\|}. \tag{2.3}$$

where $C'(t_0), C_2'(s_0)$ denote the tangential vectors of the curves C, C_2 at point P respectively. Then S_1 and S_2 meet with G^1 continuity along C.

Proof. The tangent plane of S_1 at P is $\{Q_1 \mid (Q_1 - P) \times N(t_0) = 0\}$, and the tangent plane of S_2 at P is $\{Q_2 \mid (Q_2 - P) \times (C'(t_0) \times C_2'(s_0)) = 0\}$. The

[a]either parametric or implicit, but here we only consider the parametric curve case.

two planes are obviously the same plane when (2.3) holds, as is shown in Figure 1. This means that the two surfaces have tangent plane continuity. So the theorem holds. □

Corollary 2.4. *Let S_1, S_2 be regular surfaces, let $C = C(t) = S_1 \bigcap S_2$ be a regular space curve. Suppose for each point $P = C(t_0) \in C$, there exist regular space curves $C_1 = C_1(s) \subset S_1$, $C_2 = C_1(s') \subset S_2$ with $C_1(s_0) = P$, $C_2(s'_0) = P$. Furthermore, suppose that the tangent vectors of $C_1(s)$ and $C_2(s')$ at the point P are parallel, but neither is parallel to the tangent vector of $C(t)$ at P. Then S_1 and S_2 meet with G^1-continuity along C* (see Figure 1).

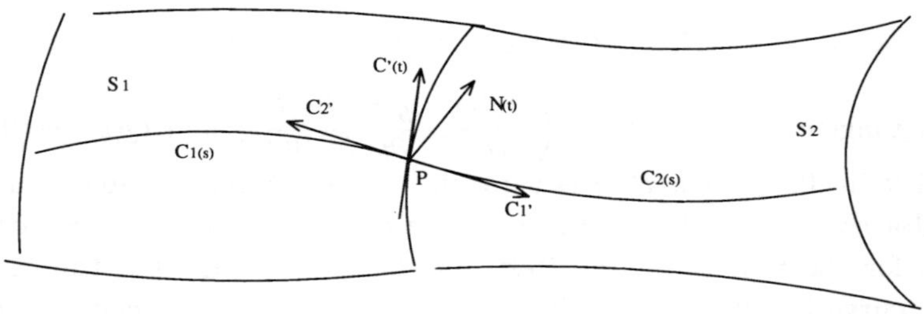

Figure 1. Two surfaces jointed with G^1-continuity.

Making use of the definition of geometric continuity, the theorem and the corollary tell us a constructive method to construct blending surfaces.

A quadric surface is a surface defined by a polynomial with degree two. Here we don't discuss surfaces or other graphs defined by quadric polynomials such as $x^2 = \pm a^2$, $x^2 + b^2y^2 = 0$. We mainly discuss the "*closed*" surfaces, whose planar sections are either ellipses or circles if the planes intersect with the surfaces appropriately. It is easy to show that there are six such surfaces: elliptic cylinder, elliptic cone, elliptic paraboloid, elliptic sphere, hyperboloid of one sheet and hyperboloid of two sheets. There are three quadric surfaces which are not closed. They are the hyperbolic paraboloid, the hyperbolic cylinder, and the parabolic cylinder.

Problem 2.5. Let S_1, S_2 be closed regular quadric surfaces, and let h_1, h_2 be two planes perpendicular to the axes of the surfaces respectively. We need to construct a blending surface which will intersect S_1 and S_2 along the intersection curves $S_1 \bigcap h_1$, $S_2 \bigcap h_2$ with G^1-continuity.

We now show how to construct the blending surface.

2.1. *Constructing the Base Curve*

Let A, B be the intersection points between the axes of quadric surfaces S_1, S_2 and h_1, h_2 respectively. Here the axis of a closed quadric surface is a straight line enclosed by the surface. For example, the X-axis is the axis of the surface $y^2 + z^2 = r^2$. Let the vertical distance and the angle between two axes be d_0 and α. Here the base curve is the curve connecting the two axes at A, B with G^1-continuity. The first step is to construct the base curve. We let one of the axes, say the one containing the point B, be the X-axis, and let the line giving the shortest distance between the two axes be the Z-axis. This line is perpendicular to both axes, meeting them at O and O', with O on the X-axis, which will be the origin. The Y-axis is perpendicular to both the X-axis and Z-axis. Let $\overline{O'A} = d_1$, $\overline{OB} = d_2$, and $OF \parallel OA$. Then $\overline{OO'} = d_0$ and $\angle BOF = \alpha$. See Figure 2.

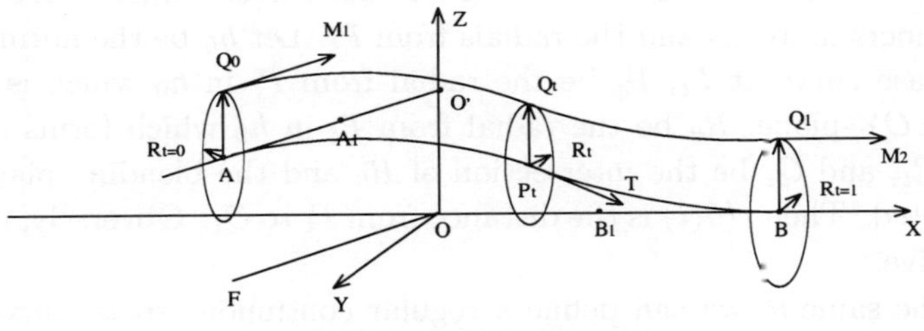

Figure 2. Position in coordinate system

We can use many methods to construct the base curve, for example, using a Bézier curve, or Hermite interpolation and so on. Here we use the Bézier's method. We use A, B and two other points A_1, B_1 as the control points to construct the base curve. Let A_1 be on the same axis as A, and B_1 be on the same axis as B. Let $\overline{OB_1} = \ell_2 d_2$, $\overline{AA_1} = (1 - \ell_1)d_1$. Here $\ell_1 \in (0,1)$, $\ell_2 \in (0,1)$. As it is shown in Figure 2, the curve from $R_{t=0}$ to P_t to $R_{t=1}$ is the base curve. As we know, the base curve has G^1-contact with the two axes of the surfaces at points A and B respectively.

Then we can get the base curve defined by the following equation.

$$P(t) = A \cdot B_{3,0}(t) + A_1 \cdot B_{3,1}(t) + B_1 \cdot B_{3,2}(t) + B \cdot B_{3,3}(t) \qquad (2.6)$$

In fact, the base curve need not always be a Bézier curve with 4 control points for our problem. For example, we can use an arc of an ellipse to contact the two axes when $d_0 = 0$. To avoid the blending surface intersecting itself, the radius of curvature of the base curve at every point should be larger than the maximal radius (in the normal plane of the base curve) at the point, which means that the following inequality should hold:

$$1/\sqrt{(\frac{\partial^2 x(t)}{\partial t^2})^2 + (\frac{\partial^2 y(t)}{\partial t^2})^2 + (\frac{\partial^2 z(t)}{\partial t^2})^2} \geq \max_{\theta\in[0,2\pi)} r(\theta,t), \ \forall t \in [0,1] \qquad (2.7)$$

Here $r(\theta, t)$ is defined by (2.8). We can adjust the value of ℓ_1 and ℓ_2 to satisfy this inequality for all t in [0,1]. Changing the value of ℓ_1 and ℓ_2 can also rectify the shape of the base curve.

2.2. *Designing the Radius Function*

Now we have a base curve $(x(t), y(t), z(t))$, $t \in [0,1]$. The second step is to construct the radius function. We will define a radius function: $r(\theta, t)$, $\theta \in [0, 2\pi)$, $t \in [0,1]$. In the normal plane of the base curve at every point $P_t = (x(t), y(t), z(t))$, there exists a one-to-one correspondence between the real numbers in $[0,2\pi)$ and the radials from P_t. Let h_0 be the normal plane of the base curve at P_t, R_t be the radial from P_t in h_0 which is parallel to the XOY-plane, R_θ be the radial from P_t in h_0 which forms an angle θ with R_t, and Q_t be the intersection of R_θ and the blending plane to be constructed. Then $r(\theta, t)$ is the distance from P_t to Q_t. Obversely, it should be positive.

To the same θ, we can define a regular continuous space curve by Q_t when t changes from 0 to 1. Let $S_\theta(t)$ denote the curve (this is the curve $\widetilde{Q_0Q_tQ_1}$ as shown in Figure 2). In order to connect the given surfaces smoothly, the tangential line of the curve at the extreme points should be in the tangential plane of the given surfaces, as shown in the theorem. Let θ change in $[0, 2\pi)$. We can get a collection of curves. All these curves form the blending surface. Fix $i = 1, 2$. Each point on the intersection curve $S_i \bigcap h_i$ and the axis of the surface S_i defines a plane, which intersects S_i at a planar curve $C_{si}(s)$. Note that $C'_{si}(s)$ is equal to M_i at the point Q_{i-1}. The tangent of the angle between the tangential line of the curve at the point and the axis is $\tan \alpha_i$, where α_i is a function of θ. We can use the planar vector $\overrightarrow{M'_i} = (1, M'_i(\theta)) = (1, \tan \alpha_i)$ to denote it; if the tangential line is perpendicular to the axis, then $\overrightarrow{M'_i} = (0, \pm 1)$. From what we state above, we know $C_{si}(s)$ connects $S_\theta(t)$ with G^1-continuity. When θ changes from 0 to 2π, $S_\theta(t)$ forms the blending surface $S(\theta, t)$. The corollary ensures that $S(\theta, t)$ connects S_1 and S_2 with G^1-continuity.

We can get the following representation for the radius function: $r(\theta, t) = \overline{P_tQ_t}$, $t \in [0,1], \theta \in [0, 2\pi)$. Let $r_1(\theta)$ (resp. $r_2(\theta)$) be the distance from

A (resp. B) to the point on the intersection curve that corresponds to $\theta \in [0, 2\pi)$. The radius function to be constructed should satisfy

$$r(\theta,0) = r_1(\theta),\ r(\theta,1) = r_2(\theta),\ \frac{\partial r(\theta,0)}{\partial t} = M_1'(\theta),\ \frac{\partial r(\theta,1)}{\partial t} = M_2'(\theta).$$

We can use Hermite interpolation method to get the radius function:

$$r(\theta,t) = (M_1' - M_2' - 2r_2 + 2r_1)t^3 + (3r_2 - 3r_1 - 2M_1' + M_2')t^2 + M_1't + r_1. \quad (2.8)$$

Here $M_i' = M_i'(\theta)$, $r_i = r_i(\theta)$, $t \in [0,1]$, and t increases along the base curve from A to B.

Example 2.9. We show how to compute the radius function of an elliptic cone and a cylinder. Let their equations in the usual coordinate system be:

$$\frac{y^2}{a^2} + \frac{z^2}{b^2} - cx^2 = 0, \quad y^2 + z^2 - 0.25 = 0.$$

In order to simplify the calculation, we consider the parametric form of the first surface: $(x(\theta,t), y(\theta,t), z(\theta,t)) = (t, a\cos\theta f(t), b\sin\theta f(t))$, where $f(t) = \sqrt{c}, t\ (t > 0)$. Then

$$r_1(\theta) = \sqrt{c}\, d_1 \sqrt{a^2 \sin^2\theta + b^2 \cos^2\theta},$$
$$M_1'(\theta) = \sqrt{c}\sqrt{a^2 \sin^2\theta + b^2 \cos^2\theta}.$$

In the same way, we can get the following result: $r_2(\theta) = 0.5$, $M_2'(\theta) = 0$.

2.3. *Getting the Parametric Blending Surface*

From the discussion above, we can get the expression of the blending surface. One can prove that it connects the given surfaces with tangent plane continuity. The expression $(x(t), y(t), z(t))$ in the formula is the base curve which can be defined by (2.6).

$$\begin{cases} x(\theta,t) = r(\theta,t)\left(\dfrac{-\frac{\partial x(t)}{\partial t}\frac{\partial z(t)}{\partial t}}{\sqrt{\left(\left(\frac{\partial x(t)}{\partial t}\right)^2+\left(\frac{\partial y(t)}{\partial t}\right)^2+\left(\frac{\partial z(t)}{\partial t}\right)^2\right)\left(\left(\frac{\partial x(t)}{\partial t}\right)^2+\left(\frac{\partial y(t)}{\partial t}\right)^2\right)}}\sin\theta \right. \\ \qquad \left. + \dfrac{-\frac{\partial y(t)}{\partial t}}{\sqrt{\left(\frac{\partial x(t)}{\partial t}\right)^2+\left(\frac{\partial y(t)}{\partial t}\right)^2}}\cos\theta\right) + x(t) \\ y(\theta,t) = r(\theta,t)\left(\dfrac{-\frac{\partial y(t)}{\partial t}\frac{\partial z(t)}{\partial t}}{\sqrt{\left(\left(\frac{\partial x(t)}{\partial t}\right)^2+\left(\frac{\partial y(t)}{\partial t}\right)^2+\left(\frac{\partial z(t)}{\partial t}\right)^2\right)\left(\left(\frac{\partial x(t)}{\partial t}\right)^2+\left(\frac{\partial y(t)}{\partial t}\right)^2\right)}}\sin\theta \right. \\ \qquad \left. + \dfrac{\frac{\partial x(t)}{\partial t}}{\sqrt{\left(\frac{\partial x(t)}{\partial t}\right)^2+\left(\frac{\partial y(t)}{\partial t}\right)^2}}\cos\theta\right) + y(t) \\ z(\theta,t) = r(\theta,t)\dfrac{\left(\frac{\partial x(t)}{\partial t}\right)^2+\left(\frac{\partial y(t)}{\partial t}\right)^2}{\sqrt{\left(\left(\frac{\partial x(t)}{\partial t}\right)^2+\left(\frac{\partial y(t)}{\partial t}\right)^2+\left(\frac{\partial z(t)}{\partial t}\right)^2\right)\left(\left(\frac{\partial x(t)}{\partial t}\right)^2+\left(\frac{\partial y(t)}{\partial t}\right)^2\right)}}\sin\theta + z(t) \end{cases} \quad (2.10)$$

3. Examples

Example 3.1. In this example, we consider connecting an elliptic cylinder $(\frac{y^2}{a_2^2}+\frac{z^2}{b_2^2}-c_2x=0)$ and an elliptic paraboloid $(\frac{y^2}{a_1^2}+\frac{z^2}{b_1^2}-1=0)$. Then we can get the blending surface defined by (2.10). Here

$$r_2(\theta)=\sqrt{c_2d_2}\sqrt{a_2^2\sin^2\theta+b_2^2\cos^2\theta},\quad r_1(\theta)=\sqrt{a_1^2\sin^2\theta+b_1^2\cos^2\theta}$$
$$M_2'(\theta)=\sqrt{a_2^2\sin^2\theta+b_2^2\cos^2\theta}\frac{c}{2\sqrt{cd_2}},\quad M_1'(\theta)=0$$

The parameters of the blending surface shown in Figure 3 are given below: $\ell_1=\ell_2=0.5$, $d_0=0.3$, $d_1=0.5$, $d_2=0.6$, $\alpha=5\pi/6$, $a_1=0.25$, $b_1=0.3$, $a_2=0.3$, $b_2=0.35$, $c_2=0.3$.

Example 3.2. Let us assume that the surfaces to be blended are two cylinders with intersecting axes. The two axes form an angle α. The radii of the cylinders are r_1 and r_2 respectively. Using the method introduced in section 2, we construct the blending surface as the following form by (2.10):

$$\begin{cases} x(\theta,t)=r(\theta,t)(\frac{-\frac{\partial y(t)}{\partial t}}{\sqrt{(\frac{\partial x(t)}{\partial t})^2+(\frac{\partial y(t)}{\partial t})^2}}\cos\theta)+x(t)\\ y(\theta,t)=r(\theta,t)(\frac{\frac{\partial x(t)}{\partial t}}{\sqrt{(\frac{\partial x(t)}{\partial t})^2+(\frac{\partial y(t)}{\partial t})^2}}\cos\theta)+y(t)\\ z(\theta,t)=r(\theta,t)\sin\theta+z(t)\end{cases}$$

We have $r(\theta,t)=2(r_1-r_2)t^3-3(r_1-r_2)t^2+r_1$, and $(x(t),y(t),z(t))=(d_1\cos\alpha,d_1\sin\alpha,0)\cdot B_{2,0}(t)+(0,0,0)\cdot B_{2,1}(t)+(d_2,0,0)\cdot B_{2,2}(t)$.

In order to get a "good" blending surface, we should think about the inequality (2.7). The problem is easily transformed to the following form: $1/\sqrt{d_1^2+d_2^2+2d_1d_2\cos\alpha}\geq 2\max\{r_1,r_2\}$. The parameters of the graph shown in Figure 4 are : $r_1=0.2$, $r_2=0.3$, $d_1=0.4$, $d_2=0.3$, $\alpha=5\pi/6$.

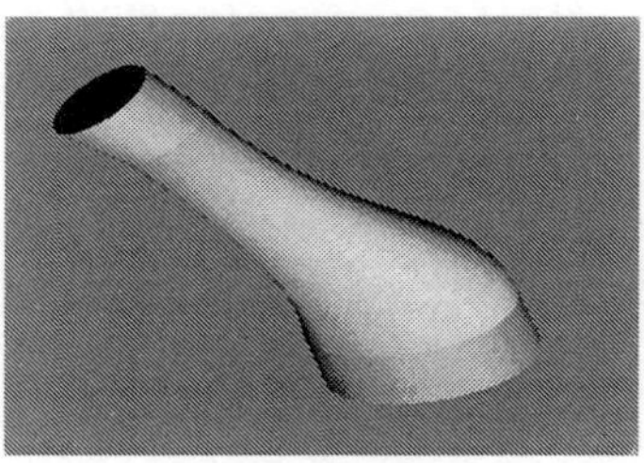

Figure 3. Blending of an elliptic cylinder to an elliptic paraboloid.

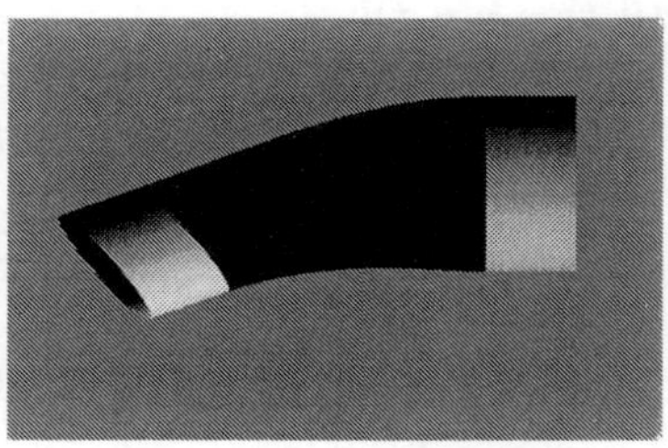

Figure 4. Smooth blending of two cylinders.

Example 3.3. In this example, we will show that our method can be modified to construct blending surfaces along non-planar blending boundaries. The axes of two cylinders are perpendicular and in the same plane. The first cylinder ($y^2 + z^2 - r_1^2 = 0$) cuts into the second cylinder ($x^2 + z^2 - r_2^2 = 0$). The cylinder ($y^2 + z^2 - r^2 = 0, r < r_2$) intersects the second cylinder at a space curve. The following representation gives the blending surface:

$$S(\theta, t) = (d_0 - d(\theta, t), R(\theta, t)\cos\theta, R(\theta, t)\sin\theta) \quad t \in [0, b(\theta)],$$

where

$$b(\theta) = d_0 - \sqrt{r_2^2 - r^2 \sin^2\theta},$$

$$R(\theta, t) = (r_1 - r)(2t^3 - 3t^2) + 3\,\ell_2(t^3 - t^2)\sqrt{\frac{r_2^2 - r^2\sin^2\theta}{r_2^2 - r^2\sin^2\theta\cos^2\theta}} + r_1$$

$$d(\theta, t) = t^3(3\ell_1 + 3\ell_2\frac{r\sin^2\theta}{\sqrt{r_2^2 - r^2\sin^2\theta\cos^2\theta}} - 2b(\theta))$$
$$-3t^2(2\ell_1 + \ell_2\frac{r\sin^2\theta}{\sqrt{r_2^2 - r^2\sin^2\theta\cos^2\theta}} - b(\theta)) + 3\ell_1 t$$

The parameters ℓ_1, ℓ_2 are used to adjust the shape of the blending surface. The parameters of the figure shown in Figure 5 are the following: $d_0 = 0.6$, $r_1 = 0.2$, $r_2 = 0.3$, $r = 0.2$, $\ell_1 = 0.1$, $\ell_2 = 0.1$.

Example 3.4. Five cylinders whose axes are jointing at the same point connecting a sphere with G^1-continuity are shown in this example (see Figure 6).

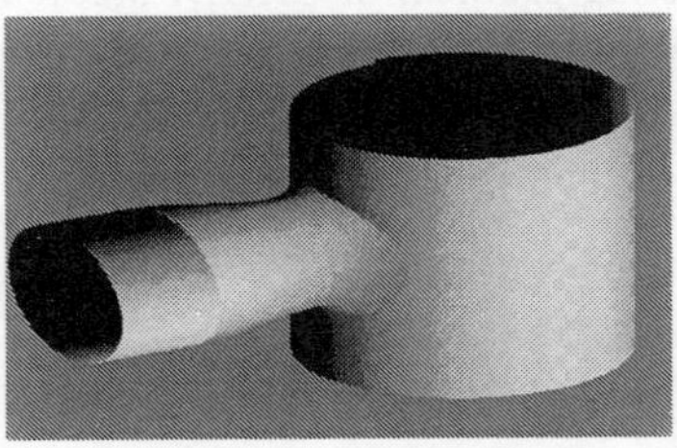

Figure 5. Smooth blending of two intersecting cylinders.

Figure 6. Smooth blending of five cylinders with a sphere.

4. Conclusion

A method for connecting two surfaces G^1-continuously is introduced. It is based on a G^1-continuous parametric regular curve. Obviously, this method can be extended to connecting general regular surfaces.

Acknowledgments

We would like to thank Professor Xiao-shan Gao for his good advices on this paper. We would also like to thank the anonymous referees for their helpful comments.

References

1. C. L. Bajaj, I. Ihm. Algebraic surface design with Hermite interpolation. *ACM Transactions on Graphics* **11**(1) (1992), 61–91.
2. M. I. G. Bloor, M. J. Wilson. Generating blending surfaces using partial differential equation. *CAD* **21**(3) (1989), 165–171.
3. F. L. Chen, C. S. Chen, J. S. Deng. Blending pipe surfaces with piecewise algebraic surfaces. *Chinese J. Computers* **23**(9) (2000), 911–916
4. E. Hartmann. Blending an implicit with a parametric surface. *CAGD* **12** (1995), 825–835.
5. E. Hartmann. G^n-continuous connections between normal ringed surfaces. *CAGD* **18** (2001), 751–770.
6. E. Hoffmann, J. Hopcroft. Quadratic blending surfaces. *CAD* **18** (1986), 301–307.
7. J. R. Rossignac, A. A. G. Requicha. Constant-radius blending in solid modeling. *Comput. Mech. Eng.* 1984.3, 65–73.
8. J. Warren. Blending algebraic surfaces. *ACM Transactions on Graphics* **8**(4) (1989), 263–278.
9. T. R. Wu, Y. S. Zhou. On blending of several quadratic algebraic surfaces. *CAGD* **17**(2000), 759–766.
10. W. T. Wu, D. K. Wang. On surface-fitting problem in CAGD. *Mathematics in Practice and Theory* **3** (1994), 26–31.
11. H. D. Zhu, T. G. Jin. Blending surface via generatrix. *Journal of engineering graphics* **3** (1998), 45–48.

BIVARIATE HERMITE INTERPOLATION AND LINEAR SYSTEMS OF PLANE CURVES WITH BASE FAT POINTS

CIRO CILIBERTO,[1] FRANCESCA CIOFFI,[2] RICK MIRANDA,[3] FERRUCCIO ORECCHIA[2]

[1] *Dip. di Matematica, Univ. di Roma II, Roma, Italy*
E-mail: cilibert@axp.mat.utovrm.it

[2] *Dip. di Matematica e Appl., Univ. di Napoli "Federico II," Napoli, Italy*
E-mail: cioffifr@unina.it,   E-mail: orecchia@unina.it

[3] *Dept. of Mathematics, Colorado State University, Ft. Collins, CO 80523*
E-mail: miranda@math.colostate.edu

It is still an open question to determine in general the dimension of the vector space of bivariate polynomials of degree at most d which have all partial derivatives up through order $m_i - 1$ vanish at each point p_i ($i = 1, \ldots, n$), for some fixed integer m_i called the multiplicity at p_i. When the multiplicities are all equal, to m say, this problem has been attacked by a number of authors (Lorentz and Lorentz, Ciliberto and Miranda, Hirschowitz) and there are several good conjectures (Hirschowitz, Ciliberto and Miranda) on the dimension of these interpolating spaces. The determination of the dimension has been already solved for $m \le 12$ and all d and n by a degeneration technique and some ad hoc geometric arguments. Here this technique is applied up through $m = 20$; since it fails in some cases, we resort (in these exceptional cases) to the bivariate Hermite interpolation with the support of a simple idea suggested by Gröbner bases computation. In summary we are able to prove that the dimension of the vector space is the expected one for $13 \le m \le 20$.

1. Introduction

In this article we work over the field of complex numbers. Fix n distinct general points $p_1, \ldots, p_n$ in the affine plane and let $m_1, \ldots, m_n$ be non-negative integers. Also fix a degree d, and consider the vector space of all polynomials in two variables of degree at most d having multiplicity at least m_i at p_i for each i. This is the space of polynomials $P(x, y)$ such that for each i, and for each (a, b) with $a + b < m_i$, $\frac{\partial^{a+b} P}{\partial x^a \partial y^b}(p_i) = 0$. Removing the identically zero polynomial and identifying polynomials which are scalar multiples, we denote the projective space of such (nonzero) polynomials by $\mathcal{L} = \mathcal{L}(m_1, \ldots, m_n) = \mathcal{L}_d(-\sum_{i=1}^n m_i p_i)$. We refer to this, in accordance with the language of algebraic geometry, as a *linear system*. Because the zero sets of a polynomial in two variables forms a plane curve, this is a linear system of plane curves. This is the same as considering the linear system of projective curves of degree d with multiplicities $m_1, \ldots, m_n$ on the projective points corresponding respectively to $p_1, \ldots, p_n$. The vanishing of a polynomial at a point p is exactly the multiplicity one condition. For this reason higher multiplicity conditions expressed as the additional partial derivative vanishings have been referred to in the algebraic geometry literature as "fat point" vanishing. The dimension of the (projective) space of all nonzero polynomials of degree at most d is $d(d+3)/2$. (This is one less than the vector space dimension.) The number of partial derivatives being asked to be zero at a point of multiplicity m is $m(m+1)/2$, and each of these conditions is a linear condition on the coefficients of the polynomial. Therefore we define the *virtual dimension* of $\mathcal{L}$ to be

$$\nu = \nu_d\Big(-\sum_{i=1}^{n} m_i p_i\Big) = \frac{d(d+3)}{2} - \sum_{i=1}^{n} \frac{m_i(m_i+1)}{2}.$$

The actual dimension of the linear system cannot be less than -1 and so we define the *expected dimension* to be

$$e = e_d\Big(-\sum_{i=1}^{n} m_i p_i\Big) = \max\{-1, \nu\}.$$

(A projective space of dimension zero is simply one point; a projective space of dimension -1 is empty.) Here we are interested in the *general dimensionality problem* for points in the plane, i.e. we want to know if the dimension of $\mathcal{L}$ is equal to the expected dimension. If this happens, we say that the linear system $\mathcal{L}$ is *non-special*. Equivalently, we are interested in the classification of all linear systems $\mathcal{L}(m_1, \ldots, m_n)$ which are *special*, that is, whose dimension exceeds the expected dimension. In this article we focus on the case in which all of the multiplicities are equal, to m say. In this case, in which we denote the system by $\mathcal{L} = \mathcal{L}_d(m^n)$ and call it *homogeneous*, the general dimensionality problem has been attacked by a number of authors [19, 7, 8, 16], and there are several good conjectures [17, 9] on the dimension and general elements of the interpolating spaces. Whether $\mathcal{L}$ has the expected dimension or not certainly depends on the position of the points, even if all the multiplicities are one. Indeed, if the points are in some very special position, the dimension of $\mathcal{L}$ can be very large. However it is elementary that for an open dense set in the parameter space of the set of n points, the dimension achieves a minimum value, and it is this dimension that is to be compared with the expected dimension. If the points are such that the dimension is this minimum possible dimension, we say that the points are *in general position.* If the points are in general position and the multiplicities are one, then the dimension of the linear system $\mathcal{L}$ is always equal to the expected dimension (for example, see [14, 23]). In [8] the authors reformulate a conjecture of Harbourne [15] and Hirschowitz [17] about the dimension of $\mathcal{L}_d(m^n)$ when the points are in general position and verify this conjecture for all $m \leq 12$. They use a degeneration technique that was developed in [7] and that produces an algorithm which has been implemented in the language C [8]. In this range ($m \leq 12$), the degeneration technique failed in some cases, and various ad hoc geometric arguments were used to complete the computational verification of the dimension expected.

The computation of the dimension of $\mathcal{L}$ is equivalent to the computation of the Hilbert function of the algebra S/I where $S = K[x, y]$ and I is the ideal of n general fat points of multiplicities m (see Section 3). So, here we replace the geometric arguments used in [8] when the degeneration technique fails by a computer algebra computation of the Hilbert function. For computing the Hilbert function of fat points we rework the Hermite polynomial interpolation problem [20, chapter 4], [21], [12, 24]. As a result we can produce an algorithm, based on both, the recursion provided by the degeneration technique of Ciliberto and Miranda and the interpolation, which, given an integer $m > 0$, verifies whether the Harbourne-Hirschowitz conjecture holds for all linear systems of the type $\mathcal{L}_d(m^n)$.

The related computer program has been implemented, tested, and gave an affirmative answer, for all m between 13 and 20. This way we have been able to prove the Harbourne-Hirschowitz conjecture for $m \leq 20$, where the bound 20 is due to a matter of computational time. Note that in [19, Theorem 8] the general dimensionality problem for points in the plane is solved (for $m \leq 4$) using a detailed study of the ranks of the relevant matrices. The computer algebra computation presented below has its origins in 1982, when Buchberger and Möller [4] described an algorithm for computing reduced Gröbner bases of ideals of n points on $\mathbb{A}^r$ in a time that is polynomial in n and in r. For many authors this algorithm has been the starting point for making computations with zero-dimensional varieties in polynomial time [1, 2, 10, 22, 25, 27, 28]. The algorithm that we describe for computing the Hilbert function of fat points is a very natural consequence of the original idea of Buchberger and Möller for simple points. The applied method has been implemented using the object oriented language C++ in a software called Points [26] which, using the arithmetic on $K = \mathbb{Z}_p$, where

p is a prime, of the NTL library of V. Shoup [29], provides over $K = \mathbb{Z}_p$ the computation of the Hilbert function for fat points with given multiplicities in any dimension. A generalization of the method of Buchberger and Möller to affine points with differential conditions has been also described in [22, 2]. A projective version of our algorithm which produces also a minimal set of generators of the ideal of projective fat points has been given in [11].

For the purpose of this paper we need to prove that the Hilbert function of the ideal of n general fat points in the plane (with the same multiplicity m) is the maximum possible. Since this is an open condition, it is enough to find an example of fat points with integer coordinates that have maximal Hilbert function. This turns out to be a condition on the maximality of the rank of certain matrices, and all the computations can be done over $K = \mathbb{Z}_p$ for a suitable prime p. In practice we have used $p = 32003$, on an Intel Pentium IV 1.6 GHz with 512 MB RAM + 240 MB swap, running Linux (kernel 2.4.3).

In Section 2 we recall some basic facts and present a summary of the main results for the application of the degeneration technique developed by Ciliberto and Miranda together with a direct computation on fat points. In Section 3 we introduce the algebraic approach on which is based the algorithm described in Section 4 for computing on fat points together with the results of its performance. We have also considered an alternative computational approach to this problem, described briefly below at the end of Section 4; but a direct computation with fat points turns out to be more efficient in the cases we are considering. For a survey of this problem from the point of view of approximation and interpolation theory, the article [3] and the more recent [21] are excellent. For a survey of the problem from an algebraic geometry viewpoint, and its connections with other geometric and algebraic problems, such as the Waring problem for forms, see [6], [23] and the references therein.

2. Background

Let us begin by first developing the notation necessary to precisely state the conjecture of Harbourne and Hirschowitz that we will then verify (up through multiplicity 20). For a nice collection of geometric explanatory examples we refer the readers to [23].

Definition 2.1. Given two linear systems $\mathcal{L} = \mathcal{L}_d(m_1, \ldots, m_n)$ and $\mathcal{L}' = \mathcal{L}'_{d'}(m'_1, \ldots, m'_n)$ of plane curves, their *extra-intersection number* is $\mathcal{L} \cdot \mathcal{L}' =$

$dd' - \sum_i m_i m_i'$. The *self-intersection* of $\mathcal{L}$ is the integer $\mathcal{L}^2 = \mathcal{L} \cdot \mathcal{L} = d^2 - \sum_i m_i^2$ obtained by extra-intersecting $\mathcal{L}$ with itself.

The reader familiar with algebraic geometry methods will recognize this as the intersection numbers of the corresponding linear systems on the blowup of the plane at the n base points.

Definition 2.2. An irreducible rational curve A, which is a member of a linear system $\mathcal{L} = \mathcal{L}_d(m_1, \ldots, m_n)$ and such that the proper transform of A on the blow-up $\widetilde{\mathbb{P}}^2$ is smooth, of self-intersection -1, is called a (-1)-*curve*.

Definition 2.3. A linear system $\mathcal{L}$ is (-1)-*special* if it is nonempty and there is a (-1)-curve A such that $\mathcal{L} \cdot A \leq -2$.

Conjecture 2.4. (***Harbourne-Hirschowitz***) *Let* $\mathcal{L} = \mathcal{L}_d(m_1, \ldots, m_n)$ *be a linear system of plane curves with general multiple base points. Then* $\mathcal{L}$ *is special if and only if it is* (-1)*-special.*

It is not hard to see that every (-1)-special system is special [7, Lemma 4.1]; hence the real content of the conjecture is that every special system is (-1)-special [8, the Main Conjecture]. Moreover, the only (-1)-special homogeneous systems $\mathcal{L}_d(m^n)$ occur when $n \leq 8$ [8, Theorem 2.4], and it is known that in this range the conjecture holds. Therefore the conjecture can be reformulated, in the homogeneous case, by saying that all homogeneous systems $\mathcal{L}_d(m^n)$ for $n \geq 9$ are non-special. Note that for every degree d there is a critical number n_0 such that the virtual dimension of $\mathcal{L}_d(m^{n_0})$ is positive while the dimension of $\mathcal{L}_d(m^{n_0+1})$ is negative. In addition, if $\mathcal{L}_d(m^n)$ is empty, then $\mathcal{L}_d(m^q)$ is empty for all $q \geq n$; and if $\mathcal{L}_d(m^n)$ is nonempty and non-special, then $\mathcal{L}_d(m^q)$ is nonempty and non-special for all $q \leq n$. Therefore, if one can show that the critical system $\mathcal{L}_d(m^{n_0})$ is non-special and that the system $\mathcal{L}_d(m^{n_0+1})$ is empty, then $\mathcal{L}_d(m^n)$ will be non-special for all n. Hence for fixed multiplicity m, and fixed degree d, we have *a priori* only two cases to check in order to prove the conjecture. The degeneration technique of Ciliberto and Miranda (used in [7, 8] to prove the conjecture for $m \leq 12$) provides a recursion in the degree d for fixed multiplicity m. As is the case with many such arguments, there are various technical difficulties in applying the recursion for low values of the degree d. Ciliberto and Miranda define the function

$$dlow(\gamma, h) = \left(\binom{m}{2} + \binom{\gamma+1}{2} + (2h+1)\binom{m+1}{2} - m^{\sim} - 1\right)/(m+1-\gamma)$$

and set

$$D(m) = max\{\lfloor (23m+16)/6 \rfloor, \lceil dlow(-1, \lceil (m^2-1)/(3m+4) \rceil) \rceil\}.$$

Ciliberto and Miranda are able to prove that the recursion always works when the degrees are larger than $D(m)$. Specifically, by applying the degeneration technique, they obtain the following result.

Theorem 2.5. *Fix $m \geq 2$ and let $D = D(m)$ as defined above. Suppose that the conjecture holds for all linear systems $\mathcal{L}_d(m^n)$ with $d < D$. Then the conjecture holds for all linear systems $\mathcal{L}_d(m^n)$.*

For low degrees more standard methods based on the theory of Cremona transformations gives the following result.

Proposition 2.6. *Fix m and $d \leq 3m$. Then for all n the conjecture holds for the homogeneous linear system $\mathcal{L}_d(m^n)$.*

For the proofs of Theorem 2.5 and Proposition 2.6, see [8, Theorem 4.1, Proposition 5.1] respectively.

From the above considerations we see that the conjecture for a fixed m and all d and n will follow if one can show that for all d in the range $[3m+1, D(m)-1]$ the system $\mathcal{L}_d(m^{n_0})$ is non-special and the system $\mathcal{L}_d(m^{n_0+1})$ is empty. For fixed m this then reduces the problem to a finite computation. The degeneration method does not fail for all d in this interval, and can further be used to reduce the number of cases one has to handle by other methods. By their method Ciliberto and Miranda investigated the dimension of $\mathcal{L}_d(m^n)$ up through $m = 12$ and proved the conjecture in this range. Here we go further, reaching the value $m = 20$.

3. Hilbert Function and Computation of $dim_K(\mathcal{L}_d(m^n))$

Let $S = K[x_1, \ldots, x_r]$ be the ring of polynomials in r variables over a field K and $S_{\leq d}$ be the set of polynomials of S of degree at most d. We are interested in computing the dimension of the linear system $\mathcal{L} = \mathcal{L}_d(m^n)$ consisting of plane curves of degree d with multiplicity at least m at n general points. This is equivalent to computing the Hilbert function of S/I when I is the ideal of fat points in the plane. Indeed, more generally, let $p_1, \ldots, p_n$ be n points of the affine space $\mathbb{A}^r$ of dimension r over a field K and $\mathcal{M}_1, \ldots, \mathcal{M}_n$ be their maximal ideals. These points are called "fat points" with multiplicities at least $m_1, \ldots, m_n$ respectively when we consider the ideal $I = \mathcal{M}_1^{m_1} \cap \ldots \cap \mathcal{M}_n^{m_n}$. Note that, if $m = \max\{m_i\}_{i=1,\ldots,n}$ is the maximum of the multiplicities, then a polynomial F belongs to I if and only if F has degree at least m and its derivatives of order up to $m_i - 1$ vanish at the point p_i for each $i = 1, \ldots, n$. Hence, the subset $I_{\leq d}$ consisting of the polynomials of I of degree at most d is isomorphic to $\mathcal{L}_d(m_1, \ldots, m_n)$ as a K-vector space.

The Hilbert function of S/I is the function $H_{S/I} : \mathbb{N}_0 \to \mathbb{N}$ such that $H_{S/I}(d) = dim_K S_{\leq d}/I_{\leq d}$. Hence, since $I_{\leq d}$ is isomorphic to $\mathcal{L}_d$ and $dim_K S_{\leq d}$ is known, our aim is equivalent to computing the Hilbert function of S/I when the fat points are general. We implemented an ad hoc algorithm for our purposes.

Since S/I is an Artinian Cohen-Macaulay ring, the defined Hilbert function of S/I is strictly increasing until it assumes the value of the Hilbert polynomial $HP = \sum_{i=1}^{n} \binom{m_i-1+r}{r}$ of S/I. If we let

$$\begin{aligned}\sigma &= min\{d \in \mathbb{N} \mid H_{S/I}(d) = H_{S/I}(d-1)\} \\ &= min\{d \in \mathbb{N} \mid H_{S/I}(d-1) = HP\},\end{aligned}$$

this means that $I_{\leq d} = S_{\leq d}$ for every $d \geq \sigma$.

Let $\mathcal{T} = \{x_1^{\alpha_1} \ldots x_r^{\alpha_n} \mid (\alpha_1, \ldots, \alpha_r) \in \mathbb{N}\}$ be the monoid of the terms of S and suppose that $\mathcal{T}$ is ordered with respect to a graded term order $<$, that is, a graded well-ordering of monoids. To interpolate the fat points $p_1, \ldots, p_n$ up to the degree d, by the Hermite interpolation, one can consider the matrix G_d^n such that the j-th column is the vector of the evaluations at the given points p_i of the derivatives of order up to $m_i - 1$ of the j-th term (in the fixed graded term order) T of degree $\leq d$. Hence,

$$G_d^n = (b_{k,j}), \quad \text{where} \quad b_{k,j} = \frac{\partial^{i_1+\ldots+i_r} T}{\partial x_1^{i_1} \ldots \partial x_r^{i_r}}(p_i) = 0,$$

with $i_1 + \ldots + i_r \leq m_i - 1$ and $\sum_{i=1}^{i-1} \binom{m_i-1+r}{r} < k \leq \sum_{i=1}^{i+1} \binom{m_i-1+r}{r}$. To know the dimension of $\mathcal{L}_d(m_1, \ldots, m_n)$ it is enough to compute the rank of such matrices. The algorithm of [4] suggests a simple idea by which it is possible to replace G_d^n with a submatrix. To single out this idea now, we are going to collect what we think are the main results for a generalization of the algorithm of [4] to affine fat points.

Since the algorithm computes a (reduced) Gröbner basis of the ideal I, we briefly review what a Gröbner basis is. If F is a polynomial of S, denote by $Lt(F)$ its leading term with respect to the fixed graded term order $<$ and define $Lt(I) = \langle Lt(F) \mid F \in I - \{0\}\rangle$ to be the ideal generated by the leading terms of the non null polynomials of I. More in general, if B is a set of non null polynomials, let $Lt(B) = \langle\{Lt(F) \mid F \in B\}\rangle$ denote the ideal generated by the leading terms of the polynomials of B. Recall that a Gröbner basis of the ideal I with respect to the fixed graded term order is a subset G of non null polynomials of I such that $Lt(G) = Lt(I)$. It is noteworthy that a Gröbner basis generates I.

Let $T'_1 < \ldots < T'_{q_d}$ be the terms of degree up to d corresponding to the first columns of G^n_d that form a maximal set of linearly independent columns. We say that such terms are *independent* and that the others are *dependent.* Now, suppose that the j-th term T of $\mathcal{T}$ is dependent and let $a_1, \ldots, a_{q_d}$ be the scalars of the linear combination of the columns corresponding to $T'_1, \ldots, T'_{q_t}$ which is equal to the column of T. With a notation similar to that in [22], we consider the following polynomial

$$b(T) = T - \sum_{i=1}^{q_d} a_i T'_i. \tag{3.1}$$

For each dependent term T we obtain the scalars $a_1, \ldots, a_{q_d}$ by computing the reduced row-echelon form of the matrix G^n_d. Note that, for each i with $a_i \neq 0$, we have $T'_i < T$.

Proposition 3.2. *The set $\mathcal{B}_{\leq\sigma}$ of the polynomials of degree $\leq \sigma$ of $\mathcal{B}$ is a Gröbner basis for I with respect to the fixed graded term order $<$.*

Proof. By construction $\mathcal{B}_{\leq d}$ is a K-basis of the K-vector space $I_{\leq d}$, for every d. Then, in particular, since $S_{\leq\sigma} = I_{\leq\sigma}$, then $\mathcal{B}_{\leq\sigma}$ is a K-basis of $S_{\leq\sigma}$. Now we show that $Lt(I)_{\leq d} = Lt(\mathcal{B})_{\leq d}$. Since $\mathcal{B} \subseteq I$, obviously $Lt(\mathcal{B})_{\leq d} \subseteq Lt(I)_{\leq d}$. Let F be a polynomial of $I_{\leq d}$. By construction, all polynomials of $\mathcal{B}_{\leq d}$ have different leading terms. Furthermore, each leading term does not appear (i.e. the corresponding coefficient vanishes) in the polynomials where it is not the leading term. Thus, since every polynomial F of $I_{\leq d}$ is a linear combination of polynomials of $\mathcal{B}_{\leq d}$, $Lt(F)$ must be the leading term of a polynomial of $\mathcal{B}_{\leq d}$. Finally, $Lt(\mathcal{B}_{\leq\sigma}) = Lt(I_{\leq\sigma}) = Lt(S_{\leq\sigma}) = (x_1, \ldots, x_r)^\sigma$ and so we are done. □

By the previous proposition it follows that the number of dependent terms of degree up to d is the dimension of $I_{\leq d}$ and, hence, that the number of independent terms of degree up to d is the dimension of $S_{\leq d}/I_{\leq d}$. This is just another way to say that $\text{rank}(G^n_d) = dim_K(S_{\leq d}/I_{\leq d})$. Moreover, there are no independent terms of degree higher than σ.

Proposition 3.3. *If T is a dependent term of degree d then every term $T\overline{T}$ that is a multiple of T is a dependent term too and the polynomial $b(T\overline{T})$ depends on $\mathcal{B}_\sigma - b(T\overline{T})$.*

Proof. By the hypothesis the vector of the evaluations of T depends on the vectors of the evaluations of $T'_1, \dots, T'_{q_d}$ and so the vector of the evaluations of $T\overline{T}$ depends on the vectors of the evaluations of $T'_1\overline{T}, \dots, T'_{q_d}\overline{T}$. Thus $T\overline{T}$ is a dependent term too and the polynomial $b(T\overline{T})$ exists. By construction, the leading term of the polynomial $F = b(T\overline{T}) - \overline{T}b(T)$ is lower than $T\overline{T}$. Since F belongs to the ideal I and since $\mathcal{B}_{\leq\sigma}$ is a Gröbner basis, there is a polynomial F_1 of $\mathcal{S}_{\leq\sigma}$ such that $Lt(F) = Lt(F_1)$. Let $F_2 = F - F_1$ and note that $Lt(F_2) < Lt(F) = Lt(F_1)$. Since F_2 belongs to I, we can repeat for F_2 the same procedure applied to F_1. This procedure ends because a term order is a well-order. □

The matrix G^n_d has already been used to compute the Hilbert function of ideals of points in [27, 25, 22]. We observe that, by Proposition 3.3, when we compute the matrix G^n_d of the evaluations of the terms of degree up to d, we can eliminate each term $T\overline{T}$ that is a multiple of a dependent term T. Hence, we obtain the submatrix $\overline{G}^n_d$ of G^n_d whose columns correspond to the terms of degree up to d that are divided *only* by the independent terms of degree $d-1$. At the same time, if we want to compute the Gröbner basis, we avoid to construct each corresponding polynomial $b(T\overline{T})$ obtaining a subset of $\mathcal{B}_{\leq\sigma}$ which consists exactly of the polynomials of degree $\leq \sigma$ of the so-called reduced Gröbner basis of I with respect to the given term order. In conclusion, the simple idea suggested by this construction concerns how to reduce the number of terms which we have to consider for computing $\mathrm{rank}(G^d_n)$. To this aim the first observation of the proof of Proposition 3.3 is sufficient.

Since at each degree d we can single out the terms of degree $d+1$ which are multiples only of the independent terms of degree d, it is more convenient to compute the Hilbert function degree by degree and, so, to consider the transpose H^n_d of G^n_d and the transpose $\overline{H}^n_d$ of $\overline{G}^n_d$.

4. Verifying the Conjecture for $\mathcal{L}_d(m^n)$ Up Through $m \leq 20$

Based on Section 3, we propose the following algorithm implemented using the object oriented language C++ in a software called Points [26] .

Algorithm 4.1.
Input: n distinct points of $\mathbb{A}^r$ and their multiplicities $m_1, \dots, m_n$.
Output: The Hilbert function of S/I at each degree d.

begin

1. set $HP = \sum_{i=1}^{n} \binom{m_i-1+r}{r}$, $m = \max\{m_i\}_{i=1,\dots,n}$ and $d = m-1$;
2. **for** $i = 0, \dots, m-1$ **do** $H(S/I, d) = H(S, d)$ **endfor**
3. computation and Gauss reduction of $\overline{H}_d^n = H_n^d$
4. **while** $H(S/I, d) < HP$ **do**

   4.1 set $d = d + 1$;

   4.2 computation of the set T_d of terms of degree d in the r variables $x_1, \dots, x_r$ which are divided only by the independent terms in r variables of degree $\leq d-1$;

   4.3 computation of the rows of $\overline{H}_d^n$ corresponding to the terms of degree d that have been computed in Step 4.2; Gauss reduction of these rows with respected to the already reduced rows of $\overline{H}_{d-1}^n$;

   4.4 set $H(S/I, d) = \text{rank}(\overline{H}_d^n)$;

   endwhile

end

Note that $H_{S/I}(d) \leq \min\{H_S(d), HP\} = \min\{\binom{d+r}{r}, \sum_{i=1}^{n} \binom{m_i-1+r}{r}\}$ and, moreover, $H_{S/I}(d) = \min\{H_S(d), HP\}$ for some d if and only if the linear system $\mathcal{L} = \mathcal{L}(m_1, \dots, m_n)$ has the expected dimension for that d. This last is an open condition [23] and is equivalent to the maximality of $\text{rank}(\overline{H}_d^n)$. Now consider a linear system $\mathcal{L} = \mathcal{L}(m_1, \dots, m_n)$ whose points $p_1, \dots, p_n$ have integer coordinates and let $\overline{p}_1, \dots, \overline{p}_n$ be the points whose coordinates are the coordinates of $p_1, \dots, p_n$ modulo a prime p. Let $\overline{I} = \overline{\mathcal{P}}_1^{m_1} \cap \dots \cap \overline{\mathcal{P}}_n^{m_n} \subset \overline{S} = \mathbb{Z}_p[X_0, ..., X_r]$ where $\overline{\mathcal{P}}_i$ is the ideal of the point $\overline{p}_i$. By previous considerations, if $H(\overline{S}/\overline{I}, d) = \min\{\binom{d+r}{r}, \sum_{i=1}^{n} \binom{m_i-1+r}{r}\}$, a generic linear system $\mathcal{L} = \mathcal{L}(m_1, \dots, m_n)$ over the complex numbers will have the expected dimension. So, since to have the expected dimension is an open condition, we implemented our algorithm for points with randomly generated coordinates over $K = \mathbb{Z}_p$. In practice it turned out that for all computations it was enough to consider $p = 32003$.

Remark 4.2. The described algorithm consists of iterated Gaussian eliminations. Hence, for evaluating its computational cost it is enough to look at the dimensions of the matrices to be reduced. To this aim note that the number of rows of the matrix $\overline{H}_d^n$ is no higher than $HP{\cdot}r$ (where recall that $HP = \sum_{i=1}^{n} \binom{m_i-1+r}{r}$) because the number of independent terms of degree up to d is lower than HP for each d. Moreover, note that the number of columns is equal to HP because there are $\binom{m_i-1+r}{r}$ columns for the points p_i. We consider the multiplication as the significative operation and so the computational cost of the rank of $\overline{H}_d^n$ is of order $O(HP^3{\cdot}r)$. Hence our method is an $O(HP^3{\cdot}r)$ algorithm. In particular, if the points are in the plane ($r = 2$) and of the same multiplicity m, the computational cost of the algorithm becomes of order $O(n^3m^6)$.

In [8] there is developed a method which can be used in a deterministic algorithm that for a fixed positive integer m gives all the finitely many triples (d, n, m) for which the degeneration method of Ciliberto and Miranda fails to compute the dimension of the linear system of plane curves of degree d having n points of multiplicity m, and therefore fails to give an answer to Conjecture 2.4. In Table 2 (next page) we list such exceptional triples (d, n, m) with m in the range $13 \leq m \leq 20$, d in the corresponding range $3m+1 \leq d \leq D(m)-1$, and n either given by the critical value n_0 or by $n_0 + 1$. The corresponding virtual dimension ν is also listed. The cases for which $m \leq 12$ have been treated by geometric arguments in [8]. The intervals $[3m + 1, D(m) - 1]$ for $13 \leq m \leq 20$ are given in Table 1 below.

Table 1. Intervals $[3m + 1, D(m) - 1]$ for $13 \leq m \leq 20$

m	13	14	15	16	17	18	19	20
3m+1	40	43	46	49	52	55	58	61
D(m)-1	60	78	84	100	112	119	144	193

Using the software Points (in which the fixed graded term order is the degree reverse lexicographic order) we have investigated the exceptional cases listed in Table 2 and we have found that in these cases the Hilbert function S/I at the degree d is always maximal (recall that I is the ideal of n general fat points of multiplicity m). Thus we complete the proof of the Harbourne-Hirschowitz conjecture up to $m = 20$. Recall that the Hilbert function of S/I is strictly increasing until it assumes constantly its maximum value HP. Since our software Points computes the Hilbert function of S/I until this function reaches the maximum value HP, we prove the cases related to the same number n of points by only one performance of Points. We ran Points over the field K_p, with $p = 32003$, on an Intel Pentium IV 1.6 GHz with 512 MB RAM + 240 MB swap, running Linux (kernel 2.4.3). So in Tables 3–10, for each m with $13 \leq m \leq 20$, we collect the exceptional couples (d, n) and list the minimum degree α of the generators of the corresponding ideal I, the Hilbert polynomial HP, the timing "Time" of the computation in seconds and the memory space "size" used during the computation in megabytes. Note that α is the minimum degree at which the virtual dimension is non negative.

Table 2. Exceptional triples (d, n, m) with m in the range $13 \le m \le 20$

d	n	m	ν	d	n	m	ν	d	n	m	ν	d	n	m	ν
40	10	13	−50	51	10	16	17	57	10	18	0	88	21	19	14
41	10	13	−8	52	10	16	70	58	10	18	59	89	21	19	104
42	10	13	35	53	10	16	124	59	10	18	119	90	22	19	5
43	10	13	79	54	11	16	43	60	11	18	9	107	31	19	−5
45	12	13	−12	56	12	16	20	61	11	18	71	108	31	19	104
46	12	13	35	57	12	16	78	63	12	18	27	125	42	19	20
47	13	13	−8	58	13	16	1	64	12	18	92	61	10	20	−148
48	13	13	41	59	13	16	61	65	13	18	−13	62	10	20	−85
59	20	13	9	60	14	16	−14	66	13	18	54	63	10	20	−21
43	10	14	−61	60	13	16	122	67	13	18	122	64	10	20	44
44	10	14	−16	73	20	16	54	68	14	18	20	65	10	20	110
45	10	14	30	74	21	16	−7	81	20	18	−18	66	11	20	−33
46	10	14	77	75	21	16	69	82	20	18	65	66	10	20	177
47	11	14	20	89	30	16	14	84	21	18	63	67	11	20	35
49	12	14	14	52	10	17	−100	100	30	18	20	68	11	20	104
51	13	14	12	53	10	17	−46	102	31	18	54	70	12	20	35
52	13	14	65	54	10	17	9	58	10	19	−131	71	12	20	107
65	21	14	5	55	10	17	65	59	10	19	−71	72	13	20	−30
78	30	14	9	56	10	17	122	60	10	19	−10	73	13	20	44
46	10	15	−73	57	11	17	27	61	10	19	52	74	13	20	119
47	10	15	−25	59	12	17	−7	62	10	19	115	75	14	20	−15
48	10	15	24	60	12	17	54	63	11	19	−11	75	13	20	195
49	10	15	74	61	13	17	−37	63	10	19	179	76	14	20	62
50	11	15	5	62	13	17	26	64	11	19	54	88	19	20	14
52	12	15	−10	63	13	17	90	66	12	19	−3	90	20	20	−15
53	12	15	44	64	14	17	2	67	12	19	65	91	20	20	77
54	13	15	−21	75	19	17	18	68	12	19	134	92	21	20	−40
55	13	15	35	77	20	17	20	69	13	19	14	93	21	20	54
56	13	15	92	79	21	17	26	70	13	19	85	94	21	20	149
66	19	15	−3	80	21	17	107	71	14	19	−33	95	22	20	35
68	20	15	14	96	31	17	9	71	13	19	157	109	29	20	14
70	21	15	35	112	42	17	14	72	14	19	40	111	30	20	27
49	10	16	−86	55	10	18	−115	84	19	19	44	113	31	20	44
50	10	16	−35	56	10	18	−58	86	20	19	27	133	43	20	14

Table 3. Multiplicity $m = 13$

(d,n)	Time	α	HP	size
(40,10) (41,10) (42,10) (43,10)	20.64	42	910	2.336
(45,12) (46,12)	36.01	46	1092	3.052
(47,13) (48,13)	45.49	48	1183	3.460
(59,20)	166.86	59	1820	7.216

Table 4, Multiplicity $m = 14$

(d,n)	Time	α	HP	size
(43,10) (44,10) (45,10) (46,10)	31.15	45	1050	2.876
(47,11)	41.64	47	1155	3.332
(49,12)	54.24	49	1260	3.828
(51,13) (52,13)	69.15	51	1365	4.372
(65,21)	293.51	65	2205	10.252
(78,30)	856.71	78	3150	20.160

Table 5. Multiplicity $m = 15$

(d,n)	Time	α	HP	size
(46,10)	46.01	48	1200	3.540
(47,10)				
(48,10)				
(49,10)				
(50,11)	61.51	50	1320	4.784
(52,12)	80.25	53	1440	5.488
(53,12)				
(54,13)	102.20	55	1560	5.492
(55,13)				
(56,13)				
(66,19)	321.58	67	2280	10.912
(68,20)	377.65	68	2400	12.012
(70,21)	436.06	70	2520	13.168

Table 6. Multiplicity $m = 16$

(d,n)	Time	α	HP	size
(49,10)	66.24	51	1360	4.344
(50,10)				
(51,10)				
(52,10)				
(53,10)				
(54,11)	88.61	54	1496	51.08
(56,12)	115.76	56	1632	5.944
(57,12)				
(58,13)	147.65	58	1768	6.848
(59,13)				
(60,13)				
(60,14)	184.86	61	1904	7.828
(73,20)	545.14	73	2720	15.220
(74,21)	631.74	75	2856	16.708
(75,21)				
(89,30)	1849.82	89	4080	33.324

Table 7. Multiplicity $m = 17$

(d,n)	Time	α	HP	size
(52,10)	93.55	54	1530	5.312
(53,10)				
(54,10)				
(55,10)				
(56,10)				
(57,11)	125.39	57	1683	6.276
(59,12)	163.43	60	1836	7.332
(60,12)				
(61,13)	212.17	62	1989	8.476
(62,13)				
(63,13)				
(64,14)	261.83	64	2142	9.716
(75,19)	661.29	75	2907	17.284
(77,20)	772.53	77	3060	19.068
(79,21)	917.57	79	3213	20.948
(80,21)				
(96,31)	2911.52	96	4743	44.764
(112,42)	7210.47	112	6426	82.176

Table 8. Multiplicity $m = 18$

(d,n)	Time	α	HP	size
(55,10)	129.56	57	1710	6.456
(56,10)				
(57,10)				
(58,10)				
(59,10)				
(60,11)	174.13	60	1881	7.660
(61,11)				
(63,12)	227.18	63	2052	8.976
(64,12)				
(65,13)	290.10	66	2223	10.412
(66,13)				
(67,13)				
(68,14)	727.55	68	2394	11.956
(81,20)	1103.41	82	3420	24.288
(82,20)				
(84,21)	1245.52	84	3591	25.984
(100,30)	3719.80	100	5130	52.236
(102,31)	4041.43	102	5301	55.688

Remark 4.3. Applying the theory of linear inverse systems (see, for example, [13, 18]), for a general choice of the points, the Hilbert function of S/I at a degree d is equal to the Hilbert function at the same degree d of an ideal $J = (L_1^{d-m_1+1}, \ldots, L_n^{d-m_n+1})$ generated by powers of n general linear forms from S_1. So, for example, if $m_i = m$, $r = 2$ and $n \leq 3$, setting $L_1 = x_0$, $L_2 = x_1$ and $L_3 = x_2$, it is not difficult to single out the cases in

which the linear systems $\mathcal{L}_d(m^n)$ are special. So, instead of computing with fat points, one could try to compute the Hilbert function of the ideal J by a computer algebra software package. But it turns out that this approach is less suitable for this problem. In fact, for example with CoCoA [5] it is impossible to get the expected results already for very few points.

Table 9. Multiplicity $m = 19$

(d,n)	Time	α	HP	size
(58,10)	176.52	61	1900	7.800
(59,10)				
(60,10)				
(61,10)				
(62,10)				
(63,10)				
(63,11)	236.98	64	2090	9.284
(64,11)				
(66,12)	310.18	67	2280	10.912
(67,12)				
(68,12)				
(69,13)	396.83	69	2470	12.680
(70,13)				
(71,13)				
(71,14)	496.95	72	2660	14.592
(72,14)				
(84,19)	1258.82	84	3610	26.248
(86,20)	1472.44	86	3800	29.008
(88,21)	1718.56	88	3990	31.900
(89,21)				
(90,22)	1964.52	90	4180	34.940
(107,31)	5523.52	108	5890	68.616
(108,31)				
(125,42)	13776.85	125	7980	121

Table 10. Multiplicity $m = 20$

(d,n)	Time	α	HP	size
(61,10)	237.33	64	2100	9.368
(62,10)				
(63,10)				
(64,10)				
(65,10)				
(66,10)				
(66,11)	319.44	67	2310	11.180
(67,11)				
(68,11)				
(70,12)	417.19	70	2520	13.168
(71,12)				
(72,13)	532.61	73	2730	15.332
(73,13)				
(74,13)				
(75,13)				
(75,14)	668.68	76	2940	17.664
(76,14)				
(88,19)	1709.68	88	3990	31.900
(90,20)	1983.66	91	4200	35.268
(91,20)				
(92,21)	2339.43	93	4410	38.804
(93,21)				
(94,21)				
(95,22)	2696.26	95	4620	42.516
(109,29)	6119.20	109	6090	73.300
(111,30)	6747.25	111	6300	78.348
(113,31)	7439.53	113	6510	83.652
(133,43)	20006.34	133	9030	155

References

1. J. Abbott, A. Bigatti, M. Kreuzer, L. Robbiano. Computing ideals of points. *J. Symbolic Computation*, 30(4) (2000), 351–356.
2. J. Abbott, M. Kreuzer, L. Robbiano. Computing zero-dimensional schemes. Preprint available at `http://cocoa.dima.unige.it/research/publications.html`, 2001.
3. A. A. Akopyan, A. A. Saakyan. Multivariate splines and polynomial interpolation. *Russian Mathematical surveys* 48(5) (1994), 1–72.

4. B. Buchberger, H. M. Möller. The construction of multivariate polynomials with preassigned zeros. In *EUROCAM 82*, *LNCS* **144**, 24–31. Springer-Verlag, 1982.
5. A. Capani, G. Niesi, L. Robbiano. *CoCoA, a system for doing Computations in Commutative Algebra*, 4.1 ed., 2001. Available via anonymous ftp from `cocoa.dima.unige.it`.
6. C. Ciliberto. Geometric Aspects of Polynomial Interpolation in More Variables and Waring's problem. In *European Congress of Mathematics, I (Barcelona, 2000)*, 289–316. *Progress in Mathematics* **201**, Birkhäuser, 2001.
7. C. Ciliberto, R. Miranda. Degeneration of planar linear systems. *Journal Reine Angew. Math* **501** (1998), 191–220.
8. C. Ciliberto, R. Miranda. Linear systems of plane curves with base points of equal multiplicity. *Trans. Amer. Math. Soc.* **352** (2000), 4037–4050.
9. C. Ciliberto, R. Miranda. The Segre and Harbourne-Hirschowitz conjectures. In *Applications of algebraic geometry to coding theory, physics and computation (Eilat, 2001)*, 37–51. *NATO Sci. Ser. II Math. Phys. Chem.* **36**. Kluwer Acad. Publ., 2001.
10. F. Cioffi. Minimally generating ideals of points in polynomial time using linear algebra. *Ricerche di Matematica*, **XLVIII**(1) (1999), 55–63.
11. F. Cioffi, F. Orecchia. Computation of minimal generators of ideals of fat points. In *Proceedings, ISSAC 2001*, 72–76. ACM Press, 2001.
12. M. Gasca, J. I. Maeztu. On Lagrange and Hermite interpolation in $\mathbf{R}^k$. *Numer. Math.* **39**(1) (1992), 1–14.
13. A. V. Geramita. Inverse system of fat points: Waring's problem, secant varieties of veronese varieties and parameter spaces for gorenstein ideals. *The Curves Seminar at Queen's* **10** (1996), 2–129.
14. A. V. Geramita, F. Orecchia. Minimally generating ideals defining certain tangent cones. *J. Algebra* **78**(1) (1982), 36–57.
15. B. Harbourne. The geometry of rational surfaces and Hilbert function of points in the plane. *Can. Math. Soc. Conf. Pro.* **6** (1986), 95–111.
16. A. Hirschowitz. La method d'horace pour l'interpolation a plusieurs variables. *Man. Math.* **50** (1985), 337–388.
17. A. Hirschowitz. Une conjecture puor la cohomologie des diviseurs sur les surfaces rationelles génériques. *J. Reine Angew. Math* **397** (1989), 208–213.
18. A. Iarrobino. Inverse system of a symbolic power, II: the Waring Problem for forms. *J. Algebra*, **174** (1995), 1091–1110.

19. G. G. Lorentz, R. A. Lorentz. Bivariate Hermite interpolation and application to algebraic geometry. *Numer. Math.* **57** (1990), 669–680.
20. R. A. Lorentz. *Multivariate Birkhoff Interpolation. LNM* **1516**. Springer Verlag, 1992.
21. R. A. Lorentz. Multivariate Hermite interpolation by algebraic polynomials: a survey. *Numerical Analysis 2000, Vol. II: Interpolation and Extrapolation. J. Comput. Appl. Math* **122**(1–2) (2000), 167–201.
22. M. G. Marinari, H. M. Moeller, T. Mora. Gröbner bases of ideals defined by functionals with an application to ideals of projective points. *AAECC* **4** (1993), 103–145.
23. R. Miranda. Linear systems of planes curves. *Notices of the AMS* **46**(2) (1999), 192–202.
24. H. Michael Möller, Thomas Sauer. H-bases for polynomial interpolation and system solving. In *Multivariate polynomial interpolation. Adv. Comput. Math.* **12**(4) (2000), 335–362.
25. F. Orecchia. A polynomial algorithm for computing the conductor of points and of a curve with reduced tangent cone. *The Curves Seminar at Queens's, Vol. VIII* (*Kingston, ON, 1990–1991*), Exp. G. *Queen's Papers in Pure and Applied Mathematics* **88**, Univ. Kingston, ON, 1991.
26. F. Orecchia, F. Cioffi, I. Ramella. *Points (software for computations on points).* Preprint, 2001. Available for the Linux platform at `http://cds.unina.it/~orecchia/gruppo/EPoints.html`.
27. I. Ramella. *Algoritmi di Computer Algebra relativi agli ideali di punti dello spazio proiettivo.* PhD thesis, Univ. di Napoli "Federico II", 1990. *Preprint* n. **30**, Dip. di Mat. e Applic. "R. Caccioppoli", 1990.
28. I. Ramella. Ideals of points in generic position: a polynomial algortithm for computing a minimal set of generators. *Ricerche di Matematica* **43** (1994), 205–217.
29. V. Shoup. *NTL: a Library for doing Number Theory.* Open source software distributed under the GNU General Public License, 2001. Available at `http://www.shoup.net/ntl`.

A SERIES OF EXACT SOLUTIONS OF TWO EXTENDED COUPLED ITO SYSTEMS

ENGUI FAN
Institute of Mathematics and Key Laboratory for Nonlinear Mathematical Models and Methods, Fudan University, Shanghai 200433, P. R. China
E-mail: faneg@fu.dan.edu.cn

A direct algebraic method, which can be implemented on a computer with the help of symbolic computation software like *Mathematica* or *Maple*, is used to construct a series of travelling wave solutions of two extended coupled Ito systems. The obtained solutions include solitary wave solutions, rational solutions, triangular periodic solutions, Jacobi and Weierstrass doubly periodic wave solutions. Among them, the Jacobi elliptic periodic wave solutions exactly degenerate to the solitary wave solutions at a certain limit condition. Compared with most existing tanh methods and elliptic function method, the proposed method gives new and more general solutions. More importantly, the method provides a guideline to classify the various types of the solution according to some parameters.

Keywords: extended coupled Ito system; travelling wave solution; algebraic method with symbolic computation.

1. Introduction

A central and very active topic in the study of soliton theory is to find as many new solutions with physical interests for nonlinear equations as possible. Much work has been dedicated to this subject and various techniques have been developed (see, for example, [1,2,3,4,6,9,11,13 17,18,20,23]). Recently, we proposed a simple and effective algebraic method for constructing many new families of travelling wave solutions with physical interests in a unified way that simply proceeds as follows [7].

Consider the general form of a nonlinear equation

$$H(u, u_t, u_x, u_{xx}, \cdots) = 0. \tag{1.1}$$

We use the solution φ of an ordinary differential equation

$$\varphi' = \varepsilon\sqrt{\sum_{j=0}^{r} c_j\varphi^j} \tag{1.2}$$

where r is some positive integer and $\varepsilon = \pm 1$, as a new variable, and propose the following series expansion as a solution of Equation (1.1)

$$u(x,t) = U(\xi) = \sum_{i=0}^{n} a_i \varphi^i, \tag{1.3}$$

where $\varphi = \varphi(\xi)$ with $\xi = x + ct$. Normally, balancing the highest derivative term with the nonlinear terms in Equation (1.1) will give a special relation between the positive integers n and r, from which we can make different choices for the integers n and r and use them in different series expansions or ansatzes to find the solutions of Equation (1.1). For example, in the case of the KdV equation

$$u_t + 6uu_x + u_{xxx} = 0,$$

we have

$$r = n + 2. \tag{1.4}$$

If we take $n = 1$ and $r = 3$ in (1.4), we may use the following series expansion

$$u = a_0 + a_1\varphi, \quad \varphi' = \varepsilon\sqrt{c_0 + c_1\varphi + c_2\varphi^2 + c_3\varphi^3}$$

as a solution of the KdV equation. Similarly, if we take $n = 2$, $r = 4$ in (1.4), we have

$$u = a_0 + a_1\varphi + a_2\varphi^2, \quad \varphi' = \varepsilon\sqrt{c_0 + c_1\varphi + c_2\varphi^2 + c_3\varphi^3 + c_4\varphi^4}.$$

The constants $c, a_i, c_j \ \ (i = 0, 1, \cdots, n, j = 0, 1, \cdots, r)$ are obtained from a system of algebraic equations that is generated by substituting the ansatz (1.3) into Equation (1.1), and then setting the coefficients of all powers like φ^i and $\varphi^i\sqrt{\sum_{j=0}^{r} c_j\varphi^j}$ to zero.

The algorithm presented here is effective, although the generation of the algebraic system from Equation (1.1) and solving it are two key procedures that are laborious to do by hand. But they can be implemented on a computer with the help of computer algebra software like *Mathematica* or *Maple*. The output solutions from the algebraic system comprise a list of the form $\{c, a_i, c_j\}$. In general, if the values of some parameters are left unspecified, then they are regarded to be arbitrary in the solution of Equation (1.1).

We see that the travelling wave solutions of Equation (1.1) depend on the explicit solvability of Equation (1.2) subject to constant parameters

c, a_i, c_j satisfying a system of algebraic equations. Such a system will require considerable computation resources, when n and r increase, and may become too difficult to solve even for a computer. But since in the case when $r = 4$, Equation (1.3) will give us a series of nonlinear waves with physical interest, for example, solitary wave solutions, rational solutions, triangular periodic solutions, Jacobi, and Weierstrass doubly periodic solutions, we only consider the case when $r = 4$ in this paper and hence

$$\varphi' = \varepsilon\sqrt{c_0 + c_1\varphi + c_2\varphi^2 + c_3\varphi^3 + c_4\varphi^4}. \tag{1.5}$$

Theorem 1.6. *Suppose that φ is a solution of the Equation (1.5). We have the following results:*

(i) *If $c_3 = c_1 = c_0 = 0$, Equation (1.5) possesses a bell shaped solitary wave solution*

$$\varphi = \sqrt{-\frac{c_2}{c_4}}\ \mathrm{sech}(\sqrt{c_2}\xi) \quad \text{if } c_2 > 0,\ c_4 < 0, \tag{1.7}$$

a triangular solution

$$\varphi = \sqrt{-\frac{c_2}{c_4}}\ \sec(\sqrt{-c_2}\xi) \quad \text{if } c_2 < 0,\ c_4 > 0,$$

and a rational solution

$$\varphi = -\frac{\varepsilon}{\sqrt{c_4}\xi} \quad \text{if } c_2 = 0,\ c_4 > 0.$$

(ii) *If $c_3 = c_1 = 0$, and $c_0 = \frac{c_2^2}{4c_4}$, Equation (1.5) possesses a kink shaped solitary wave solution*

$$\varphi = \varepsilon\sqrt{-\frac{c_2}{2c_4}}\ \tanh(\sqrt{-\frac{c_2}{2}}\xi) \quad \text{if } c_2 < 0,\ c_4 > 0 \tag{1.8}$$

and a triangular solution

$$\varphi = \varepsilon\sqrt{\frac{c_2}{c_4}}\ \tan(\sqrt{\frac{c_2}{2}}\ xi) \quad \text{if } c_2 > 0,\ c_4 > 0.$$

(iii) *If $c_3 = c_1 = 0$, Equation (1.5) admits two Jacobi elliptic function solutions*

$$\varphi = \sqrt{\frac{-c_2m^2}{c_4(2m^2-1)}}\ \mathrm{cn}(\sqrt{\frac{c_2}{2m^2-1}}\xi), \quad \text{if } c_2 > 0,\ c_0 = \frac{c_2^2m^2(m^2-1)}{c_4(2m^2-1)^2}, \tag{1.9}$$

$$\varphi = \sqrt{\frac{-c_2}{c_4(2-m^2)}}\,\mathrm{dn}(\sqrt{\frac{c_2}{2-m^2}}\xi), \tag{1.10}$$

$$\text{if } c_2 > 0,\ c_0 = \frac{c_2^2(1-m^2)}{c_4(2-m^2)^2},$$

and

$$\varphi = \varepsilon\sqrt{\frac{-c_2 m^2}{c_4(m^2+1)}}\,\mathrm{sn}(\sqrt{\frac{-c_2}{m^2+1}}\xi), \tag{1.11}$$

$$\text{if } c_2 < 0,\ c_0 = \frac{c_2^2 m^2}{c_4(m^2+1)^2}.$$

As $m \to 1$, the Jacobi doubly periodic solutions (1.9) and (1.11) degenerates to the solitary wave solutions (1.7) and (1.8) respectively.

(iv) *If $c_4 = c_0 = c_1 = 0$, Equation (1.5) possesses a bell shaped solitary wave solution*

$$\varphi = -\frac{c_2}{c_3}\,\mathrm{sech}^2(\frac{\sqrt{c_2}}{2}\xi), \quad \text{if } c_2 > 0, \tag{1.12}$$

a triangular solution

$$\varphi = -\frac{c_2}{c_3}\sec^2(\frac{\sqrt{-c_2}}{2}\xi), \quad \text{if } c_2 < 0,$$

and a rational solution

$$\varphi = \frac{1}{c_3\xi^2}, \quad \text{if } c_2 = 0.$$

(v) *If $c_4 = c_2 = 0, c_3 > 0$, Equation (1.5) admits a Weierstrass elliptic function solution*

$$\varphi = \wp\left(\frac{\sqrt{c_3}}{2}\xi, g_2, g_3\right), \tag{1.13}$$

where $g_2 = -4c_1/c_3$, and $g_3 = -4c_0/c_3$.

Remark 1.14. In fact, other types of travelling wave solutions such as $\csc\xi$, $\cot\xi$, $\mathrm{csch}\,\xi$ and $\coth\xi$ can be obtained in Theorem 1.6. We omit them here since they appear in pairs with the corresponding $\sec\xi$, $\tan\xi$, $\mathrm{sech}\,\xi$ and $\tanh\xi$. For simplicity, we also omit the solutions in terms of functions $\sec\xi$, $\tan\xi$, $1/\xi$ in the following.

Remark 1.15. Let's consider two special cases of Theorem 1.6. When $c_1 = c_3 = 0, c_0 = 1, c_2 = -2, c_4 = 1$, Equation (1.5) has a solution $\tanh\xi$ and our method reduces to the tanh method [19,16]. When$c_1 = c_3 = 0, c_0 =$

$b^2, c_2 = 2b, c_4 = 1$, Equation (1.5) degenerates to a Riccati equation. In this case our proposed method becomes the extended tanh method [10,16]. In conclusion, our proposed method is a generalization of both the tanh and the extended hyperbolic-function method.

In the present paper, we consider two new extended coupled Ito systems

$$\begin{aligned} u_t &= v_x, \\ v_t &= -2(v_{xxx} + 3uv_x + 3vu_x) - 12ww_x + 6p_x \\ w_t &= w_{xxx} + 3uw_x, \\ p_t &= p_{xxx} + 3up_x. \end{aligned} \tag{1.17}$$

and

$$\begin{aligned} u_t &= v_x, \\ v_t &= -2(v_{xxx} + 3uv_x + 3vu_x) - 6(wp)_x \\ w_t &= w_{xxx} + 3uw_x, \\ p_t &= p_{xxx} + 3up_x. \end{aligned} \tag{1.18}$$

These systems were proposed and investigated by Tam, Hu and Wang [21]. Obviously the second equation in system (1.17) is different from that of system (1.18), while others are identical correspondingly. If we set $p = 0$ in system (1.17) and $w = p$ in system (1.18) respectively, then the two systems all become the coupled Ito system [21,22]. If we set $w = p = 0$ in systems (1.17) and (1.18) respectively, the two systems all reduce to the following system

$$\begin{aligned} u_t &= v_x, \\ v_t &= -2(v_{xxx} + 3uv_x + 3vu_x), \end{aligned}$$

which is just equivalent to the Ito equation [12]:

$$u_{tt} = -2u_{xxt} - 6(uu_t)_x - 6(u_x \partial^{-1} u_t)_x.$$

Starting from their Hirota bilinear equations and using symbolic computation software *Mathematica*, Tam et al. further constructed the 3- and 4-solitary waves of the systems (1.17) and (1.18) [21]. But to our knowledge, there have been no investigations on other kinds of solutions. In this paper, we will apply the above proposed method to establish new families of travelling wave solutions of the systems (1.17) and (1.18) including solitary wave solutions, rational solutions, triangular periodic solutions, Jacobi and Weierstrass doubly periodic wave solutions. It is shown that the Jacobi elliptic periodic wave solutions exactly degenerate to the solitary wave solutions under a certain limit condition.

2. The Solutions of Ito System (1.17)

In this section, we shall apply the technique developed in Section 1 to the system (1.17) and search for its multiple travelling wave solutions. For this purpose we introduce the transformations

$$\begin{aligned} u(x,t) &= U(\xi), \\ v(x,t) &= V(\xi), \\ w(x,t) &= W(\xi), \\ p(x,t) &= P(\xi), \quad \xi = x + ct \end{aligned}$$

and change system (1.17) into the form

$$\begin{aligned} cU' - V' &= 0, \\ cV' + 2(V''' + 3UV' + 3U'V) + 12WW' - 6P' &= 0, \\ cW' - W''' - 3UW' &= 0, \\ cP' - P''' - 3UP' &= 0, \end{aligned} \tag{2.1}$$

According to (1.2) and (1.3), we propose the following series expansion as a solutions of system (2.1)

$$U = \sum_{i=0}^{n_1} a_i \varphi^i, \quad V = \sum_{i=0}^{n_2} b_i \varphi^i, \quad W = \sum_{i=0}^{n_3} d_i \varphi^i, \quad P = \sum_{i=0}^{n_4} e_i \varphi^i,$$

where φ satisfies Equation (1.2).

Balancing the linear terms of the highest order with nonlinear terms in system (2.1) gives

$$n_1 = n_2 = r - 2, \quad n_3 \le n_1, \quad n_4 \le n_1 + 1.$$

Therefore we may choose $r = 4, n_1 = n_2 = n_3 = 2, n_4 = 3$ and have the following ansatz

$$\begin{aligned} U &= a_0 + a_1\varphi + a_2\varphi^2, \\ V &= b_0 + b_1\varphi + b_2\varphi^2, \\ W &= d_0 + d_1\varphi + d_2\varphi^2, \\ P &= e_0 + e_1\varphi + e_2\varphi^2 + e_3\varphi^3, \end{aligned} \tag{2.2}$$

where φ satisfies (1.5).

With the help of the symbolic software *Mathematica*, by substituting (2.2) into (2.1) and setting the coefficients of

$$\varphi^i \sqrt{c_0 + c_1\varphi + c_2\varphi^2 + c_3\varphi^3 + c_4\varphi^4} \quad (i = 0, 1, \cdots)$$

to zero, we further obtain a system of algebraic equations

$$
\begin{aligned}
0 &= ca_1 - b_1,\\
0 &= c_1a_2 - b_2,\\
0 &= 6\varepsilon a_1b_0 + c\varepsilon b_1 + 6\varepsilon a_0b_1 + 6\varepsilon^3b_2c_1 + 2\varepsilon^3b_1c_2 + 12\varepsilon d_0d_1 - 6\varepsilon e_1,\\
0 &= 12\varepsilon a_2b_0 + 12\varepsilon a_1b_1 + 2\varepsilon cb_2 + 12\varepsilon a_0b_2 + 16\varepsilon^3b_2c_2\\
&\quad + 6\varepsilon^3b_1c_3 + 12\varepsilon d_1^2 + 24\varepsilon d_0d_2 - 12\varepsilon e_2,\\
0 &= 18\varepsilon a_2b_1 + 18\varepsilon a_1b_2 + 30\varepsilon^3b_2c_3 + 12\varepsilon^3b_1c_4 + 36\varepsilon d_1d_2 - 18\varepsilon e_3,\\
0 &= 24\varepsilon a_2b_2 + 48\varepsilon^3b_2c_4 + 24\varepsilon d_2^2,\\
0 &= -\varepsilon cd_1 + 3\varepsilon a_0d_1 + \varepsilon^3c_2d_1 + 3\varepsilon^3c_1d_2,\\
0 &= 3\varepsilon a_1d_1 + 3\varepsilon^3c_3d_1 - 2\varepsilon cd_2 + 6\varepsilon a_0d_2 + 8\varepsilon^3c_2d_2,\\
0 &= 3\varepsilon a_2d_1 + 6\varepsilon^3c_4d_1 + 6\varepsilon a_1d_2 + 15\varepsilon^3c_3d_2,\\
0 &= 6\varepsilon a_2d_2 + 24\varepsilon^3c_4d_2,\\
0 &= -\varepsilon ce_1 + 3\varepsilon a_0e_1 + \varepsilon^3c_2e_1 + 3\varepsilon^3c_1e_2 + 6\varepsilon^3c_0e_3,\\
0 &= 3\varepsilon a_1e_1 + 3\varepsilon^3c_3e_1 - 2\varepsilon ce_2 + 6\varepsilon a_0e_2 + 8\varepsilon^3c_2e_2 + 15\varepsilon^3c_1e_3,\\
0 &= 3\varepsilon a_2e_1 + 6\varepsilon^3c_4e_1 + 6\varepsilon a_1e_2 + 15\varepsilon^3c_3e_2 - 3\varepsilon ce_3 + 9\varepsilon a_0e_3 + 27\varepsilon^3c_2e_3,\\
0 &= 6\varepsilon a_2e_2 + 24\varepsilon^3c_4e_2 + 9\varepsilon a_1e_3 + 42\varepsilon^3c_3e_3,\\
0 &= 9\varepsilon a_2e_3 + 6\varepsilon^3c_4e_3.
\end{aligned}
$$

Note that $\varepsilon = \pm 1$ and hence $\varepsilon^3 = \varepsilon$. We may eliminate ε from above system, such that corresponding system does not involve ε. From the output of *Mathematica* (here we may use command: Reduce[{system of the above algebraic equations}, {List of variables}]), we find two kinds of solutions, namely,

$$
\begin{aligned}
&c_3 = c_1 = a_1 = b_1 = d_2 = e_2 = e_3 = 0, a_0 = \frac{1}{3}(c - c_2),\\
&b_0 = -\frac{1}{2a_2}(a_2c^2 + 2a_2c_0c + 2d_1^2), e_1 = 2d_0d_1, c_4 = -\frac{1}{2}a_2, b_2 = a_2c,
\end{aligned} \tag{2.3}
$$

with $c_0, c_2, c, d_0, d_1, e_0, b_2, a_2 \neq 0$ being arbitrary constants, and,

$$
\begin{aligned}
&c_4 = a_2 = b_2 = d_2 = e_2 = e_3 = 0, a_0 = -\frac{1}{3a_1^2}(c_2a_1^2 + 2d_1^2),\\
&b_0 = -\frac{1}{a_1^4}(2d_0d_1a_1^3 + 2d_1^4 - e_1a_1^3), c_3 = -a_1, b_1 = -\frac{2d_1^2}{a_1},
\end{aligned} \tag{2.4}
$$

with $c_0, c_1, c_2, c, d_0, e_0, e_1, a_1 \neq 0$ being arbitrary constants.

Now all possible explicit solutions of the coupled Ito system (1.17) are discussed as follows:

(A) Making use of (1.7)–(1.11) and (2.3), we obtain two solitary wave solutions and three Jacobi doubly periodic solutions

$$u_i = a_0 + a_2\varphi_i^2, \quad v_i = b_0 + b_2\varphi_i^2,$$
$$w_i = d_0 + d_1\varphi_i, \quad p_i = e_0 + e_1\varphi_i, \quad i = 1, 2, 3, 4, 5;$$

where

$$\varphi_1 = \sqrt{\frac{2c_2}{a_2}}\ \mathrm{sech}(\sqrt{c_2}\xi),$$

$$\varphi_2 = \sqrt{\frac{c_2}{a_2}}\ \tanh(\sqrt{-\frac{c_2}{2}}\xi),$$

$$\varphi_3 = \sqrt{\frac{2c_2m^2}{a_2(2m^2-1)}}\ \mathrm{cn}(\sqrt{\frac{c_2}{2m^2-1}}\xi),$$

$$\varphi_4 = \sqrt{\frac{2c_2}{a_2(2-m^2)}}\ \mathrm{dn}(\sqrt{\frac{c_2}{2-m^2}}\xi),$$

$$\varphi_5 = \sqrt{\frac{2c_2m^2}{a_2(m^2+1)}}\ \mathrm{sn}(\sqrt{-\frac{c_2}{m^2+1}}\xi),$$

where $\xi = x + ct$, the constants a_0, b_0, e_1, b_2 are given by (2.3), and the constants $c_0, c_2, c, d_0, d_1, b_2, a_2$ are arbitrary. As $m \to 1$, the Jacobi periodic solutions (u_3, v_3, w_3, p_3) and (u_4, v_4, w_4, p_4) degenerate to the solitary wave solution (u_1, v_1, w_1, p_1) and (u_2, v_2, w_2, p_2) respectively.

(B) Making use of (1.12), (1.13) and (2.4), we obtain a solitary wave solution ($i = 6$) and a Weierstrass periodic solution ($i = 7$):

$$u_i = a_0 + a_1\varphi_i, \quad v_i = b_0 + b_1\varphi_i,$$
$$w_i = d_0 + d_1\varphi_i, \quad p_i = e_0 + e_1\varphi_i, \quad (i = 6, 7),$$

and

$$\varphi_6 = \frac{c_2}{a_1}\ \mathrm{sech}^2(\frac{\sqrt{c_2}}{2}\xi), \quad \varphi_7 = \wp\left(\frac{\sqrt{-a_1}}{2}\xi\right),$$

$$g_2 = 4c_1/a_1, g_3 = 4c_0/a_1,$$

where $\xi = x+ct$, the constants a_0, b_0, b_1 are given by (2.4), and the constants $c_0, c_1, c_2, c, d_0, e_0, e_1, a_1$ are arbitrary. It is easy to verify that these solutions satisfy the Ito system (1.17) with help of *Mathematica*.

3. The Solutions of Ito System (1.18)

In a similar way to the system (1.17), the solitary wave, Jacobi and Weierstrass periodic solutions of system (1.18) are obtained as follows:

$$u_1 = \frac{1}{3}(c - 4c_2) + 4c_2\varphi^2,$$

$$v_1 = -\frac{1}{2}(c^2 - \frac{a_2 e_0 c}{e_2} + \frac{2d_0 e_2}{a_2}) + 4cc_2\varphi^2$$

$$w_1 = d_0 - \frac{2a_2 c_2 c}{e_2}\varphi^2,$$

$$p_1 = e_0 + \frac{4e_2 c_2}{a_2}\varphi^2,$$

$$\varphi = \operatorname{sech}^2(\sqrt{c_2}\xi);$$

$$u_2 = \frac{1}{3}(c - 4c_2) - 4c_4\varphi^2,$$

$$v_2 = -\frac{1}{2}(c^2 - \frac{a_2 e_0 c}{e_2} + \frac{2d_0 e_2}{a_2}) + 4cc_4\varphi^2.$$

$$w_2 = d_0 - \frac{2a_2 c_4 c}{e_2}\varphi^2,$$

$$p_2 = e_0 + \frac{4e_2}{a_2}\varphi^2,$$

$$\varphi = \sqrt{-\frac{c_2 m^2}{c_4(2m^2 - 1)}} \operatorname{cn}\left(\sqrt{\frac{c_2}{2m^2 - 1}}\xi\right);$$

$$u_3 = \frac{1}{3}c + a_1\varphi,$$

$$v_3 = -\frac{1}{2}(c^2 - \frac{a_1 e_0 c}{e_1} + \frac{2d_0 e_2}{a_1}) + a_1 c\varphi$$

$$w_3 = d_0 - \frac{a_1^2 c}{2e_1}\varphi,$$

$$p_3 = e_0 + e_1\varphi,$$

$$\varphi = \wp\left(\frac{\sqrt{-a_1}}{2}\xi, g_2, g_3\right),$$

where $g_2 = 4c_1/a_1, g_3 = 4c_0/a_1$ and $\xi = x + ct$. As $m \to 1$, the Jacobi periodic solutions (u_2, v_2, w_2, p_2) degenerates to the solitary wave solution

(u_1, v_1, w_1, p_1). It is easy to verify that these solutions satisfy the Ito system (1.18) with help of *Mathematica*.

In summary, we have applied a unified algebraic method with symbolic computation to construct a series of travelling wave solutions of two extended coupled Ito systems. However we still have no way to get all travelling wave solutions of the systems due to complication of nonlinear system. More general solutions are expected to be obtained by studying the case when $r > 4$. In addition, a large number of nonlinear equations may be studied and solved in this simple and systematic way, including the classical KdV, MdV, KP, Jaulent-Miodek, BBM, Kawachra, variant Boussinesq, Schrödinger, sine-Gordon, sinh-Gordon, Dodd-Bullough-Mikhailov, coupled KdV and coupled Schrödinger-Boussinesq equations.

Acknowledgments

The author is mostly grateful to the referees and editors for helpful suggestions and timely help. This work has been supported by the Chinese Basic Research Plan "Mathematics Mechanization and a Platform for Automated Reasoning" and Shanghai "Shuguang" Project.

References

1. M. J. Ablowitz, H. Segur. Soliton and the inverse scattering transformation. In *Studies in Applied Mathematics*, 4. SIAM, Philadelphia, Pa., 1981
2. V. G. Dubrousky, B. G. Konopelchenko. Delta-dressing and exact solutions for the $(2+1)$-dimensional Harry Dym equation. *J. Phys. A.* **27** (1994), 4719–4721.
3. P. G. Estevez. Darboux transformation and solutions for an equation in 2+1 dimensions. *J. Math. Phys.* **40** (1999), 1406–1419.
4. E. G. Fan. Darboux transformation and soliton-like solutions for the Gerdjikov-Ivanov equation, *J. Phys. A* **33** (2000), 6925–6933.
5. E. G. Fan. Extended tanh-function method and its applications to nonlinear equations. *Phys. Lett. A* **277** (2000), 212–218.
6. E. G. Fan. A family of completely integrable multi-Hamiltonian systems explicitly related to some celebrated equations. *J. Math. Phys.* **42** (2001), 4327–4344.
7. E. G. Fan. Multiple travelling wave solutions of nonlinear evolution equations using a unified algebraic method. *J. Phys. A* **35** (2002), 6853–6872.
8. E. G. Fan. Travelling wave solutions for nonlinear equations using symbolic computation. *Comput. Math. Appl.* **43** (2002), 671–680.
9. C. H. Gu, H. S. Hu, and Z. X. Zhou. Darboux transformations in soliton theory and its geometric applications. *Shanghai Sci. Tech. Publ.*, 1999.

10. W. Hereman. Exact solitary wave solutions of coupled nonlinear evolution equations using *MACSYMA*. *Comput. Phys. Commun.* **65** (1991), 143–150.
11. R. Hirota, J. Satsuma. Soliton solution of a coupled KdV equation. *Phys. Lett. A* **85** (1981), 407–408.
12. M. Ito. An extension of nonlinear evolution equations of the KdV (MKdV) type to higher orders. *J. Phys. Soc. Jpn.* **49** (1980), 771–778.
13. S. B. Leble, N. V. Ustinov. Darboux transforms, deep reductions and solitons. *J. Phys. A* **26** (1993), 5007–5016.
14. Z. B. Li, Y. P. Liu. Exact solitary wave and soliton solutions of the fifth order model equation. *Acta Math. Sin.* **22** (2002), 138–144.
15. Z. B. Li, S. Q. Zhang. Symbolic computation for exact solutions of nonlinear evolution equations. *Acta Math. Sin.* **17** (1997), 81–90.
16. W. Malfliet. Solitary wave solutions of nonlinear wave equations. *Am. J. Phys.* **60** (1992), 650–654.
17. V. B. Matveev, M. A. Salle, *Darboux transformation and solitons.* Springer-Verlag, Berlin 1991.
18. G. Neugebauer, D. Kramerl. Einstein-Maxwell solutions. *J. Phys. A* **16** (1983), 1927–1936.
19. E. J. Parkes, B.R. Duffy. An automated tanh-function method for finding solitary wave solutions to nonlinear evolution equations. *Comput. Phys. Commun.* **98** (1996), 288–300.
20. J. Satsuma, R. Hirota. A coupled KdV equation is one case of the 4-reduction of the KP hierarchy. *J. Phys. Soc. Jpn.* **51** (1982), 3390–3397.
21. H. T. Tam, X. B. Hu, D. L. Wang. Two integrable coupled nonlinear systems. *J. Phys. Soc. Jpn.* **68** (1999), 369–379.
22. H. W. Tam, W. X. Ma, X. B. Hu. The Hirota-Satsuma coupled KdV equation and a coupled Ito system revisited. *J. Phys. Soc. Jpn.*, **69** (2000), 45–51.
23. M. L. Wang. Exact solutions for a compund KdV-Burgers equation. *Phys. Lett. A* **215** (1996), 279–287.
24. Y. X. Yao, Z. B. Li. New exact solutions for three evolution equations. *Phys. Lett. A* **297** (2002), 196–204.

CORNER POINT PASTING AND DIXON $\mathcal{A}$-RESULTANT QUOTIENTS

MAO-CHING FOO, ENG-WEE CHIONH

School of Computing
National University of Singapore
Singapore 117543
E-mail: {foomaoch,chionhew}@comp.nus.edu.sg

The Dixon determinant is the exact $\mathcal{A}$-resultant when $\mathcal{A}$ is a complete rectangle [7] or a corner-cut rectangle [3]. The Dixon determinant becomes a multiple of the $\mathcal{A}$-resultant when $\mathcal{A}$ is a complete rectangle made smaller with corner edge cutting, and the extraneous factors are brackets determined by the vertices of $\mathcal{A}$. Consequently the $\mathcal{A}$-resultant can be expressed explicitly as a quotient [9]. This paper reports a corresponding quotient formula when $\mathcal{A}$ is a corner-cut rectangle made larger with corner point pasting. Besides being useful in itself as an $\mathcal{A}$-resultant quotient formula, the result also shows that for $\mathcal{A}$-resultants, there is a duality relationship between edge cutting a complete rectangle and point pasting a corner-cut rectangle.

Keywords: rectangular corner cutting, corner edge cutting, corner point pasting, Dixon resultant, $\mathcal{A}$-resultant

1. Introduction

The Dixon determinant for three bi-degree (m, n) polynomials gives the exact $\mathcal{A}$-resultant (the sparse resultant customed for the monomial support $\mathcal{A}$) for two types of unmixed $\mathcal{A}$: a complete rectangle [7] or a corner-cut rectangle [3]. For other monomial supports, it is known that a maximal minor of the Dixon determinant is a multiple of the $\mathcal{A}$-resultant [8], [12]. To recover the $\mathcal{A}$-resultant in these situations, both a maximal minor and its extraneous factors have to be identified. When $\mathcal{A}$ is a complete rectangle made smaller with corner chipping called edge cutting, the Dixon determinant is the only maximal minor and the extraneous factors are brackets determined by the vertices of $\mathcal{A}$ [9]. This paper reports a dual phenomenon: when $\mathcal{A}$ is a corner-cut rectangle made larger with corner pasting called point pasting, the Dixon determinant is also the only maximal minor and similarly the extraneous factors are brackets determined by the vertices of $\mathcal{A}$. For example, when $(m, n) = (5, 4)$, the following edge cutting and point pasting monomial supports (0 and 1 indicate the absence and presence of a monomial respectively) have similar quotient formulas for their $\mathcal{A}$-resultants.

$$\begin{matrix} 1 & 1 & 1 & 1 & 1 & 1 \\ 1 & 1 & 1 & 1 & 1 & 1 \\ 0 & 1 & 1 & 1 & 1 & 1 \\ 0 & 1 & 1 & 1 & 1 & 1 \\ 0 & 0 & 0 & 0 & 1 & 1 \end{matrix} \qquad \begin{matrix} 1 & 1 & 1 & 1 & 1 & 1 \\ 1 & 1 & 1 & 1 & 1 & 1 \\ 0 & 0 & 0 & 1 & 1 & 1 \\ 0 & 0 & 0 & 0 & 1 & 1 \\ 0 & 0 & 0 & 0 & 1 & 1 \end{matrix}$$

The general situation of this duality phenomenon for $\mathcal{A}$-resultants between corner edge cutting a complete rectangle and corner point pasting a corner-cut rectangle is depicted in Figure 1.

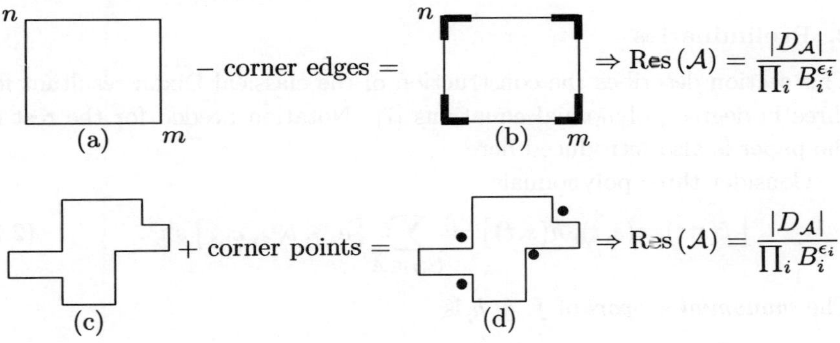

Figure 1. (a) A complete rectangle. (b) A complete rectangle with corner edge cutting; the thick lines are the corner edges cut. (c) A corner-cut rectangle. (d) A corner-cut rectangle with corner point pasting.

As the results can be applied to one or more of the four corners independently, a quotient formula for the $\mathcal{A}$-resultant can be obtained readily for an edge cutting or a point pasting unmixed monomial support $\mathcal{A}$. An immediate applications of these formulas will be the implicitization of toric patches [4], [15]. In particular, the quotient formulas can implicitize any general bi-degree $(3,3)$ toric patches as defined in [11] except the following monomial support

$$\begin{matrix} 1 & 1 & 1 & 1 \\ 0 & 1 & 1 & 1 \\ 0 & 0 & 1 & 1 \\ 0 & 0 & 0 & 1 \end{matrix}$$

or its equivalent. This is important since bi-degree $(3,3)$ is the most common bi-degree for surface patches in computer aided geometric design.

For a general discussion of resultants we refer the reader to [6,13]. The $\mathcal{A}$-resultant quotient formula reported in this paper involves pure bracket expressions. Recent results on $\mathcal{A}$-resultants which involve Macaulay style quotient formulas or hybrid resultants are reported in [5,1,10,14].

The rest of the paper is organized in four sections. Section 2 describes the Dixon method and defines row and column supports of the Dixon matrix. Section 3 states the result as a theorem and illustrates the theorem with an example. Section 4 proves the theorem by identifying the extraneous factors, showing that the Dixon determinant has the right degree, and establishing that the Dixon determinant is non-zero in general. Section 5 summarizes the paper and reports some related observations.

2. Preliminaries

This section describes the construction of the classical Dixon resultant for three bi-degree polynomial equations [7]. Notation needed for the rest of the paper is also introduced here.

Consider three polynomials

$$\left[f(s,t),\, g(s,t),\, h(s,t)\right] = \sum_{(i,j)\in\mathcal{A}} \left[a_{i,j},\, b_{i,j},\, c_{i,j}\right] s^i t^j. \tag{2.1}$$

The *monomial support* of f, g, h is

$$\mathcal{A} = \{(i,j) \mid a_{i,j}, b_{i,j}, c_{i,j} \neq 0\} \subseteq \mathcal{A}_{m,n} \tag{2.2}$$

where

$$\mathcal{A}_{m,n} = \{(i,j) \mid i = 0,\ldots,m; j = 0,\ldots,n\} = 0..m \times 0..n \tag{2.3}$$

is the monomial support of a general bi-degree (m,n) polynomial. Note that $a..b$ denotes the set of consecutive integers from a to b and $a..b \times c..d$ is the cartesian product of $a..b$ and $c..d$.

The *Dixon polynomial* of f, g, h (2.1) is

$$\Delta_{\mathcal{A}}(f(s,t), g(\alpha,t), h(\alpha,\beta)) = \frac{\begin{vmatrix} f(s,t) & g(s,t) & h(s,t) \\ f(\alpha,t) & g(\alpha,t) & h(\alpha,t) \\ f(\alpha,\beta) & g(\alpha,\beta) & h(\alpha,\beta) \end{vmatrix}}{(s-\alpha)(t-\beta)}. \tag{2.4}$$

Our aim is to investigate the matrix form

$$\Delta_{\mathcal{A}} = \left[\cdots\, s^\sigma t^\tau \cdots\right] D_{\mathcal{A}} \left[\cdots\, \alpha^a \beta^b \cdots\right]^T \tag{2.5}$$

where the coefficient matrix $D_{\mathcal{A}}$ is called the Dixon matrix of (2.1).

The monomials $s^\sigma t^\tau$ (resp. $\alpha^a\beta^b$) that occur in $\Delta_\mathcal{A}$ are called the row (resp. column) indices of $D_\mathcal{A}$. The monomial support $\mathcal{R}_A$ (resp. $\mathcal{C}_A$) of $\Delta_\mathcal{A}$ considered as a polynomial in s, t (resp. α, β) is called the row (resp. column) support of $D_\mathcal{A}$. That is, we have

$$\mathcal{R}_A = \{(\sigma,\tau) \mid cs^\sigma t^\tau \alpha^a\beta^b \text{ is a term in } \Delta_\mathcal{A} \text{ for some } a, b \text{ with } c \neq 0\},$$
$$\mathcal{C}_A = \{(a,b) \mid cs^\sigma t^\tau \alpha^a\beta^b \text{ is a term in } \Delta_\mathcal{A} \text{ for some } \sigma,\tau \text{ with } c \neq 0\}.$$

The ordered pair notation is overloaded to also mean a coefficient row vector: $(i,j) = \left[a_{i,j}\ b_{i,j}\ c_{i,j}\right]$. A *bracket* is a vector triple product

$$(i,j)\cdot(k,l)\times(p,q) = \begin{vmatrix} a_{i,j} & b_{i,j} & c_{i,j} \\ a_{k,l} & b_{k,l} & c_{k,l} \\ a_{p,q} & b_{p,q} & c_{p,q} \end{vmatrix}. \tag{2.6}$$

Using brackets, we can write the numerator in (2.4) as

$$\sum_{(i,j),(k,l),(p,q)\in\mathcal{A}} (i,j)\cdot(k,l)\times(p,q)s^i t^{j+l}\alpha^{k+p}\beta^q \tag{2.7}$$

and thus the entries of $D_\mathcal{A}$ are linear in the coefficients of each of f, g, h. Clearly $\Delta_{\mathcal{A}_{m,n}}$ is of degree $m-1$ in s, $2n-1$ in t, $2m-1$ in α, and $n-1$ in β. Consequently,

$$\begin{aligned} \mathcal{R}_{\mathcal{A}_{m,n}} &= 0\mathinner{..}(m-1)\times 0\mathinner{..}(2n-1), \\ \mathcal{C}_{\mathcal{A}_{m,n}} &= 0\mathinner{..}(2m-1)\times 0\mathinner{..}(n-1), \end{aligned} \tag{2.8}$$

and the set cardinalities $\#\mathcal{R}_{\mathcal{A}_{m,n}} = \#\mathcal{C}_{\mathcal{A}_{m,n}} = 2mn$. Thus $D_{\mathcal{A}_{m,n}}$ is a square matrix of order $2mn$. The determinant $|D_{\mathcal{A}_{m,n}}|$ is the classical Dixon resultant of f, g, h.

We shall use the abbreviation $ijklpq = (i,j)\cdot(k,l)\times(p,q)$ in the examples. For instance, $\left|D_{\mathcal{A}_{1,1}}\right| = \begin{vmatrix} 100100 & 101100 \\ 110100 & 110110 \end{vmatrix}$.

3. The Dixon $\mathcal{A}$-Resultant Quotients

The main result of the paper is an explicit quotient expression for the $\mathcal{A}$-resultant for a monomial support $\mathcal{A}$ obtained by rectangular corner cutting followed by corner point pasting. (See Theorem 3.1 for the precise definition of $\mathcal{A}$.)

In the following discussions, we define for any set $\mathcal{S}$: $\mathcal{S}^1 = \mathcal{S}$ and $\mathcal{S}^0 = \emptyset$. Also, we define the indeterminate 0^0 to be 1.

Theorem 3.1. *Let the monomial support of f, g, h be $\mathcal{A} = \mathcal{A}_{m,n} - \cup_{i=1}^{4} \mathcal{S}_i$ where*

$$
\begin{aligned}
\mathcal{S}_1 &= \ 0 \mathinner{..} b_1 \times 0 \mathinner{..} l_1 \ - \ \{(b_1, l_1)\}^{\delta_1}, \\
\mathcal{S}_2 &= b_2 \mathinner{..} m \times 0 \mathinner{..} r_1 \ - \ \{(b_2, r_1)\}^{\delta_2}, \\
\mathcal{S}_3 &= t_2 \mathinner{..} m \times r_2 \mathinner{..} n \ - \ \{(t_2, r_2)\}^{\delta_3}, \\
\mathcal{S}_4 &= \ 0 \mathinner{..} t_1 \times l_2 \mathinner{..} n \ - \ \{(t_1, l_2)\}^{\delta_4},
\end{aligned}
\tag{3.2}
$$

with

$$
\begin{aligned}
&-1 \le b_1 < b_1 + 1 < b_2 \le m + 1, \ \ -1 \le r_1 < r_1 + 1 < r_2 \le n + 1, \\
&-1 \le t_1 < t_1 + 1 < t_2 \le m + 1, \ \ -1 \le l_1 < l_1 + 1 < l_2 \le n + 1;
\end{aligned}
\tag{3.3}
$$

and $\delta_i = 0, 1$. The $\mathcal{A}$-resultant of f, g, h is

$$
\operatorname{Res}(\mathcal{A}) = \frac{|D_{\mathcal{A}}|}{B_1^{\delta_1 \epsilon_1} B_2^{\delta_2 \epsilon_2} B_3^{\delta_3 \epsilon_3} B_4^{\delta_4 \epsilon_4}}
\tag{3.4}
$$

where $D_{\mathcal{A}}$ is the Dixon matrix for the monomial support $\mathcal{A}$,

$$
\begin{aligned}
B_1 &= \ (0, l_1 + 1) \cdot (b_1, l_1) \times (b_1 + 1, 0), \\
B_2 &= (b_2 - 1, 0) \cdot (b_2, r_1) \times (m, r_1 + 1), \\
B_3 &= (t_2 - 1, n) \cdot (t_2, r_2) \times (m, r_2 - 1), \\
B_4 &= \ (0, l_2 - 1) \cdot (t_1, l_2) \times (t_1 + 1, n);
\end{aligned}
\tag{3.5}
$$

and

$$
\begin{aligned}
&\epsilon_1 = 0 \text{ if } b_1 \le 0 \text{ or } l_1 \le 0; \qquad \epsilon_2 = 0 \text{ if } b_2 \ge m \text{ or } r_1 \le 0; \\
&\epsilon_3 = 0 \text{ if } t_2 \ge m \text{ or } r_2 \ge n; \qquad \epsilon_4 = 0 \text{ if } t_1 \le 0 \text{ or } l_2 \ge n.
\end{aligned}
\tag{3.6}
$$

Remark 3.7. Note that $\delta_i = 1$ (or 0) when a point is pasted (or not pasted). Condition (3.6) states when $\mathcal{S}_i$ is a vertical line, a horizontal line, a point, or null, we have $\epsilon_i = 0$ because there is no extraneous factor and the value of δ_i does not matter.

The theorem statement may look complicated but its application is actually straightforward. This is illustrated in the following example.

Example 3.8. Consider the monomial support $\mathcal{A}$ and the corresponding δ_i, ϵ_i:

t^5	4	4	4	•	•	3	3	3
t^4	4	4	4	•	•	•	3	3
t^3	4	4	•	•	•	•	•	•
t^2	•	•	•	•	2	2	2	2
t^1	•	•	•	•	2	2	2	2
1	•	•	•	•	2	2	2	2
$\mathcal{A}$	1	s	s^2	s^3	s^4	s^5	s^6	s^7

$i =$	1	2	3	4
δ_i	–	0	1	1
ϵ_i	0	1	1	1
$\epsilon_i \delta_i$	0	0	1	1

where elements of the monomial support are marked "•" and elements of $\mathcal{S}_i$ are marked i. By Theorem 3.1, the $\mathcal{A}$-resultant is

$$\text{Res}\,(\mathcal{A}) = \frac{|D_{\mathcal{A}}|}{0^0 \cdot 304273^0 \cdot 455473^1 \cdot 022335^1} = \frac{|D_{\mathcal{A}}|}{455473 \cdot 022335}.$$

4. Proof of the Main Theorem

In this section, we present important facts and develop intermediate results needed to prove Theorem 3.1 at the end of the section. Here, we define the Minkowski sum of an element and a set as

$$(a,b) \oplus \{(x,y) \mid \cdots\} = \{(a+x, b+y) \mid \cdots\}. \tag{4.1}$$

Entry Formula for the Dixon Matrix

Theorem 4.2. *The Dixon matrix entry indexed by* $(s^\sigma t^\tau, \alpha^a \beta^b)$ *is*

$$\begin{aligned} &D(s^\sigma t^\tau, \alpha^a \beta^b) \\ &= \sum_{u=0}^{\min(a,m-1-\sigma)} \sum_{v=0}^{\min(b,2n-1-\tau)} \sum_{k=\max(0,a-u-\sigma)}^{\min(m,a-u)} \sum_{l=\max(b+1,\tau+1+v-b)}^{\min(n,\tau+1+v)} B \\ &+ \sum_{u=0}^{\min(a,m-1-\sigma)} \sum_{v=0}^{\min(b,2n-1-\tau)} \sum_{k=\max(0,a-u-m)}^{\min(\sigma,a-u)} \sum_{l=\max(b+1,\tau+1+v-n)}^{\min(n,\tau+v-b)} B \end{aligned} \tag{4.3}$$

where $B = (\sigma+1+u, \tau+1+v-l) \cdot (k,l) \times (a-u-k, b-v)$.

The entry formula proved in [2] is needed to derive a simple formula in Proposition 4.5 for three special columns of the Dixon matrix after bottom-left corner point pasting. The entry formula is also needed in the automated proof of column independence in Proposition 4.15.

The Row and Column Supports of $\mathcal{A} = \mathcal{A}_{m,n} - \cup_{i=1}^4 \mathcal{S}_i$

Proposition 4.4. *Let* $\mathcal{A} = \mathcal{A}_{m,n} - \cup_{i=1}^4 \mathcal{S}_i$. *The row support of* $\mathcal{A}$ *is*

$$\mathcal{R}_A = \mathcal{R}_{m,n} - \mathcal{S}_1 - [(-1,0) \oplus \mathcal{S}_2] - [(-1,n-1) \oplus \mathcal{S}_3] - [(0,n-1) \oplus \mathcal{S}_4]$$

and the column support of $\mathcal{A}$ *is*

$$\mathcal{C}_A = \mathcal{C}_{m,n} - \mathcal{S}_1 - [(m-1,0) \oplus \mathcal{S}_2] - [(m-1,-1) \oplus \mathcal{S}_3] - [(0,-1) \oplus \mathcal{S}_4].$$

The row and column supports proposition is proved in [3]. Briefly, the proposition says that if $\mathcal{A}$ is obtained by cutting $\mathcal{S}_1$, $\mathcal{S}_2$, $\mathcal{S}_3$, $\mathcal{S}_4$ from $\mathcal{A}_{m,n}$, then the row and column supports of $\mathcal{A}$ are obtained simply by cutting the corresponding corner sub-supports from the row and column supports $\mathcal{R}_{m,n}$ and $\mathcal{C}_{m,n}$.

Entries of the Dixon Matrix After Bottom-Left Corner Point Pasting

The following proposition shows that rectangular corner cutting followed by corner point pasting simplifies the entries of the Dixon Matrix $D_{\mathcal{A}}$.

Proposition 4.5. *Let $\mathcal{A} = \mathcal{A}_{m,n} - \mathcal{S}_1$, $1 \le b_1 \le m-1$, $1 \le l_1 \le n-1$. The entries in the three columns of $D_{\mathcal{A}}$ indexed by*

$$(a,b) = (0, l_1+1), (b_1, l_1), (b_1+1, 0)$$

are as follows

$$D(s^\sigma t^\tau, \alpha^a \beta^b) = \sum_{l=\max(l_1+1,\tau+1-n)}^{\min(n,\tau+1)} (\sigma+1, \tau+1-l) \times (0,l) \cdot (a,b),$$

where $(\sigma, \tau) \in \mathcal{R}_A$.

Proof. Substitute $a = 0$, $b = l_1 + 1$ into Equation (4.3). The summation ranges require $0 \le u \le a$, $0 \le k \le a - u$, thus forcing $u = k = 0$. For $(a-u-k, b-v) = (0, l_1+1-v) \notin \mathcal{S}_1$, we need $v = 0$. Hence $D(s^\sigma t^\tau, \alpha^0 \beta^{l_1+1})$ simplifies to

$$\sum_{l=\max(l_1+2,\tau-l_1)}^{\min(n,\tau+1)} (\sigma+1, \tau+1-l) \times (0,l) \cdot (0, l_1+1)$$

$$+ \sum_{l=\max(l_1+2,\tau+1-n)}^{\min(n,\tau-l_1-1)} (\sigma+1, \tau+1-l) \times (0,l) \cdot (0, l_1+1).$$

When $\tau - l_1 - 1 \le n$ and $\tau - l_1 \ge l_1 + 2$, it is obvious that the sums can be combined as

$$D(s^\sigma t^\tau, \alpha^0 \beta^{l_1+1}) = \sum_{l=\max(l_1+2,\tau+1-n)}^{\min(n,\tau+1)} (\sigma+1, \tau+1-l) \times (0,l) \cdot (0, l_1+1).$$

It can be checked that when $\tau - l_1 - 1 > n$ or $\tau - l_1 < l_1 + 2$, the combination remains valid. Finally, since the bracket vanishes when $l = l_1 + 1$, we can enlarge the lower bound as $l = \max(l_1 + 1, \tau + 1 - n)$. (See Remark 4.7.) This completes the proof for the entry formula for $D(s^\sigma t^\tau, \alpha^0 \beta^{l_1+1})$.

To prove the other two formulas, we note that both formulas demand $(a - u - k, b - v) \notin \mathcal{S}_1$, thus forcing $k = u = v = 0$. The rest of the proof is then done similarly. To get the summation range of $D(s^\sigma t^\tau, \alpha^{b_1+1} \beta^0)$, note that $(0, l) \in \mathcal{S}_1$ when $l \leq l_1$. □

Remark 4.6. The summation range is null when $\tau < l_1$, thus the column entries are zero whenever $\tau < l_1$.

Remark 4.7. The proof of Proposition 4.9 requires the summation ranges for all three formulas to be alike.

Remark 4.8. When $l_1 = n - 1$, the column indexed by $\alpha^0 \beta^{l_1+1}$ does not exist. The summation range of the other two formulas is either null or $n \leq l \leq n$. Thus the entries simplify further to become

$$D(s^\sigma t^\tau, \alpha^{b_1} \beta^{n-1}) = (\sigma + 1, \tau + 1 - n) \times (0, n) \cdot (b_1, n - 1),$$
$$D(s^\sigma t^\tau, \alpha^{b_1+1} \beta^0) = (\sigma + 1, \tau + 1 - n) \times (0, n) \cdot (b_1 + 1, 0).$$

Divisibility by the Brackets B_1, B_2, B_3, B_4

Proposition 4.9. *$|D_{\mathcal{A}_{m,n} - \mathcal{S}_1}|$ is divisible by the bracket B_1.*

Proof. We examine two cases: $l_1 < n - 1$ and $l_1 = n - 1$. When $l_1 < n - 1$, it suffices to show that the determinant of any 3×3 submatrix of the columns indexed by $\alpha^0 \beta^{l_1+1}$, $\alpha^{b_1} \beta^{l_1}$, $\alpha^{b_1+1} \beta^0$ contains the factor B_1. By Proposition 4.5 we have

$$\begin{vmatrix} (\sum_{l \in l_A} F_l \times P_l) \cdot X & (\sum_{l \in l_A} F_l \times P_l) \cdot Y & (\sum_{l \in l_A} F_l \times P_l) \cdot Z \\ (\sum_{l \in l_B} G_l \times P_l) \cdot X & (\sum_{l \in l_B} G_l \times P_l) \cdot Y & (\sum_{l \in l_B} G_l \times P_l) \cdot Z \\ (\sum_{l \in l_C} H_l \times P_l) \cdot X & (\sum_{l \in l_C} H_l \times P_l) \cdot Y & (\sum_{l \in l_C} H_l \times P_l) \cdot Z \end{vmatrix}$$

$$= \begin{vmatrix} \sum_{l \in l_F} F_l \times P_l \\ \sum_{l \in l_G} G_l \times P_l \\ \sum_{l \in l_H} H_l \times P_l \end{vmatrix} |X^T Y^T Z^T| \tag{4.10}$$

where the range $l_F = \max(l_1 + 1, \tau_F + 1 - n) .. \min(n, \tau_F + 1)$, the ordered pair $F_l = (\sigma_F + 1, \tau_F + 1 - l)$; l_G, G_l, l_H, and H_l are similarly defined;

$P_l = (0, l)$; $X = (0, l_1+1)$, $Y = (b_1, l_1)$, $Z = (b_1+1, 0)$. The proposition is proved by observing that $|X^T Y^T Z^T| = B_1$.

When $l_1 = n-1$, we consider the determinant of any 2×2 submatrix of the columns indexed by $\alpha^{b_1}\beta^{l_1}$, $\alpha^{b_1+1}\beta^0$. By Remark 4.8 of Proposition 4.5, we have by direct computation

$$\begin{vmatrix} (\sigma_1+1, \tau_1+1-n) \times (0,n) \cdot (b_1, n-1) & (\sigma_1+1, \tau_1+1-n) \times (0,n) \cdot (b_1+1, 0) \\ (\sigma_2+1, \tau_2+1-n) \times (0,n) \cdot (b_1, n-1) & (\sigma_2+1, \tau_2+1-n) \times (0,n) \cdot (b_1+1, 0) \end{vmatrix} =$$

$$\Big((\sigma_1+1, \tau_1+1-n) \times (\sigma_2+1, \tau_2+1-n) \cdot (0,n)\Big)\Big((b_1, n-1) \times (b_1+1, 0) \cdot (0,n)\Big).$$

Thus B_1 divides the determinant of these 2×2 submatrices. □

Proposition 4.11. $|D_{\mathcal{A}_{m,n}-\mathcal{S}_i}|$ *is divisible by the bracket* B_i, $i = 2, 3, 4$.

Proof. To show that $|D_{\mathcal{A}_{m,n}-\mathcal{S}_3}|$ is divisible by the bracket B_3, let $L_{i,j} = [a_{i,j}\ b_{i,j}\ c_{i,j}]$, $\overline{L}_{i,j} = [a_{m-i,n-j}\ b_{m-i,n-j}\ c_{m-i,n-j}]$, $\overline{s} = s^{-1}$, $\overline{t} = t^{-1}$, $\overline{\alpha} = \alpha^{-1}$ and $\overline{\beta} = \beta^{-1}$. We have,

$$\frac{\begin{vmatrix} \sum_{i=0}^m \sum_{j=0}^n L_{i,j} s^i t^j \\ \sum_{i=0}^m \sum_{j=0}^n L_{i,j} \alpha^i t^j \\ \sum_{i=0}^m \sum_{j=0}^n L_{i,j} \alpha^i \beta^j \end{vmatrix}}{(s-\alpha)(t-\beta)} = s^{m-1} t^{2n-1} \alpha^{2m-1} \beta^{n-1} \frac{\begin{vmatrix} \sum_{i=0}^m \sum_{j=0}^n \overline{L}_{i,j} \overline{s}^i \overline{t}^j \\ \sum_{i=0}^m \sum_{j=0}^n \overline{L}_{i,j} \overline{\alpha}^i \overline{t}^j \\ \sum_{i=0}^m \sum_{j=0}^n \overline{L}_{i,j} \overline{\alpha}^i \overline{\beta}^j \end{vmatrix}}{(\overline{s}-\overline{\alpha})(\overline{t}-\overline{\beta})}.$$

The monomial support with respect to $\overline{L}_{i,j}$ is $\overline{\mathcal{A}} = \mathcal{A}_{m,n} - \overline{\mathcal{S}_1}$ where $\overline{\mathcal{S}_1} = 0 \,..\, b_1' \times 0 \,..\, l_1' - (b_1', l_1')\}$, $b_1' = m - t_2$, $l_1' = n - r_2$. By Proposition 4.9, we know that the determinant is divisible by

$$\begin{aligned} \overline{B_1} &= (\overline{0}, \overline{l_1'+1}) \times (\overline{b_1'}, \overline{l_1'}) \cdot (\overline{b_1'+1}, \overline{0}) \\ &= (m, n - l_1' - 1) \times (m - b_1', n - l_1') \cdot (m - b_1' - 1, n) \\ &= (m, r_2 - 1) \times (t_2, r_2) \cdot (t_2 - 1, n) = B_3. \end{aligned}$$

Furthermore, the three columns responsible for this bracket are

$$\begin{aligned} &\alpha^{2m-1}\beta^{n-1}\big[\overline{\alpha}^0\overline{\beta}^{l_1'+1}, \overline{\alpha}^{b_1'}\overline{\beta}^{l_1'}, \overline{\alpha}^{b_1'+1}\overline{\beta}^0\big] \\ &= \big[\alpha^{2m-1}\beta^{n-l_1'-2}, \alpha^{2m-1-b_1'}\beta^{n-l_1'-1}, \alpha^{2m-2-b_1'}\beta^{n-1}\big] \\ &= \big[\alpha^{2m-1}\beta^{r_2-2}, \alpha^{m+t_2-1}\beta^{r_2-1}, \alpha^{m+t_2-2}\beta^{n-1}\big]. \end{aligned}$$

To show that $|D_{\mathcal{A}_{m,n}-\mathcal{S}_2}|$ is divisible by the bracket B_2, let $\overline{s} = s^{-1}$, $\overline{\alpha} = \alpha^{-1}$, and $\overline{L}_{i,j} = [a_{m-i,j} b_{m-i,j} c_{m-i,j}]$. The proof is similar and the three columns responsible are similarly found to be

$$\big[\alpha^{2m-1}\beta^{r_1+1}, \alpha^{m+b_2-1}\beta^{r_1}, \alpha^{m+b_2-2}\beta^0\big].$$

To show that $|D_{\mathcal{A}_{m,n}-\mathcal{S}_4}|$ is divisible by the bracket B_4, let $\bar{t} = t^{-1}$, $\overline{\beta} = \beta^{-1}$, and $\overline{L}_{i,j} = [a_{i,n-j}b_{i,n-j}c_{i,n-j}]$. The proof is similar and the three columns responsible are similarly found to be

$$\left[\alpha^0\beta^{l_2-2}, \alpha^{t_1}\beta^{l_2-1}, \alpha^{t_1+1}\beta^{n-1}\right]. \qquad \square$$

Theorem 4.12. *The Dixon determinant* $|D_{\mathcal{A}_{m,n}-\cup_{i=1}^4 \mathcal{S}_i}|$ *is divisible by the product* $B_1^{\delta_1\epsilon_1}B_2^{\delta_2\epsilon_2}B_3^{\delta_3\epsilon_3}B_4^{\delta_4\epsilon_4}$.

Proof. The result follows from the proofs above because the divisibility proof at each corner is independent of the situations at the other corners except when $l_2 - 1 = l_1 + 1$ or $r_2 - 1 = r_1 + 1$. When this happens, we note that only two columns are needed to produce the divisor B_i. To see this, we need only examine the bottom left corner as the situation at the other corners can be reduced to that of the bottom left corner.

When $l_2 - 1 = l_1 + 1$, the determinant of every 2×2 submatrix of the two columns indexed by $\alpha^{b_1}\beta^{l_1}, \alpha^{b_1+1}\beta^0$ is of the form

$$\begin{vmatrix} (F_{l_1+1} \times P_{l_1+1}) \cdot Y & (F_{l_1+1} \times P_{l_1+1}) \cdot Z \\ (G_{l_1+1} \times P_{l_1+1}) \cdot Y & (G_{l_1+1} \times P_{l_1+1}) \cdot Z \end{vmatrix}, \tag{4.13}$$

where $F_{l_1+1}, G_{l_1+1}, P_{l_1+1}, Y, Z$ are defined in Proposition 4.9, because the brackets $(\sigma+1, \tau+1-l) \times (0,l) \cdot (b_1, l_1)$ and $(\sigma+1, \tau+1-l) \times (0,l) \cdot (b_1+1, 0)$ are non-zero only when $l = l_1 + 1$. By direct computation, the above 2×2 determinant is found to be $|(F_{l_1+1} \times G_{l_1+1}) \cdot P_{l_1+1}|\,|YZP_{l_1+1}|$. Since $|YZP_{l_1+1}| = B_1$, the proof is complete. $\square$

Proof of Theorem 3.1

Finally, we are ready to prove Theorem 3.1. By the theory of $\mathcal{A}$-resultants [6], we only have to show that (1) $\frac{|D_{\mathcal{A}}|}{B_1^{\delta_1\epsilon_1}B_2^{\delta_2\epsilon_2}B_3^{\delta_3\epsilon_3}B_4^{\delta_4\epsilon_4}}$ has the correct degree in the polynomial coefficients and (2) the determinant $|D_{\mathcal{A}}| \neq 0$.

Proposition 4.14. *If* $\frac{|D_{\mathcal{A}}|}{B_1^{\delta_1\epsilon_1}B_2^{\delta_2\epsilon_2}B_3^{\delta_3\epsilon_3}B_4^{\delta_4\epsilon_4}} \neq 0$, *it has the correct degree in the coefficients.*

Proof. Note that the entries of $D_{\mathcal{A}}$ and of the brackets B_i are linear in each of the coefficients of f, g, h. Thus we need only to show that the order of $D_{\mathcal{A}}$ minus $\sum_{i=1}^4 \delta_i\epsilon_i$ is equal to twice the area of the Newton polygon of $\mathcal{A}$.

By Proposition 4.4, the order of $D_{\mathcal{A}}$ is $2mn - \sum_{i=1}^4 \#\mathcal{S}_i$. By direct calculation, we see that when $\mathcal{S}_i$ is cut, an area of size $\frac{\#\mathcal{S}+\delta_i\epsilon_i}{2}$ is chipped away from the rectangular Newton polygon of $\mathcal{A}_{m,n}$. Thus the area of the Newton polygon is $mn - \sum_{i=1}^{4} \frac{\#\mathcal{S}_i + \delta_i\epsilon_i}{2}$. Since clearly, we have

$$\text{order of } D_{\mathcal{A}} - \sum_{i=1}^{4} \delta_i\epsilon_i = 2 \times \text{ area of Newton polygon,}$$

and thus, $\frac{|D_{\mathcal{A}}|}{B_1^{\delta_1\epsilon_1} B_2^{\delta_2\epsilon_2} B_3^{\delta_3\epsilon_3} B_4^{\delta_4\epsilon_4}}$ has the expected degree. □

Proposition 4.15. *The columns responsible for the factor $B_1^{\delta_1\epsilon_1}$, $B_2^{\delta_2\epsilon_2}$, $B_3^{\delta_3\epsilon_3}$, $B_4^{\delta_4\epsilon_4}$ are linearly independent.*

Proof. When $\sum_{i=1}^{4} \epsilon_i = 0$, there are no extraneous factors. So it suffices to consider the cases when $\sum_{i=1}^{4} \epsilon_i \geq 1$. Here, we show that $k = 3\sum_{i=1}^{4} \epsilon_i = 3$, 6, 9 or 12 columns responsible for the extraneous factors are independent because any $k \times k$ submatrix of the k columns is lower triangular with non-zero diagonal entries. The proof is done mechanically by a Maple program with the assume facility for symbolic m, n, l_1, l_2, r_1, r_2, b_1, b_2, t_1 and t_2. We first select the monomial support $\mathcal{A}$ depending on the type of cutting. Next, we select the columns and row indices for the Maple program to generate the $k \times k$ submatrix. Finally, we observe that the $k \times k$ submatrix is lower triangular with non-zero diagonal entries, hence the columns responsible for the factors are independent.

Monomials. We need only consider a monomial support $\mathcal{A}$ whose monomials are the vertices. This is because if the k columns of this special $\mathcal{A}$ are independent then they will remain independent for any $\mathcal{A}$ containing these vertices.

There are three possibilities at each corner i of $\mathcal{A}$, $i = 1, 2, 3, 4$: (1) $\mathcal{S}_i = \emptyset$; (2) $\mathcal{S}_i$ is a rectangle; and (3) $\mathcal{S}_i$ is the union of two rectangles. Consider the bottom-left corner. We have respectively (1) $(0, 0) \in \mathcal{A}$; (2) $(0, l_1 + 1)$, $(b_1 + 1, 0) \in \mathcal{A}$; (3) $(0, l_1 + 1)$, (b_1, l_1), $(b_1 + 1, 0) \in \mathcal{A}$. The same is done at the other corners. Hence $\mathcal{A}$ has at most twelve monomials.

Column indices. Consider the twelve ordered columns that generate all possible extraneous factors B_i for the cases when $\delta_i\epsilon_i = 1$:

$$\begin{aligned}
&\{\alpha^{b_1+1}, \alpha^{b_1}\beta^{l_1}, \beta^{l_1+1}\}^{\delta_1\epsilon_1} \\
&\bigcup \{\alpha^{m+b_2-2}, \alpha^{m+b_2-1}\beta^{r_1}, \alpha^{2m-1}\beta^{r_1+1}\}^{\delta_2\epsilon_2} \\
&\bigcup \{\alpha^{m+t_2-2}\beta^{n-1}, \alpha^{m+t_2-1}\beta^{r_2-1}, \alpha^{2m-1}\beta^{r_2-2}\}^{\delta_3\epsilon_3} \\
&\bigcup \{\alpha^{t_1+1}\beta^{n-1}, \alpha^{t_1}\beta^{l_2-1}, \beta^{l_2-2}\}^{\delta_4\epsilon_4}.
\end{aligned}$$

We select 3, 6, 9 or 12 of the 12 columns based on the above formula.

Row indices. Corresponding to the k columns, we have to select k rows to form a $k \times k$ submatrix. Let the variables γ_1 (resp. $\gamma_2, \gamma_3, \gamma_4$) be 1 if $(0,0)$ (resp. $(m,0)$, (m,n), $(0,n)$) belongs to $\mathcal{A}$ and 0 otherwise. The set of rows to choose are as follows

$$\begin{aligned}
&\{s^{b_1-1}t^{2l_1}, s^{b_1}t^{l_1}\}^{\delta_1\epsilon_1} \cup \{s^{b_1}t^{n-1}\}^{(1-\gamma_4)\delta_1\epsilon_1} \cup \{s^{t_1}t^{n+l_2-2}\}^{\gamma_4\delta_1\epsilon_1}\\
\bigcup &\{s^{b_2}t^{2r_1}, s^{b_2-1}t^{r_1}\}^{\delta_2\epsilon_2} \cup \{s^{b_2-1}t^{n-1}\}^{(1-\gamma_3)\delta_2\epsilon_2} \cup \{s^{t_2-1}t^{r_1+r_2-2}\}^{\gamma_3\delta_2\epsilon_2}\\
\bigcup &\{s^{t_2}t^{2r_2-1}, s^{t_2-1}t^{n+r_2-1}\}^{\delta_3\epsilon_3} \cup \{s^{t_2-1}t^{n}\}^{(1-\gamma_2)\delta_3\epsilon_3} \cup \{s^{b_2-1}t^{r_1+1}\}^{\gamma_2\delta_3\epsilon_3}\\
\bigcup &\{s^{t_1-1}t^{2l_2-1}, s^{t_1}t^{n+l_2-1}\}^{\delta_4\epsilon_4} \cup \{s^{t_1}t^{n}\}^{(1-\gamma_1)\delta_4\epsilon_4} \cup \{s^{b_1}t^{l_1+1}\}^{\gamma_1\delta_4\epsilon_4}.
\end{aligned}$$

The expression looks complicated, but it simply means that for each corner, two of the three rows needed are given in $\{\cdots,\cdots\}^{\delta_i\epsilon_i}$, $i = 1, 2, 3, 4$. The choice of the third row depends on the cutting configuration of the vertically opposite corner and is given in $\{\cdots\}^{(1-\gamma_{5-i})\delta_i\epsilon_i} \cup \{\cdots\}^{\gamma_{5-i}\delta_i\epsilon_i}$. The vertically opposite corner of the top right (top left, bottom left, bottom right) corner is the bottom right (bottom left, top left, top right) corner respectively.

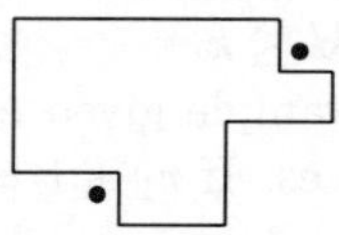

Figure 2. A rectangular bi-degree (m, n) monomial support with rectangular corner cutting, without corner point pasting at the bottom right corner and with corner point pasting at the top right and bottom left corners.

For instance, consider a monomial support with rectangular corner cutting without corner point pasting at the bottom right corner, corner point pasting at the top right, bottom left corners (see Figure 2). We want to prove that the six columns responsible for the brackets B_1 and B_3 are independent. The six columns in $D_{\mathcal{A}}$ indexed by

$$\alpha^{b_1+1}, \alpha^{b_1}\beta^{l_1}, \beta^{l_1+1}, \alpha^{m+t_2-2}\beta^{n-1}, \alpha^{m+t_2-1}\beta^{r_2-1}, \alpha^{2m-1}\beta^{r_2-2},$$

and the six rows in $D_{\mathcal{A}}$ indexed by

$$s^{b_1-1}t^{2l_1}, s^{b_1}t^{l_1}, s^{b_1}t^{n-1}, s^{t_2}t^{2r_2-1}, s^{t_2-1}t^{n+r_2-1}, s^{t_2-1}t^{r_1+1},$$

form a 6×6 lower triangular matrix with non-zero diagonal entries $-B_1$, B_1, $-(0, l_1+1) \times (0, n) \cdot (b_1+1, 0)$, B_3, $-B_3$, $(b_2-1, 0) \times (m, r_1+1) \cdot (m, r_2-1)$.

Given the $3^4 = 81$ combinations of $\mathcal{A}$, and its respective row and column indices, our mechanical prover produces a lower triangular $k \times k$ submatrix with non-zero diagonal entries for each of the cases.

Mechanical Proving. We describe the mechanical prover for the case when all the corners of the monomial support undergo rectangular cutting and point pasting. The other cutting and pasting configurations are proved similarly. The program proves that the 12×12 submatrix is a lower triangular matrix with non-zero diagonal entries by actually finding the matrix entries using Equation (4.3) of Theorem 4.2. Since $\mathcal{A}$ has twelve monomials, there are $12 \times 11 \times 10 = 1320$ (but only $\binom{12}{3} = 220$ of which are distinct modulo sign) non-zero brackets $(e_1, e_2) \times (e_3, e_4) \cdot (e_5, e_6)$ to consider. For each of the 12×12 matrix entries, the value of (σ, τ) and (a, b) are known. The program then solves for u, v, k, l by equating $\sigma + 1 + u = e_1$, $\tau + 1 + v - l = e_2$, $k = e_3$, $l = e_4$, $a - u - k = e_5$, and $b - v = e_6$. A bracket is in the entry if and only if u, v, k, l satisfy the summation bounds of Equation (4.3).

Degenerate cases. When $l_1 + 1 = l_2 - 1$, or $r_1 + 1 = r_2 - 1$ or both occur, we may not need all three columns in a corner to generate an extraneous factor (see Theorem 4.12). In such cases, we observe that the original $k \times k$ non-singular matrix would shrink to a lower triangular $k' \times k'$ submatrix with non-zero diagonal entries, $k' \leq k$.

For instance, consider the example given earlier in the proof when we discuss the selection of row indices. If $r_1 + 1 = r_2 - 1$, the five columns in $D_{\mathcal{A}}$ indexed by

$$\alpha^{b_1+1}, \alpha^{b_1}\beta^{l_1}, \beta^{l_1+1}, \alpha^{m+t_2-2}\beta^{n-1}, \alpha^{m+t_2-1}\beta^{r_2-1},$$

and the five rows in $D_{\mathcal{A}}$ indexed by

$$s^{b_1-1}t^{2l_1}, s^{b_1}t^{l_1}, s^{b_1}t^{n-1}, s^{t_2}t^{2r_2-1}, s^{t_2-1}t^{n+r_2-1},$$

form a 5×5 lower triangular matrix with non-zero diagonal entries $-B_1$, B_1, $-(0, l_1 + 1) \times (0, n) \cdot (b_1 + 1, 0)$, B_3, $-B_3$. □

Proposition 4.16. $|D_{\mathcal{A}}| \neq 0$.

Proof. It is known that any maximal minor of $D_{\mathcal{A}}$ is a multiple of the $\mathcal{A}$-resultant [12,8]. Since the columns in Proposition 4.15 are independent, there is a maximal minor M containing these columns. By Proposition 4.15, the maximal minor M has the factors $B_1^{\delta_1\epsilon_1} B_2^{\delta_2\epsilon_2} B_3^{\delta_3\epsilon_3} B_4^{\delta_4\epsilon_4}$. Thus $M = B_1^{\delta_1\epsilon_1} B_2^{\delta_2\epsilon_2} B_3^{\delta_3\epsilon_3} B_4^{\delta_4\epsilon_4} N$ for some polynomial N and N is a multiple of the $\mathcal{A}$-resultant. By Proposition 4.14, the degree of N in the coefficients of the polynomials f, g, h is at least

$$2mn - \sum_{i=1}^{4} \#\mathcal{S}_i - \sum_{i=1}^{4} \delta_i\epsilon_i.$$

That means the degree of M in the coefficients of each of the polynomials is at least

$$2mn - \sum_{i=1}^{4} \#\mathcal{S}_i$$

which is the order of $D_{\mathcal{A}}$. This means M and $|D_{\mathcal{A}}|$ differ by a constant factor and thus $|D_{\mathcal{A}}|$ is non-zero since M is not. □

5. Conclusion

We have shown that when the monomial support $\mathcal{A}$ is obtained by rectangular corner cutting followed by corner point pasting, the Dixon formulation produces a multiple of the $\mathcal{A}$-resultant with at most four extraneous factors, each of which is a bracket. We have identified precisely these brackets and thereby determined that the $\mathcal{A}$-resultant is $\frac{|D_{\mathcal{A}}|}{B_1^{\delta_1\epsilon_1} B_2^{\delta_2\epsilon_2} B_3^{\delta_3\epsilon_3} B_4^{\delta_4\epsilon_4}}$.

The proof consists of three components. Focusing on the bottom left corner, we used the Dixon matrix entry formula to reveal that there are three columns such that the determinant of any 3×3 submatrix of these three columns has a factor B_1. Transformations were used to prove similar results in the other three corners. Following that, we checked that $\frac{|D_{\mathcal{A}}|}{B_1^{\delta_1\epsilon_1} B_2^{\delta_2\epsilon_2} B_3^{\delta_3\epsilon_3} B_4^{\delta_4\epsilon_4}}$ has the right degree if $|D_{\mathcal{A}}| \neq 0$. Finally, we confirmed that indeed $|D_{\mathcal{A}}| \neq 0$. Hence $\frac{|D_{\mathcal{A}}|}{B_1^{\delta_1\epsilon_1} B_2^{\delta_2\epsilon_2} B_3^{\delta_3\epsilon_3} B_4^{\delta_4\epsilon_4}}$ is the $\mathcal{A}$-resultant.

With corner point pasting, the Dixon determinant is the maximal minor and the extraneous factors can be identified explicitly. We have observed that with other types of corner cutting (for example isosceles triangle cutting), the Dixon determinant is no longer a maximal minor. To obtain an $\mathcal{A}$-resultant, we need to identify dependent rows and columns to remove in order to retrieve the maximal minor. For this case and other new cases, the problem becomes more complex, involving both the locating of extraneous factors and the retrieval of a maximal minor. Such an endeavor seems worth pursuing.

References

1. C. D'Andrea. Macaulay style formulas for the sparse resultant. *Trans. Amer. Math. Soc.* **354** (2002), 2579–2594.
2. E. W. Chionh. Concise parallel Dixon determinant. *Computer Aided Geometric Design* **14** (1997), 561–570.
3. E. W. Chionh. Rectangular corner cutting and Dixon $\mathcal{A}$-resultants. *J. Symbolic Computation*, **31** (2001), 651–669.

4. E. W. Chionh, M. Zhang, R. N. Goldman. Implicitization by Dixon $\mathcal{A}$-resultants. *Proceedings of Geometric Modeling and Processing*, 2000, Hong Kong, 310–318.
5. A. D. Chtcherba, D. Kapur. A complete analysis of resultants and extraneous factors for unmixed bivariate polynomial systems using the Dixon formulation. *Proc. of 8th Rhine Workshop on Computer Algebra* (RWCA'02), Mannheim, Germany, March 2002, 136–166.
6. D. Cox, J. Little, D. O'Shea. *Using Algebraic Geometry.* Springer-Verlag, New York, 1998.
7. A. L. Dixon. The eliminant of three quantics in two independent variables. *Proc. London Math. Soc.* **6** (1908) 49–69, 473–492.
8. I. Z. Emiris, B. Mourrain. Computer algebra methods for studying and computing molecular conformations. *Algorithmica* **25**(2–3) (1999), 372–402.
9. M. C. Foo, E. W. Chionh. Corner edge cutting and Dixon $\mathcal{A}$-resultant quotients. Submitted to *J. Symbolic Computation*, 2002.
10. A. Khetan. The resultant of an unmixed bivariate system. Submitted to *J. Symbolic Computation*, 2002
11. R. Krasauska. Toric surface patches. *Computational Mathematics* **17** (2002), 89–113.
12. T. Saxena. Efficient variable elimination using resultants. Ph.D. thesis, State University of New York, 1997.
13. D. Wang. *Elimination Methods.* Springer Verlag, Austria, 2000.
14. M. Zhang, R. N. Goldman. Rectangular corner cutting and Sylvester $\mathcal{A}$-resultants. *Proceedings of the 2000 International Symposium on Symbolic and Algebraic Computation*, Scotland, ACM Press, 301–308.
15. S. Zube. The n-sided toric patches and $\mathcal{A}$-resultants. *Computer Aided Geometric Design* **17** (2000), 695–714.

ZERO DECOMPOSITION THEOREMS FOR COUNTING THE NUMBER OF SOLUTIONS FOR PARAMETRIC EQUATION SYSTEMS*

XIAO-SHAN GAO, DINGKANG WANG
Institute of Systems Science, AMSS, Academia Sinica
Beijing 100080, China
Email:(xgao,dwang)@mmrc.iss.ac.cn

Reducing a set of equations in general form to triangular form is a basic idea in equation solving. In this paper, we show how to recover the information about the number of solutions for parametric equation systems by introducing two new forms of triangularization techniques: *the decomposition tree* and *the refined cover*. The decomposition tree can be used to compute the maximal number of solutions for the parametric equation system. The refined cover gives a complete classification of the number of solutions for the parametric equation system. That is, we decompose the value space for the parameters into disjoint components. On each component, the equation system has a fixed number of solutions.

Keywords: Parametric equation system, number of solutions, characteristic set method, triangularization, zero decomposition theorem.

1. Introduction

Parametric algebraic equation systems occur in many application fields such as robotics, celestial mechanics [9,21], chemical equilibrium [3], computer vision [8], etc. Actually, most equation systems raised from applications have parameters. Parametric equations are more complicated than equation systems without parameters in that their solutions vary dramatically for different values of the parameters. For instance, consider the following equation system for the P3P problem from computer vision [8]:

$$p_1 = y^2 + z^2 - 2yzp - a^2 = 0,$$
$$p_2 = z^2 + x^2 - 2zxq - b^2 = 0,$$
$$p_3 = x^2 + y^2 - 2xyr - c^2 = 0,$$

*This work is supported in part by a National Key Basic Research Project of China (NO. G1998030600)

where a, b, c, p, q, r are the parameters. The polynomials p_1, p_2, p_3 generate a prime ideal and this ideal is of zero dimension if the parameters are considered as algebraically independent indeterminates. But if the parameters satisfy certain algebraic condition, the system will have infinitely many solutions.

In [20], Weispfenning proposed the concept of comprehensive Gröbner bases (CGB) for a parametric polynomial set. The idea is to divide the parametric space into domains on which the polynomial set has a fixed form for its Gröbner basis. In [17], Suzuki and Sato gave a new algorithm for CGB. In [15], Sit proposed the concept of a cover for linear parametric equation systems, which allows an effective method of solving parametric equation systems. Sit's method was extended to general equation systems and differential equation systems in [7,18]. In [11], Kapur proposed a method for solving parametric equations with both the Gröbner basis method and the characteristic set method. In [9], Lazard proposed a method to count the number of solutions for parametric systems with parameters less than or equal to two.

The previous work mainly focuses on two problems about the parametric equation systems: (1) for what values of the parameters the equation systems have solutions, and (2) for these parametric values how to solve the equation system. In this paper, we will consider the problem of counting the number of solutions for parametric systems over the field of complex numbers.

Reducing a set of equations in general form to triangular form is a basic idea in equation solving, and there exist many methods of triangularization [1,14,18,21]. For equation systems with parameters, the information about the number of solutions is not discussed. In this paper, we show how to recover the information about the number of solutions for parametric equation systems by introducing two new forms of triangularization techniques: *the zero decomposition tree* and *the refined cover*. The decomposition tree can be used to compute the maximal number of solutions for the equation system. The idea is to modify the original zero decomposition algorithm [21] so that the number of solutions could be computed effectively. The refined cover gives a complete classification of the number of solutions for the equation systems. That is, we decompose the value space for the parameters into disjoint components. On each component, the equation system has a fixed number of solutions and these solutions can be represented by equation systems in triangular form. For instance, a complete solution classification for the simple parametric equation $ax^2 + bx + c = 0$ is as follows:

(1) If $a \neq 0$ and $b^2 - 4ac \neq 0$, the equation has 2 different solutions.
(2) If $a \neq 0$ and $b^2 - 4ac = 0$, the equation has only one solution.
(3) If $a = 0$ and $b \neq 0$, the equation has only one solution.
(4) If $a = b = 0$ and $c \neq 0$, the equation has no solution.
(5) If $a = b = c = 0$, the equation has infinitely many solutions.

The zero decomposition tree is implemented as a part of the `wsolve` package [19]. The program has been used to solve the P3P problem [8], which is quite difficult.

We only count the number of solutions over an algebraically closed field. It would be interesting to go further by counting the number of real solutions. There are other methods to count the number of solutions for equation systems without parameters, for example, [5,2]. It would be interesting to see whether these methods can be extended to parametric equations.

2. Preliminary Results

Let $\mathbf{K}$ be the field of rational numbers, let $\mathcal{B} = \mathbf{K}[u_1, \cdots, u_m]$ be the polynomial ring in the variables $\mathcal{U} = \{u_1, \cdots, u_m\}$ over $\mathbf{K}$, and let $\mathcal{B}[x_1, \cdots, x_n]$ be the polynomial ring in the variables $\mathcal{X} = \{x_1, \cdots, x_n\}$ over $\mathcal{B}$. A polynomial in $\mathcal{B}$ is called a *u-pol.* Unless explicitly mentioned otherwise, all polynomials in this paper are in $\mathcal{B}[\mathcal{X}]$.

For $P \in \mathcal{B}[\mathcal{X}] - \mathcal{B}$, we can write $P = c_d x_p^d + ... + c_1 x_p + c_0$, where $c_i \in \mathcal{B}[x_1, ..., x_{p-1}]$, $p > 0$, and $c_d \neq 0$. We define $\mathrm{cls}(P) = p$, $\mathrm{init}(P) = c_d$, $\mathrm{sep}(P) = \partial P/\partial x_p$, $\mathrm{lvar}(P) = x_p$, $\mathrm{ldeg}(P) = d$, and call them respectively, the *class*, the *initial*, the *separant*, the *leading variable*, and the *leading degree* of P.

A sequence of polynomials $\mathcal{C} = A_1, ..., A_p$ in $\mathcal{B}[\mathcal{X}] - \mathcal{B}$ is said to be an *ascending chain* (abbr. chain) or in *triangular form*, if either $p = 1$ or $0 < \mathrm{cls}(A_i) < \mathrm{cls}(A_j)$ and the degree of A_j in $\mathrm{lvar}(A_i)$ is less than $\mathrm{ldeg}(A_i)$ for $1 \leq i < j$. For a chain $\mathcal{C}$, we use $\mathrm{I}(\mathcal{C})$ to denote the product of the initials of the polynomials in $\mathcal{C}$. The *degree* of $\mathcal{C}$ is defined as $\deg(\mathcal{C}) = \prod_{i=1}^p \mathrm{ldeg}(A_i)$. We may define ascending chains in $\mathcal{B}$ and $\mathcal{B}[\mathcal{X}]$ similarly. For a chain $\mathcal{C}$ and a polynomial P, we use $\mathrm{Prem}(P, \mathcal{C})$ to denote the *pseudo-remainder* of P with respect to $\mathcal{C}$ (see [13,21]). We assume that $\mathrm{Prem}(P, \emptyset) = P$.

Let $\mathcal{S}$ be a polynomial set and D be a polynomial in $\mathbf{K}[\mathcal{U}, \mathcal{X}]$. For a universal extension field [13] $\mathbf{E}$ of $\mathbf{K}$, let

$$\mathrm{Zero}(\mathcal{S}) = \{a \in \mathbf{E}^{m+n} \mid \forall P \in \mathcal{S}, P(a) = 0\},$$
$$\mathrm{Zero}(\mathcal{S}/D) = \mathrm{Zero}(\mathcal{S}) - \mathrm{Zero}(D).$$

A *quasi-variety* is defined to be $\cup_{i=1}^{t}\mathrm{Zero}(\mathcal{S}_i/D_i)$ where $\mathcal{S}_i$ are polynomial sets and D_i are polynomials in $\mathbf{K}[\mathcal{U},\mathcal{X}]$. A *basic quasi-variety* is of the form $\mathrm{Zero}(\mathcal{C}\,/\,\mathrm{I}(\mathcal{C})\cdot D)$ where $\mathcal{C}$ is a chain and D is a polynomial.

The *saturation* for a chain $\mathcal{C}$ is defined as the ideal

$$\mathrm{Sat}(\mathcal{C}) = \{P \mid \exists\, J \text{ such that } JP \in \mathrm{Ideal}(\mathcal{C})\}$$

where J is a product of powers of initials of the polynomials in $\mathcal{C}$. For a chain $\mathcal{C} \subset \mathbf{K}[\mathcal{X}]$, we define the *dimension* (in $\mathcal{X}$) of $\mathcal{C}$ to be $\dim(\mathcal{C}) = n - |\mathcal{C}|$. The next lemma is taken from [6].

Lemma 2.1. *Let $\mathcal{C}$ be a chain, P a polynomial in $\mathbf{K}[\mathcal{X}]$. Then* $\mathrm{Zero}(\mathcal{C}/\mathrm{I}(\mathcal{C}))$ *and* $\mathrm{Zero}(\mathrm{Sat}(\mathcal{C}))$ *are either empty or un-mixed with dimension* $\dim(\mathcal{C})$.

Let P be a polynomial and $\mathcal{C} = A_1, \cdots, A_p$ be a chain in $\mathbf{K}[\mathcal{U},\mathcal{X}] - \mathbf{K}[\mathcal{U}]$. We recursively define

$$\mathrm{Res}(P,\mathcal{C}) = \mathrm{Res}(\cdots \mathrm{Res}(P, A_p, \mathrm{lvar}(A_p)), \cdots, A_1, \mathrm{lvar}(A_1))$$

where $\mathrm{Res}(P,Q,v)$ represents the resultant of P and Q with respect to the variable v. If P does not involve v, $\mathrm{Res}(P,Q,v) = P$.

A chain $\mathcal{C} = A_1, ..., A_p$ is called *regular* or *with invertible initials* if for $1 \le i \le p$ $\mathrm{Res}(\mathrm{init}(A_i),\mathcal{C}) \neq 0$ [1,14,18]. Rody *et al.* ([14], Lemma 1.2.1) proved:

Lemma 2.2. *Let $\mathcal{C}$ be a regular chain in $\mathcal{B}[\mathcal{X}] - \mathcal{B}$, P be a polynomial in $\mathcal{B}[\mathcal{X}]$, and D be a non-zero u-pol. If $DP \in \mathrm{Ideal}(\mathcal{C})$ then $\mathrm{Prem}(P,\mathcal{C}) = 0$.*

A chain $\mathcal{C}$ is called a *u-chain* if $\mathrm{I}(\mathcal{C})$ is a u-pol.

Lemma 2.3. *If $\mathcal{C}$ is a regular chain whose dimension in $\mathcal{X}$ is zero, then we can find a u-chain $\mathcal{C}'$ such that $\mathrm{Sat}(\mathcal{C}) = \mathrm{Sat}(\mathcal{C}')$. Furthermore, for any $P \in \mathrm{Sat}(\mathcal{C})$ we have $\mathrm{Prem}(P,\mathcal{C}) = \mathrm{Prem}(P,\mathcal{C}') = 0$.*

Proof. Let $\mathcal{C} = B_1, \cdots, B_q, A_1, \cdots, A_n$ be such that B_i are u-pols and $A_i \in \mathcal{B}[X] - \mathcal{B}$. Let $I_i = \mathrm{init}(A_i)$. It is clear that $R_1 = I_1$ is a u-pol. Let $A_1' = A_1$. Since $\mathcal{C}$ is regular, $R_k = \mathrm{Res}(I_k, A_1, \cdots, A_{k-1})$ is a nonzero u-pol for $k > 1$. By the properties of the resultants, there exist $C_i \in \mathbf{K}[\mathcal{U},\mathcal{X}], i = 1, \cdots, k$ such that $R_k = C_k I_k + \sum_{i=1}^{k-1} C_i A_i$. Let

$$A_k' = C_k A_k + (\sum_{i=1}^{k-1} C_i A_i) x_k^{\mathrm{ldeg}(A_k)}.$$

Then the initial of A_k' is R_k. Let $\mathcal{C}' = B_1, \cdots, B_q, A_1', A_2', \cdots, A_n'$. To prove $\mathrm{Sat}(\mathcal{C}') \subset \mathrm{Sat}(\mathcal{C})$, let $P \in \mathrm{Sat}(\mathcal{C}')$. Then there is a product J of initials of $\mathcal{C}'$ such that $JP \in \text{Ideal}(\mathcal{C}') \subset \text{Ideal}(\mathcal{C})$. Since J is a u-pol, by Lemma 2.2, $\text{Prem}(P, \mathcal{C}) = 0$. Thus $P \in \text{Sat}(\mathcal{C})$. To prove $\text{Sat}(\mathcal{C}) \subset \text{Sat}(\mathcal{C}')$, let $P \in \text{Sat}(\mathcal{C})$. Then there is a product $K = \prod_{i=1}^{n} I_i^{d_i}$ of initials of $\mathcal{C}$ such that $KP = \sum_{i=1}^{n} B_i A_i$. Without loss of generality, we assume $d_i \geq 1$. Multiplying $C_n^{d_n}$ to both sides, we have

$$\prod_{i=1}^{n-1} I_i (I_n C_n)^{d_n} P = C_n^{d_n} \sum_{i=1}^{n-1} B_i A_i + C_n^{d_n - 1} B_n (C_n A_n).$$

Substituting $C_n I_n = R_n - \sum_{i=1}^{n-1} C_i A_i$ and $C_n A_n = A_n' - \sum_{i=1}^{n-1} C_i A_i$, we have

$$\prod_{i=1}^{n-1} I_i (R_n)^{d_n} P = \sum_{i=1}^{n-1} D_i A_i + D_n A_n'.$$

Repeat the above process for $i = n-1, \cdots, 2$, we have $\prod_{i=1}^{n} R_i^{d_i} P = \sum_{i=1}^{n} E_i A_i'$. Thus $P \in \text{Sat}(\mathcal{C}')$. From Lemma 2.2, $\text{Prem}(P, \mathcal{C}') = 0$. □

A chain $\mathcal{C} = A_1, ..., A_p$ is called *simple* or *with invertible initials and separants* (see [1,14,18]) if for $1 \leq i \leq p$, $\text{Res}(\text{init}(A_i)\text{sep}(A_i), \mathcal{C}) \neq 0$. The chain $\mathcal{C}$ is called *simple with respect to a polynomial* D if either $\text{Prem}(D, \mathcal{C}) = 0$ or $\text{Res}(D, \mathcal{C}) \neq 0$.

For a quasi-variety $\mathbf{Q} \in \mathbf{E}^{m+n}$ and an $a \in \mathbf{E}^m$, let

$$K_a = \{b \in \mathbf{E}^n \mid (a, b) \in \mathbf{Q}\}$$

and let $\mathbf{M}(\mathbf{Q})$ be the *maximal cardinal number* of the set $K_a(\mathbf{Q})$ as a runs through E^m.

Lemma 2.4. *Let $\mathcal{C}$ be a chain in $\mathbf{K}[\mathcal{U}, \mathcal{X}]$ with zero dimension in $\mathcal{X}$. We have $\mathbf{M}(\text{Zero}(\mathcal{C}/\text{I}(\mathcal{C}))) \leq \deg(\mathcal{C})$, and if $\mathcal{C}$ is simple then $\mathbf{M}(\text{Zero}(\mathcal{C}/\text{I}(\mathcal{C})))$ $= \deg(\mathcal{C})$.*

Proof. Let $A_i \in \mathcal{C}$ be the polynomial with leading variable $x_i, i = 1, \cdots, n$. For specific values of the parameters, we may solve x_1 with $A_1 = 0$. Substitute each solution for $A_1 = 0$ into $A_2 = 0$ to obtain an equation $B_2 = 0$ for x_2. We may solve x_2 with $B_2 = 0$, and so on. Since $\text{I}(C) \neq 0$, the total number of solutions is less than or equal to $\deg(\mathcal{C})$. Let D be the product of $\text{I}(C)$ and the separants of A_i in x_i for $i = 1, \cdots, n$. Since $\mathcal{C}$ is simple, $R = \text{Res}(D, \mathcal{C}) \neq 0$ is a u-pol. For a set of parametric values that does not vanish R, each $A_1 = 0$ or $B_i = 0, i = 2, \cdots, n$ has $\text{ldeg}(A_i)$ distinct roots. Hence $\mathbf{M}(\text{Zero}(\mathcal{C}/\text{I}(\mathcal{C}))) = \deg(\mathcal{C})$. □

3. Zero Decomposition Tree

A *parametric equation system* is a set of equations and inequations

$$\begin{cases} P_1(u_1,\cdots,u_m,x_1,\cdots,x_n)=0,\cdots,P_s(u_1,\cdots,u_m,x_1,\cdots,x_n)=0 \\ D_1(u_1,\cdots,u_m,x_1,\cdots,x_n)\neq 0,\cdots,D_t(u_1,\cdots,u_m,x_1,\cdots,x_n)\neq 0 \end{cases}$$

where $u_1,\cdots,u_m$ are the parameters and $x_1,\cdots,x_n$ are the variables to be solved. Let $\mathcal{S}=\{P_1,\cdots,P_s\}, D=\prod_{i=1}^t D_i$. We need to consider $\mathrm{Zero}(\mathcal{S}/D)$.

The zero decomposition theorem is to reduce the zero set for a general equation system into the union of zero sets of equation systems in triangular form.

Theorem 3.1. (*Coarse Form* [21]) *For a polynomial set* $\mathcal{S}$ *and a polynomial* D, *we can find chains* $\mathcal{C}_i$ *such that*

$$\mathrm{Zero}(\mathcal{S}/D)=\cup_{i=1}^{\ell}\mathrm{Zero}(\mathcal{C}_i/D\,\mathrm{I}(\mathcal{C}_i))=\cup_{i=1}^{\ell}\mathrm{Zero}(\mathrm{Sat}(\mathcal{C}_i)/D)$$

where $\mathrm{Prem}(D,\mathcal{C}_i)\neq 0$.

Many new forms for the above theorem were proposed. In particular, we may assume (see [1,10,14,4,18,22]) that the $\mathcal{C}_i$ in the zero decomposition are regular chains or simple chains. We may further assume (see [22]) that $\mathcal{C}_i$ is simple w.r.t. D.

In the above theorem, it is easy to estimate the number of solutions for each component in the right-hand side with Lemma 2.4. In order to estimate the number of solutions for the union of these components, we will present a new form for the zero decomposition theorem. To do that, we need only to consider the steps that decompose one zero set into the union of two zero sets. By a careful analysis of the zero decomposition algorithm [21], the following two steps need to be considered.

[1] **Well-Ordering.** A chain $\mathcal{C}$ is called a *characteristic set* of a polynomial set $\mathcal{S}$ if $\mathcal{C}\subset\mathrm{Ideal}(\mathcal{S})$ and $\mathrm{Prem}(P,\mathcal{C})=0$ for each $P\in\mathcal{S}$. If $\mathcal{C}$ is a characteristic set of $\mathcal{S}$, we have

$$\mathrm{Zero}(\mathcal{S})=\mathrm{Zero}(\mathcal{C}/\mathrm{I}(\mathcal{C}))\cup\mathrm{Zero}(\mathcal{S}\cup\{\mathrm{I}(\mathcal{C})\}).$$

The well-ordering principle says that for a finite set $\mathcal{S}$ of polynomials, there is an algorithm [21] to find a characteristic set $\mathcal{C}$ of $\mathcal{S}$.

[2] **Splitting.** For a polynomial set $\mathcal{S}$ and two polynomials P_1 and P_2,

$$\mathrm{Zero}(\mathcal{S},P_1P_2)=\mathrm{Zero}(\mathcal{S}\cup\{P_1\})\cup\mathrm{Zero}(\mathcal{S}\cup\{P_2\}/P_1)$$

Theorem 3.2. *($\mathcal{U}$-Zero Decomposition) For a finite polynomial set $\mathcal{S}$ and a polynomial D, we can find regular (simple) chains $\mathcal{C}_i$ and polynomials D_i such that*

$$\mathrm{Zero}(\mathcal{S}/D) = \bigcup_{i=1}^{r} \mathrm{Zero}(\mathcal{C}_i/DD_i\,\mathrm{I}(\mathcal{C}_i))$$

where $\mathrm{Prem}(DD_i, \mathcal{C}_i) \neq 0$. *For $i = 1, \cdots, l$, $\mathcal{C}_i$ are u-chains and of zero dimension in $\mathcal{X}$. For $i = l+1 \cdots, r$, $\mathcal{C}_i$ are of positive dimensions in $\mathcal{X}$. The components* $\mathrm{Zero}(\mathcal{C}_i/DD_i\mathrm{I}(\mathcal{C}_i)), i = 1, \cdots, r$ *are disjoint.*

Proof. Basically speaking, the zero decomposition algorithm works as follows. We first use the well-ordering principle to obtain a characteristic set $\mathcal{C}$ of $\mathcal{S}$. If $\mathcal{C}$ is of positive dimension in $\mathcal{X}$ then $\mathrm{Zero}(\mathcal{S}) = \mathrm{Zero}(\mathcal{C}/\mathrm{I}(\mathcal{C})) \cup \mathrm{Zero}(\mathcal{S} \cup \mathrm{I}(\mathcal{C}))$. Repeat the process for $\mathrm{Zero}(\mathcal{S} \cup \mathrm{I}(\mathcal{C}))$. If $\mathcal{C}$ is regular and with zero dimension in $\mathcal{X}$, by Lemma 2.3 we may find a u-chain $\mathcal{C}'$ such that $\mathcal{C}'$ is still a characteristic set of $\mathcal{S}$. Then $\mathrm{Zero}(\mathcal{S}) = \mathrm{Zero}(\mathcal{C}'/\mathrm{I}(\mathcal{C}')) \cup \mathrm{Zero}(\mathcal{S} \cup \mathrm{I}(\mathcal{C}'))$. Repeat the process for $\mathrm{Zero}(\mathcal{S} \cup \mathrm{I}(\mathcal{C}'))$. If $\mathcal{C}$ is not regular, a polynomial in $\mathcal{C}$ can be factorized into the product of two polynomials [14]. Now, we use the splitting step to reduce $\mathrm{Zero}(\mathcal{S})$ into two 'simpler' equation systems. Note that in this step, a D_i in the decomposition equation will be introduced. Similar to [21], we may prove that the process will terminate and give a $\mathcal{U}$-zero decomposition. Since the two components obtained in the well-ordering step and the splitting step are disjoint, the components in the decomposition are disjoint. □

Lemma 3.3. *In the well-ordering step, if $\mathcal{C}$ is a u-chain*

$$\mathbf{M}(\mathrm{Zero}(\mathcal{S})) = \max\{\mathbf{M}(\mathrm{Zero}(\mathcal{C}/\mathrm{I}(\mathcal{S}))), \mathbf{M}(\mathrm{Zero}(\mathcal{S} \cup \{\mathrm{I}(\mathcal{S})\}))\}.$$

Proof. Since $\mathcal{C}$ is a u-chain, $J = \mathrm{I}(\mathcal{S})$ is a u-pol. For a set of specific values u_0 of the parameters, we have either $J(u_0) = 0$ or $J(u_0) \neq 0$. If $J(u_0) \neq 0$, $K_{u_0}(\mathrm{Zero}(\mathcal{S} \cup \{J\})) = \emptyset$ and $K_{u_0}(\mathrm{Zero}(\mathcal{S})) = K_{u_0}(\mathrm{Zero}(\mathcal{C}/J))$. If $J(u_0) = 0$, $K_{u_0}(\mathrm{Zero}(\mathcal{C}/J)) = \emptyset$ and $K_{u_0}(\mathrm{Zero}(\mathcal{S})) = K_{u_0}(\mathrm{Zero}(\mathcal{S} \cup \{J\}))$. It is clear that $\mathbf{M}(\mathrm{Zero}(\mathcal{S})) \geq \mathbf{M}(\mathrm{Zero}(\mathcal{C}/J))$, $\mathbf{M}(\mathrm{Zero}(\mathcal{S})) \geq \mathbf{M}(\mathrm{Zero}(\mathcal{S} \cup \{J\})$. Then the result is clearly true. □

The following example shows why it is difficult to extend the above lemma to the case when J is not a u-pol.

Example 3.4. Let $\mathcal{S} = \{x_1^2 - u_1, (x_1 - 1)x_2^2 - x_2 + 1\}$. Note that $\mathcal{S}$ itself is a chain. We have

$$\mathrm{Zero}(\mathcal{S}) = \mathrm{Zero}(\mathcal{S}/(x_1 - 1)) \cup \mathrm{Zero}(\{u_1 - 1, x_1 - 1, x_2 - 1\}).$$

If $u_1 = 1$, the first component has two solutions and the second component has one solution, which are the three solutions of the equation system. Therefore, $K_1(\text{Zero}(\mathcal{S}))$ equals

$$K_1(\text{Zero}(\mathcal{S}/(x_1 - 1))) \cup K_1(\text{Zero}(\{u_1 - 1, x_1 - 1, x_2 - 1\})).$$

Due to this, we cannot obtain Lemma 3.3 in the general case.

Lemma 3.5. *In the splitting step, if P_1 is a u-pol, then*

$$\begin{aligned}&\mathbf{M}(\text{Zero}(\mathcal{S} \cup \{P_1P_2\}))\\ &= \max\{\mathbf{M}(\text{Zero}(\mathcal{S} \cup \{P_1\})), \mathbf{M}(\text{Zero}(\mathcal{S} \cup \{P_2\}/P_1))\}.\end{aligned}$$

If both P_1 and P_2 are not u-pol, then

$$\mathbf{M}(\text{Zero}(\mathcal{S} \cup \{P_1P_2\})) \leq \mathbf{M}(\text{Zero}(\mathcal{S} \cup \{P_1\})) + \mathbf{M}(\text{Zero}(\mathcal{S} \cup \{P_2\}/P_1)).$$

Proof. The proof of the first case is similar to that of Lemma 3.3. The second case is obviously true. □

The simple example $\text{Zero}((x - 1)(x - 2)) = \text{Zero}(x - 1) \cup \text{Zero}(x - 2)$ shows that if both P_1 and P_2 are not u-pol, we cannot use the first formula in Lemma 3.5.

We will define the *decomposition tree* T. The leafs of T are disjoint quasi-varieties S_i with zero dimension in $\mathcal{X}$. A non-leaf node is either a +-node or an m-node. For each node V in T, let $\text{Zero}(V) = \cup_{i=1}^k S_i$ where S_i are the descendant leafs of V. Let V be a node and $V_i, i = 1, \cdots, k$ its direct children. If V is a +-node, $\mathbf{M}(V) \leq \sum_{i=1}^k \mathbf{M}(V_i)$. If V is an m-node, $\mathbf{M}(V) = \max_{i=1}^k \mathbf{M}(V_i)$.

Theorem 3.6. *Let $\mathcal{S}$ be a polynomial set and let D be a polynomial in $\mathbf{K}[\mathcal{U}, \mathcal{X}]$. If $\mathbf{M}(\text{Zero}(\mathcal{S}/D))$ is finite, then we have an algorithm to find a decomposition tree with root r such that $\text{Zero}(r) = \text{Zero}(\mathcal{S}/D)$.*

Proof. We first obtain a $\mathcal{U}$-Zero decomposition with Theorem 3.2. To obtain a decomposition tree, in each of the well-ordering steps or splitting steps, we will use Lemmas 3.3 or 3.5 to generate a +-node or an m-node. Since $\mathbf{M}(\text{Zero}(\mathcal{S}/D))$ is finite, all the chains occurring in the decomposition have zero dimension in $\mathcal{X}$ and hence are u-chains. This guarantees that Lemma 3.3 can always be used in the well-ordering steps. □

Example 3.7. Continuing from Example 3.4. With Theorem 3.2, we obtain a $\mathcal{U}$-Zero decomposition: $\text{Zero}(\mathcal{S}) = C_1 \cup C_2 \cup C_3$ where

$$C_1 = \text{Zero}(\{x_1^2 - u_1, (u_1-1)x_2^2 - (x_1+1)(x_2-1)\}/(u_1-1)),$$
$$C_2 = \text{Zero}(\{u_1-1, x_1-1, x_2-1\}),$$
$$C_3 = \text{Zero}(\{u_1-1, x_1+1, 2x_2^2+x_2-1\}).$$

A corresponding decomposition tree is as follows.

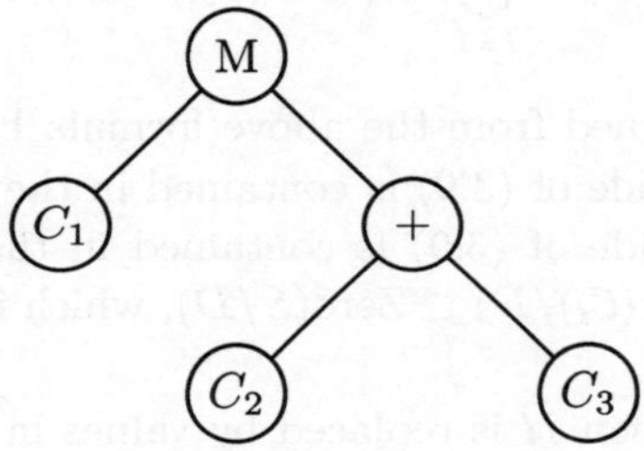

Figure 1. The decomposition tree for $\text{Zero}(\mathcal{S})$

From this tree, we may conclude that the maximal number of solutions for $\mathcal{S} = 0$ is $\max\{\mathbf{M}(C_1), \mathbf{M}(C_2)+\mathbf{M}(C_3)\} = 4$. Note that the Bezout number of the original system is six.

The decomposition tree only gives an upper bound for the maximal number of solutions.

For a polynomial set $\mathcal{S}$ and a polynomial D in $\mathbf{K}[\mathcal{X}]$, an *isolated zero* of $\text{Zero}(\mathcal{S}/D)$ is an element in $\text{Zero}(\mathcal{S}/D)$ which is not contained in any component of $\text{Zero}(\mathcal{S}/D)$ with positive dimension. For a polynomial set $\mathcal{S}$ and a polynomial D in $\mathbf{K}[\mathcal{U}, \mathcal{X}]$, the set $\mathcal{T}$ of *isolated zero* of $\text{Zero}(\mathcal{S}/D)$ is a quasi-variety contained in $\text{Zero}(\mathcal{S}/D)$ such that if $\mathcal{U}$ is replaced by any $u_0 \in \mathbf{K}^m$, resulting in corresponding sets $\mathcal{T}', \mathcal{S}', D'$, then $\mathcal{T}'$ is the set of isolated zeros of $\text{Zero}(\mathcal{S}'/D')$.

Theorem 3.8. *Let $\mathcal{S}$ be a polynomial set and let D be a polynomial in $\mathbf{K}[\mathcal{U}, \mathcal{X}]$. We can find regular chains $C_i, i = 1, \cdots, r$ with positive dimensions in $\mathcal{X}$ and a decomposition tree T such that if r is the root of T, then* $\text{Zero}(r)$ *is the set of isolated zeros of* $\text{Zero}(\mathcal{S}/D)$ *and*

$$\text{Zero}(\mathcal{S}/D) = \bigcup_{i=1}^{r} \text{Zero}(\text{Sat}(C_i)/D) \cup \text{Zero}(r). \tag{3.9}$$

Proof. In the well-ordering step, if $\text{Zero}(\mathcal{S}) = \text{Zero}(\mathcal{C}/J) \cup \text{Zero}(\mathcal{S} \cup J)$ then it is easy to show that $\text{Zero}(\mathcal{S}) = \text{Zero}(\text{Sat}(\mathcal{C})) \cup \text{Zero}(\mathcal{S} \cup J)$. If $\mathcal{C}$ is of positive dimension in $\mathcal{X}$, then we will use the latter form. Otherwise, we may assume that $\mathcal{C}$ is a u-chain and add $\text{Zero}(\mathcal{C}/J)$ to the decomposition tree. The decomposition tree may be obtained from the components with zero dimension in $\mathcal{X}$. Similar to Theorem 3.2, we get a decomposition of the following form.

$$\text{Zero}(\mathcal{S}/D) = \bigcup_{i=1}^{r} \text{Zero}(\text{Sat}(\mathcal{C}_i)/DD_i) \cup \text{Zero}(r).$$

Formula (3.9) is obtained from the above formula by removing the D_i. It is clear that the left-side of (3.9) is contained in the right-side of (3.9). To show that the right-side of (3.9) is contained in the left-side of (3.9), we need to show $\text{Zero}(\text{Sat}(\mathcal{C}_i)/D) \subset \text{Zero}(\mathcal{S}/D)$, which is a consequence of the fact $\mathcal{S} \subset \text{Sat}(\mathcal{C}_i)$.

By Lemma 2.1, when $\mathcal{U}$ is replaced by values in $\mathbf{K}^m$, $\text{Zero}(\text{Sat}(\mathcal{C}_i)/D)$ will become empty or an unmixed quasi-variety of positive dimension and a component in $\mathcal{T}$ will become empty or a zero dimensional variety. Then the isolated zeros of $\text{Zero}(\mathcal{S}/D)$ are contained in $\text{Zero}(r)$. We need to remove those zeros in $\text{Zero}(r)$ which are contained in a component $\text{Zero}(\text{Sat}(\mathcal{C}_i)/D)$ of positive dimension. Let $\mathcal{S}_i$ be a basis of $\text{Sat}(\mathcal{C}_i)$ which can be computed with Gröbner basis [6]. Let $S_i = \{P_{i,1}, \cdots, P_{i,d_i}\}$. Let the leaves of T be $L_j = \text{Zero}(\mathcal{T}_j/\text{I}(\mathcal{T}_j)DE_j), j = 1 \cdots, s$, where E_j are polynomials. Let

$$\begin{aligned} L_i' &= L_i - \bigcup_{j=1}^{r} \text{Zero}(\mathcal{S}_j/D) \\ &= \text{Zero}(\mathcal{T}_i/\text{I}(\mathcal{T}_i)DE_i) - \bigcup_{j=1}^{r} \text{Zero}(\mathcal{S}_j) \\ &= \bigcup_{k_1,\cdots,k_r} \text{Zero}(\mathcal{T}_i/DE_i\text{I}(\mathcal{T}_i)\prod_{\ell=1}^{r} P_{\ell,k_\ell}) \\ &= \bigcup_{k_1,\cdots,k_r} \text{Zero}(\mathcal{T}_i/DE_i\text{I}(\mathcal{T}_i)\prod_{\ell=1}^{r} R_{\ell,k_\ell}), \end{aligned}$$

where $1 \leq k_j \leq d_j, j = 1, \cdots, r$ and $R_{\ell,k_\ell} = \text{Prem}(P_{\ell,k_\ell}, \mathcal{T}_i)$. We may delete those components where $R_{\ell,k_\ell} = 0$. Since $L_i', i = 1, \cdots, s$ are disjoint quasi-varieties with zero dimension in $\mathcal{X}$, T' obtained by replacing L_i with L_i' is still a zero decomposition tree and (3.9) is still valid. □

Example 3.10. Let

$$\mathcal{S} = \{x_1^3 - x_1x_2^2 + u_1x_2^2 - u_1x_1^2,\ x_2^3 - x_1^2x_2 - u_2x_2^2 + u_2x_1^2\}.$$

Then $\text{Zero}(\mathcal{S}) = C_1 \cup C_2$, where

$$C_1 = \text{Zero}(x_2^2 - x_1^2), \quad C_2 = \text{Zero}(\{x_1 - u_1, x_2 - u_2\}).$$

C_1 is of positive dimension and C_2 is of zero dimension in x_1, x_2. According to Theorem 3.8, the set of isolated zeros is

$$\text{Zero}(\{x_1 - u_1, x_2 - u_2\}/x_2^2 - x_1^2) = \text{Zero}(\{x_1 - u_1, x_2 - u_2\}/u_2^2 - u_1^2).$$

4. Refined Cover for Complete Solution Classification

By a *complete solution classification* for a parametric equation system, we mean a subdivision of $\mathbf{E}^m$ into the union of disjoint components such that when the parameters $u_1, \cdots, u_m$ taking values on the same component, the parametric system has the same number of distinct solutions.

4.1. *Strong $\mathcal{U}$-Zero Decomposition Theorem*

For a set of polynomials $\mathcal{S}$, a polynomial D and a quasi-variety $\mathbf{Q} \subset \mathbf{E}^m$, if $K_a(\text{Zero}(\mathcal{S}/D))$ contains a fixed number of distinct elements for all possible $a \in \mathbf{Q}$, we say that $\text{Zero}(\mathcal{S}/D)$ is *uniform* on $\mathbf{Q}$, and use the notation $\mathbf{N}(\text{Zero}(\mathcal{S}/D))$ to denote the number $|K_a(\text{Zero}(\mathcal{S}/D))|$.

Let $\mathcal{C} = B_1, \cdots, B_q, A_1, \cdots, A_n$ be a simple chain with zero dimension in $\mathcal{X}$, and $\mathcal{A} = \{A_1, \cdots, A_n\} \subset \mathcal{B}[\mathcal{X}] - \mathcal{B}$. By Lemma 2.3, we may assume that $I = \prod_{i=1}^n \text{Res}(\text{init}(A_i), \mathcal{C})$ and $R = \prod_{i=1}^n \text{Res}(\text{sep}(A_i), \mathcal{C})$ are non-zero u-pols. We call IR the *ID-product* of $\mathcal{C}$ and let $\mathbf{Q} = \text{Zero}(B_1, \cdots, B_q/IR)$.

Lemma 4.1. *With the above notations,* $\text{Zero}(\mathcal{A}/IR)$ *is uniform over* $\mathbf{Q}$ *and* $\mathbf{N}(\text{Zero}(\mathcal{A}/IR)) = \deg(\mathcal{A}) = \prod_{i=1}^n \text{ldeg}(A_i)$.

Proof. For any $a \in \mathbf{Q}$, $I(a)R(a) \neq 0$. We will show that $K_a(\text{Zero}(\mathcal{A}/IR))$ contains exactly $\deg(\mathcal{A})$ elements. Let A_i' be obtained from A_i by replacing $\mathcal{U}$ by a. Now $A_1' = 0$ is a univariate equation in x_1 of degree $\text{ldeg}(A_1)$ since its initial is not zero, and has $\text{ldeg}(A_1)$ solutions. From the definition of R, $\text{sep}(A_1')$ is not zero for these solutions. Hence the $\text{ldeg}(A_1)$ solutions are distinct. Substituting each solution of $A_1' = 0$ into $A_2' = 0$, we obtain a univariate equation $A_2'' = 0$ of degree $\text{ldeg}(A_2)$ in x_2. Similarly, we may prove that $A_2'' = 0$ has $\text{ldeg}(A_2)$ distinct solutions. Continuing, we may prove the result. □

The well-ordering principle and the zero decomposition theorem can be modified as follows without difficulty.

Well-ordering Principle. Let $\mathcal{C}$ be a characteristic set of $\mathcal{S}$. If $\mathcal{C}$ is simple and of zero dimension in $\mathcal{X}$, we have

$$\mathrm{Zero}(\mathcal{S}) = \mathrm{Zero}(\mathcal{C}/J) \cup \mathrm{Zero}(\mathcal{S} \cup \{J\})$$

where J is the ID-product of $\mathcal{C}$.

With the above well-ordering principle, Theorem 3.1 can be extended to the following form.

Theorem 4.2. *For a polynomial set $\mathcal{S}$ and a polynomial D, we can find simple chains $\mathcal{C}_i$ and polynomials D_i such that*

$$\mathrm{Zero}(\mathcal{S}/D) = \bigcup_{i=1}^{\ell} \mathrm{Zero}(\mathcal{C}_i/DD_iJ_i)$$

where J_i is the I-product (resp. ID-product) of $\mathcal{C}_i$ if $\mathcal{C}_i$ is of positive (resp. zero) dimension in $\mathcal{X}$ and the sets $\mathrm{Zero}(\mathcal{C}_i/DD_iJ_i)$ are disjoint.

Lemma 4.3. *Let D be a polynomial, and $\mathcal{C}$ be a simple u-chain with zero dimension in $\mathcal{X}$. We assume that $\mathcal{C}$ is simple w.r.t. D. Then we can construct simple u-chains $\mathcal{C}_i, i = 1, \cdots, r$ of zero dimension in $\mathcal{X}$ and u-pols D_i such that*

$$\mathrm{Zero}(\mathcal{C}/\mathrm{I}(\mathcal{C})D) = \bigcup \mathrm{Zero}(\mathcal{C}_i/\mathrm{I}(\mathcal{C}_i)D_i)$$

and the sets $\mathrm{Zero}(\mathcal{C}_i/\mathrm{I}(\mathcal{C}_i)D_i)$ are disjoint.

Proof. If D doesn't involve any variables in $\mathcal{X}$, the theorem is obvious. We assume that $\mathrm{Prem}(D, \mathcal{C}) \neq 0$, for otherwise $\mathrm{Zero}(\mathcal{C}/JD) = \emptyset$. Since $\mathcal{C}$ is simple w.r.t. D, $R = \mathrm{Res}(D, \mathcal{C}) \neq 0$ is a u-pol. Then

$$\begin{aligned}\mathrm{Zero}(\mathcal{C}/\mathrm{I}(\mathcal{C})D) &= \mathrm{Zero}(\{\mathcal{C}, R\}/\mathrm{I}(\mathcal{C})D) \cup \mathrm{Zero}(\mathcal{C}/\mathrm{I}(\mathcal{C})DR) \\ &= \mathrm{Zero}(\{\mathcal{C}, R\}/\mathrm{I}(\mathcal{C})D) \cup \mathrm{Zero}(\mathcal{C}/\mathrm{I}(\mathcal{C})R).\end{aligned}$$

Note that the two components in the above formula are disjoint. Applying Theorem 4.2 to $\{\mathcal{C}, R\}$, we have

$$\mathrm{Zero}(\{\mathcal{C}, R\}/\mathrm{I}(\mathcal{C})) = \bigcup_j \mathrm{Zero}(\mathcal{C}'_j/E_i\, \mathrm{I}(\mathcal{C})\, \mathrm{I}(\mathcal{C}'_j))$$

where E_i are polynomials. Since $R = 0$, $\mathcal{C}$ must be reducible. Therefore, $\mathcal{C}'_j$ is of lower rank than $\mathcal{C}$. Repeat the above process for $\mathrm{Zero}(\mathcal{C}'_j/\mathrm{I}(\mathcal{C})D\mathrm{I}(\mathcal{C}_j))$. The process will terminate after a finite number of steps and the components are disjoint. □

Theorem 4.4. ***(Strong $\mathcal{U}$-Zero Decomposition Theorem)*** *For a polynomial set $\mathcal{S}$ and a polynomial D, we can find simple chains $\mathcal{C}_i$ and u-pols D_i such that*

$$\mathrm{Zero}(\mathcal{S}/D) = \cup_{i=1}^{\ell}\mathrm{Zero}(\mathcal{C}_i/D_iJ_i) \bigcup \cup_{i=l+1}^{r}\mathrm{Zero}(\mathcal{C}_i/DJ_i),$$

where for $i = 1, \cdots, \ell$, $\mathcal{C}_i$ are u-chains and of zero dimension in $\mathcal{X}$ and J_i is the ID-product of $\mathcal{C}_i$; for $i = l+1 \cdots, r$, $\mathcal{C}_i$ are of positive dimensions in $\mathcal{X}$ and J_i the I-product of $\mathcal{C}_i$; and the components $\mathrm{Zero}(\mathcal{C}_i/D_iJ_i), i = 1, \cdots, \ell$ are disjoint.

Proof. This follows directly from Theorems 3.2, 4.2 and Lemma 4.3. □

4.2. *Solution Classification*

For a polynomial set $\mathcal{S}$ and a polynomial D in $\mathbf{K}[\mathcal{U}, \mathcal{X}]$, we define the projection with $\mathcal{X}$ as follows

$$\mathrm{Proj}_{x_1,\cdots,x_n}\mathrm{Zero}(\mathcal{S}/D) = \{a \in \mathbf{E}^m | \exists e \in \mathbf{E}^n s.t. (a, e) \in \mathrm{Zero}(\mathcal{S}/D)\}$$

It is known that the projection of a quasi variety is also a quasi variety [7].

We extend the concept of cover introduced by Sit [15] as follows. A *refined cover* for parametric equation system $\mathrm{Zero}(\mathcal{S}/D)$ is a set of *solution functions*

$$(\mathbf{S}_i, \{\mathcal{C}_{i,j}, D_{i,j}\}_{j=1,\cdots,k_i}), i = 1, \cdots, d$$

such that

- $\mathbf{S}_i, i = 1, \cdots, d$ are disjoint quasi-varieties in E^m and $\mathcal{C}_{i,j}$ are simple chains in $\mathbf{K}[\mathcal{U}, \mathcal{X}] - \mathbf{K}[\mathcal{U}]$ and $D_{i,j}$ are u-pols.
- $\mathrm{Proj}_{x_1,\ldots,x_n}\mathrm{Zero}(\mathcal{S}/D) = \cup_i \mathbf{S}_i$.
- For each $a \in \mathbf{S}_i$,
$$\mathrm{Zero}(\mathcal{S}'/D') = \bigcup_j^{k_i} \mathrm{Zero}(\mathcal{C}'_{i,j}/\mathrm{I}(\mathcal{C}_{i,j}) * D'_{i,j})$$
where $\mathcal{S}', D'\mathcal{C}'_{i,j}, D'_{i,j}$ are obtained by replacing $\mathcal{U}$ with a.
- If $\mathrm{Zero}(\mathcal{S}/D)$ is of zero dimension in $\mathcal{X}$, then $\mathrm{Zero}(\mathcal{S}/D)$ is uniform over $\mathbf{S}_i$ and on $\mathbf{S}_i$, $\mathbf{N}(\mathrm{Zero}(\mathcal{S}/D)) = \sum_{j=1}^{k_i}\deg(\mathcal{C}_{i,j})$. Since $\mathbf{S}_i$ are disjoint, $\mathrm{Zero}(\mathcal{S}/D)$ could have the following numbers of solutions: $\sum_{j=1}^{k_i}\deg(\mathcal{C}_{i,j}), i = 1 \cdots, d$.

Theorem 4.5. *We can construct a refined cover for parametric equation system $\mathrm{Zero}(\mathcal{S}/D)$.*

Proof. By Theorem 4.4, we have a decomposition of the following form

$$\mathrm{Zero}(\mathcal{S}/D) = \cup_{i=1}^{\ell}\mathrm{Zero}(\mathcal{C}_i/D_iJ_i) \bigcup \cup_{i=\ell+1}^{r}\mathrm{Zero}(\mathcal{C}_i/DJ_i)$$

where $\mathcal{C}_i, i = 1, \cdots, \ell$ are u-chains and of zero dimension in $\mathcal{X}$ and J_i is the ID-product of $\mathcal{C}_i$. Let

$$\mathcal{T}_i = \mathrm{Proj}_{x_1,\cdots,x_n}\mathrm{Zero}(\mathcal{C}_i/D_iJ_i), \quad \text{for } i = 1, \cdots, \ell$$
$$\mathcal{T}_i = \mathrm{Proj}_{x_1,\cdots,x_n}\mathrm{Zero}(\mathcal{C}_i/DJ_i), \quad \text{for } i = \ell+1, \cdots, r.$$

We can compute $\mathcal{T}_i$ effectively [7]. Let $\mathcal{A}_i = \mathcal{C}_i \cap (K[\mathcal{U}, \mathcal{X}] - K[\mathcal{U}])$. We will obtain the refined cover from $\mathcal{T}_i$ and $\mathcal{A}_i$ by induction on i.

First, let $\mathbf{S}_1 = \mathcal{T}_1$, $\mathcal{C}_{1,1} = \mathcal{A}_1$, $D_{1,1} = D_1$. Let

$$(\mathbf{S}_i, \{\mathcal{C}_{i,j}, D_{i,j}\}_{j=1,\cdots,k_i}), i = 1, \cdots, d$$

be a set of solution functions obtained from $\mathcal{T}_i$ and $\mathcal{A}_i, i = 1, \cdots, d$. We will show how to obtain a set of new solution functions by adding $\mathcal{T}_{d+1}$ and $\mathcal{A}_{d+1}$. For each $\mathbf{S}_i, i = 1\cdots, d$, let $\mathbf{S}'_i = \mathbf{S}_i \cap \mathcal{T}_{d+1}$, $\mathbf{S}''_i = \mathbf{S}_i - \mathbf{S}'_i$, $\mathbf{S}'''_i = \mathcal{T}_{d+1} - \mathbf{S}'_i$. Then on $\mathbf{S}'_i$, the solutions of the equation system may be obtained from $\mathcal{C}_{d+1}$ and $\mathcal{C}_{i,j}, j = 1, \cdots, k_i$. On $\mathbf{S}''_i$, the solutions of the equation system may be obtained from $\mathcal{C}_{i,j}, j = 1, \cdots, k_i$. On $\mathbf{S}'''_i$, the solutions of the equation system may be obtained from $\mathcal{C}_{d+1}$. We replace $(\mathbf{S}_i, \{\mathcal{C}_{i,j}, D_{i,j}\}_{j=1,\cdots,k_i})$ by the three new solution functions: $(\mathbf{S}'_i, \{\mathcal{A}_{d+1}, D_{d+1}\} \cup \{\mathcal{C}_{i,j}, D_{i,j}\}_{j=1,\cdots,k_i})$, $(\mathbf{S}''_i, \{\mathcal{C}_{i,j}, D_{i,j}\}_{j=1,\cdots,k_i})$, $(\mathbf{S}'''_i, \{\mathcal{A}_{d+1}, D_{d+1}\})$ to obtain a new set of solution functions. Of course, we need only to add those solution functions whose first part is not empty.

We will prove that the set of solution functions thus obtained are a refined cover. The first and second conditions are clearly true. The third condition is also true, because the $\mathbf{S}_i$ are disjoint. For the fourth condition, we need to assume that the system is of zero dimension in $\mathcal{X}$. From Theorem 4.4 and Lemma 4.1, $\mathcal{A}_i$ is uniform over $\mathcal{T}_i$ and $\mathbf{N}(\mathrm{Zero}(\mathcal{A}_i/\mathrm{I}(\mathcal{A}_i)))$ equals to the product of the leading degrees of the polynomials in $\mathcal{A}_i$. Since the $\mathbf{S}_i$ are disjoint, the number of solutions for the equation system is the maximal number of solutions for the solution functions. From Theorem 4.4, the components $\mathrm{Zero}(\mathcal{C}_{i,j}/\mathrm{I}(\mathcal{C}_{i,j})D_{i,j})$ are disjoint for a given value for $\mathcal{U}$. Then, for each value in a specific $\mathbf{S}_i$, the number of solutions of the equation system is $\sum_{j=1}^{k_i}\deg(\mathcal{C}_{i,j})$. □

In the above theorem, we need to compute the intersection and the difference of two quasi-varieties. Algorithms for these are available from [16].

Example 4.6. Continuing from Example 3.7. With Theorem 4.4, a strong u-zero decomposition for $\mathrm{Zero}(\mathcal{S})$ is $\mathrm{Zero}(\mathcal{S}) = C_1 \cup C_2 \cup C_3 \cup C_4 \cup C_5 \cup C_6$, where

$$C_1 = \mathrm{Zero}(\mathcal{C}/u_1(u_1-1)(16u_1-25)) \text{ in which}$$
$$\mathcal{C} = x_1^2 - u_1, (u_1-1)x_2^2 - (x_1+1)(x_2-1),\ D_1 = 1$$
$$C_2 = \mathrm{Zero}(\{u_1-1, x_1-1, x_2-1\}),\ D_2 = 1$$
$$C_3 = \mathrm{Zero}(\{u_1-1, x_1+1, 2x_2^2+x_2-1\}),\ D_3 = 1$$
$$C_4 = \mathrm{Zero}(\{u_1, x_1, x_2^2+x_2-1\},\ D_4 = 1$$
$$C_5 = \mathrm{Zero}(\{16u_1-25, 4x_1-5, x_2-2\},\ D_5 = 1$$
$$C_6 = \mathrm{Zero}(\{16u_1-25, 4x_1+5, 9x_2^2-4x_2+4\},\ D_6 = 1$$

From this decomposition, we may obtain the following refined cover for $\mathcal{S} = 0$:

$$\begin{cases} \mathbf{S}_1 = \mathrm{Zero}(/u_1(u_1-1)(16u_1-25)) \\ \mathcal{C}_{1,1} = \{x_1^2 - u_1, (u_1-1)x_2^2 - (x_1+1)(x_2-1)\},\ D_{1,1} = 1 \end{cases}$$
$$\begin{cases} \mathbf{S}_2 = \mathrm{Zero}(\{u_1-1\}); \\ \mathcal{C}_{2,1} = \{x_1-1, x_2-1\}),\ D_{2,1} = 1; \\ \mathcal{C}_{2,2} = \{x_1+1, 2x_2^2+x_2-1\}),\ D_{2,2} = 1 \end{cases}$$
$$\begin{cases} \mathbf{S}_3 = \mathrm{Zero}(u_1); \\ \mathcal{C}_{3,1} = \{x_1, x_2^2+x_2-1\},\ D_{3,1} = 1 \end{cases}$$
$$\begin{cases} \mathbf{S}_4 = \mathrm{Zero}(16u_1-25); \\ \mathcal{C}_{4,1} = \{4x_1-5, x_2-2\},\ D_{4,1} = 1; \\ \mathcal{C}_{4,2} = \{4x_1+5, 9x_2^2-4x_2+4\},\ D_{4,2} = 1. \end{cases}$$

From this refined cover it is easy to see that the equation system may have 2,3,4 solutions respectively.

Acknowledgments

We want to thank the anonymous referees for valuable suggestions which lead to improvements of the paper.

References

1. P. Aubry, D. Lazard, M.M. Maza. On the theory of triangular sets. *Journal of Symbolic Computation* **25** (1999), 105–124.
2. G. Björck, R. Fröberg. A faster way to count the solutions of inhomogeneous systems of algebraic equations. *Journal of Symbolic Computation* **12** (1991), 329–336.

3. W. Boege, R. Gebauer, H. Kredel. Some examples for solving systems of algebraic equations by calculating Gröbner bases. *Journal of Symbolic Computation* **1** (1986), 83–98.
4. D. Bouziane, K. A. Rody, H. Maârouf. Unmixed-dimensional decomposition of a finitely generated perfect differential ideal. *Journal of Symbolic Computation* **31** (2001), 631–649.
5. I.Z. Emiris, J. Verschelde. How to count efficiently all affine roots of a polynomial system. *Discrete Appl. Math.* **93**(1) (1999), 21–32.
6. X.S. Gao, S.C. Chou. On the dimension for arbitrary ascending chains. *Chinese Bull. of Scis.* **38** (1993), 396–399.
7. X.S. Gao, S.C. Chou. Solving parametric algebraic systems. *Proc. of ISSAC-92*, 335–341. ACM Press, New York, 1992.
8. X.S. Gao, X. Hou, J. Tang, H. Cheng. Complete solution classification for the P3P problem. Preprint, 2001. Accepted by *IEEE T. PAMI.*
9. D. Lazard. Resolution of polynomial systems. In X.S. Gao, D. Wang eds. *Computer Mathematics, Proceedings of ASCM 2000, Chiang Mai, Thailand*, 1–8. *Lecture Notes series on Computing* **8**. World Scientific, Singapore, 2000.
10. D. Lazard, A new method for solving algebraic systems of positive dimension. *Discrete Appl. Math.* **33** (1991), 147–160.
11. D. Kapur. An Approach for solving systems of parametric polynomial equations. *Principles, Practice of Constraint Programming*, (eds. Saraswat, Van Hentenryck), MIT press, 1995.
12. E.R. Kolchin. *Differential Algebra and Algebraic Groups.* Academic Press, New York, 1973.
13. J.F. Ritt. *Differential Algebra.* Amer. Math. Soc. Colloquium, New York, 1950.
14. K.A. Rody, H. Maârouf, M. Ssafini. Triviality and dimension of a system of algebraic differential equations. *J. Automatic Reasoning* **20** (1998), 365–385.
15. W.Y. Sit. An algorithm for solving parametric linear systems. *Journal of Symbolic Computation* **13** (1992), 353–394.
16. W.Y. Sit. Computations on quasi-algebraic sets. IMACS ACA'98 Electronic Proceedings, available at `http://www-troja.fjfi.cvut.cz/aca98/sessions/geom/sit.html`.
17. A. Suzuki, Y. Sato. An alternative approach to comprehensive Gröbner Bases. *Proc. ISSAC'02*, 255–262. ACM Press, New York, 2002.
18. D. Wang. *Elimination Methods.* Springer, Berlin, 2000.
19. D. K. Wang. `http://www.mmrc.iss.ac.cn/~dwang/wsolve.txt`.
20. V. Weispfenning. Comprehensive Gröbner bases. *Journal of Symbolic Computation.* **14** (1992), 1–29.
21. W.T. Wu. *Basic Principles of Mechanical Theorem Proving in Geometries.* Springer-Verlag, Berlin, 1994.
22. L. Yang, J. Z. Zhang, X. R. Hou. *Non-linear Algebraic Equations and Automated Theorem Proving.* ShangHai Sci. and Edu. Publ., ShangHai, 1996.

AN EXPLORATION OF HOMOTOPY SOLVING IN MAPLE

K. HAZAVEH,* D.J. JEFFREY,* G.J. REID,*
S.M. WATT,* A.D. WITTKOPF†

*Ontario Research Centre for Computer Algebra,
The University of Western Ontario,
London, Ontario, Canada

†Centre for Experimental and Computational Mathematics,
Simon Fraser University,
Vancouver, British Columbia, Canada

Homotopy continuation methods find approximate solutions of a given system by a continuous deformation of the solutions of a related exactly solvable system. There has been much recent progress in the theory and implementation of such path following methods for polynomial systems. In particular, exactly solvable related systems can be given which enable the computation of all isolated roots of a given polynomial system. Extension of such methods to determine manifolds of solutions has also been recently achieved. This progress, and our own research on extending continuation methods to identifying missing constraints for systems of differential equations, motivated us to implement higher order continuation methods in the computer algebra language *Maple*. By higher order, we refer to the iterative scheme used to solve for the roots of the homotopy equation at each step. We provide examples for which the higher order iterative scheme achieves a speed up when compared with the standard second order scheme. We also demonstrate how existing *Maple* numerical ODE solvers can be used to give a predictor only continuation method for solving polynomial systems. We apply homotopy continuation to determine the missing constraints in a system of nonlinear PDE, which is to our knowledge, the first published instance of such a calculation.

1. Introduction

Newton's local method for square polynomial systems is a classical method for finding a root of systems with finitely many roots. Homotopy continuation methods [1] deform the known roots of a related system into the roots of the system of interest, and can calculate all the (isolated complex) roots of such systems.

Recent developments by Sommese, Verschelde and Wampler [20,19] include the extension of such homotopy methods to non-square (over- and under-determined systems), and characterize the components or manifolds of solutions of such systems. This is part of the rapidly developing area of Numerical Algebraic Geometry initiated in [23]. This yields new methods for problems which have been traditionally approached with symbolic methods from Computer Algebra, such as factorization [3], Gröbner bases and the completion of systems of partial differential equations. We have extended this work to systems of differential equations in [15], and initiated a study of Numerical Jet Geometry, using homotopy methods.

Despite the availability of very well developed implementations for homotopy continuation methods [26,12] surprisingly little has been implemented in the context of computer algebra systems for numerical solutions for polynomial systems. We note that, for example, Gröbner bases in *Maple* are limited to polynomials with rational coefficients. In *Maple*, the existing solvers focus on univariate equations. Even when working with a powerful Polynomial Homotopy Continuation package (in our case we have extensively used Verschelde's PHCpack [26]) we found the ability to perform experiments and try out ideas in a rich environment such as *Maple* to be a valuable asset. The work we discuss here represents a starting point for *Maple*, since many of the other standard algorithms of Numerical Algebraic Geometry (such as the computation of mixed volumes) still are not implemented in that context.

Existing Homotopy implementations in *Maple* include the univariate program of Fee [6]. In that work Fee truncates the Riemann zeta function, and uses a very efficient homotopy method he has developed for analytic functions to find roots of this truncated function in a given domain. Root counts are verified by using Cauchy's integral formula, using numerical quadrature, around the boundary of the domain. Kotsireas [11] has developed a multivariate fixed step homotopy method in *Maple*.

We have implemented a variable step homotopy continuation method in *Maple*, both for second and third orders, using the code of Smith [18] as a starting point. We compare the methods, and apply them to a variety of problems arising in polynomial system solving. For scalar functions, higher-order schemes are often called Halley methods [7], because of Halley's discovery in Newton's era. Higher-order schemes allow more rapid convergence and larger step sizes in processes such as homotopy solution techniques.

In this paper, we first present the higher order method for solving a single scalar equation. Then in the next section we apply it to systems, and extend it to a homotopy method. In the applications section we apply it to some well-known examples having finitely many roots. Finally we give the first published example of a method using homotopy continuation to identify the missing constraints in a nonlinear system of PDE.

2. Iterative Schemes

Newton's method to find solutions of a single nonlinear equation $f(x) = 0$ is well known; it is also well known that the method is second order and that higher-order methods have been derived [25,7]. Here we start by giving a uniform treatment of the higher-order scalar schemes, as a preparation for the vector case.

Consider solving the scalar equation $f(x) = 0$, given an initial estimate x_0 for the solution. We expand $f(x)$ as a Taylor series around x_0

$$f(x) = f(x_0) + (x - x_0)f'(x_0) + \tfrac{1}{2}(x - x_0)^2 f''(x_0) + \cdots . \tag{2.1}$$

Setting $\Delta = x - x_0$ and assuming $f(x) = 0$, we can solve for Δ by series reversion. Abbreviating $f(x_0)$ to f for clarity, gives

$$\Delta = -\frac{1}{f'}f - \frac{f''}{2(f')^3}f^2 + \frac{3(f'')^2 - f'f'''}{6(f')^5}f^3 + \cdots . \tag{2.2}$$

The series is written as shown to emphasize that it is a series in powers of $f(x_0)$, where $f(x_0)$ will be small in some sense when x_0 is close to the root being sought. The classical Newton iteration is obtained by taking one term of this series; taking two terms gives the third-order scheme

$$\Delta = -\frac{f}{f'} - \frac{f''f^2}{2(f')^3}, \tag{2.3}$$

which has been called Chebyshev's method. The Halley form of (2.3) is

$$\Delta = -\frac{f}{f' - \frac{1}{2}ff''/f'} . \tag{2.4}$$

One derivation of this form solves (2.1) by writing $0 = f + f'\Delta + \frac{1}{2}f''\Delta^2$ as

$$-f = (f' + \tfrac{1}{2}f''\Delta)\Delta .$$

Now assume that the Δ within the parentheses can be approximated by its Newton approximation, obtaining

$$-f = (f' + \tfrac{1}{2}(-f/f')f'')\Delta ,$$

and solve this equation for Δ. None of the methods above can be applied at a point x_0 where $f'(x_0) = 0$, and Halley's method cannot be used for a function satisfying $2(f')^2 - ff'' = 0$, which means any function of the form $f(x) = 1/(Ax + B)$.

For the vector case, we use Cartesian-tensor notation [9]. When applying these results to homotopy methods, we shall give equivalent results in vector-matrix notation. Let $f : \mathbb{R}^m \to \mathbb{R}^m$ be a vector function, with component functions f_i. Let f depend upon the vector x, which in turn has components x_j. We wish to solve $f_i(x) = 0$, starting from an initial estimate $x^{(0)}$. We direct the reader to the literature where a multivariate Halley method of the type below is given [5].

The Taylor series for f about $x^{(0)}$ can be written using $\Delta_j = x_j - x_j^{(0)}$:

$$f_i(x^{(0)}) = f_i(x^{(0)}) + f_{i,k}(x^{(0)})\Delta_k + \tfrac{1}{2} f_{i,kh}(x^{(0)})\Delta_k \Delta_h + \cdots. \tag{2.5}$$

Let $\check{f}_{ki}$ be the inverse of $f_{i,k}$, defined by $\check{f}_{ki} f_{i,j} = \delta_{kj}$, where δ_{kj} is the Kronecker delta. The inverse exists and can be readily computed. Setting the left side of (2.5) to zero and solving to first order in Δ, we obtain the standard Newton iteration: $\Delta_k \approx -\check{f}_{ki} f_i$. To obtain a third-order formula, we avoid reverting (2.5) by adopting the simplified approach used above. We write $\Delta_j \approx -\check{f}_{ji} f_i + \widetilde{\Delta}_j$, and substitute this into (2.5). Solving for $\widetilde{\Delta}$, we obtain a third-order expression for Δ, analogous to the Chebyshev form above:

$$\Delta_k \approx -\check{f}_{ki} f_i - \tfrac{1}{2} \check{f}_{ki}\, f_{i,jh}\, \check{f}_{j\ell}\, f_\ell \check{f}_{hm}\, f_m\,. \tag{2.6}$$

An equivalent form of the third-order scheme can be found that is closer to Halley's scalar form. Convert (2.5) into an equation for Δ as

$$-f_i = f_{i,k}\Delta_k + \tfrac{1}{2} f_{i,kh}\Delta_k \Delta_h = \Delta_k (f_{i,k} + \tfrac{1}{2} f_{i,kh}\Delta_h)\,.$$

We replace Δ_h with its Newton approximation and get

$$-f_i = \Delta_k (f_{i,k} - \tfrac{1}{2} f_{i,kh} \check{f}_{h,m} f_m) = \Delta_k T_{ik}\,.$$

Denoting the inverse of T by $\check{T}$ we obtain the Halley form as

$$\Delta_k = -f_i \check{T}_{ki}\,. \tag{2.7}$$

In the scalar case, if the two third-order schemes (2.3) and (2.4) are computed in a naïve way, Halley's form reduces the operation count by one multiplication, but in the vector case, we have an additional matrix inverse to compute (or linear system to factor).

In moving from a second-order method to a third-order method, there are two separate effects to consider: the speed of convergence and the basin of attraction. If an estimate $x^{(0)}$ is sufficiently close to an isolated non-singular root $x^{(e)}$, then with each iteration a second-order method will approximately double the number of digits that are correct, while a third-order method will triple that number [8]. Thus, for example, an estimate that is correct to 1 digit can be improved to 8 digits in 3 second-order steps or 2 third-order steps. Since the computation of the second derivative term is often expensive, in the scalar case, a higher-order method is usually not an advantage. However, in the vector case, an iteration requires a matrix inverse, making the iteration more expensive. In addition higher order derivatives for polynomial systems can be cheaply obtained (e.g. by automatic differentiation [4]) and this opens the possibility that the third-order method will be more efficient.

If an initial estimate $x^{(0)}$ is further away from the root, and the convergence theorems do not apply, then we must consider the basins of attraction of the root. Graphical presentations of how particular basins of attraction change with the iterative scheme have been published recently [25]. We expect the basin of attraction to be larger for higher-order methods, but have yet to investigate this.

3. Homotopy Method

Consider a system of equations $p(x) = 0$, which we wish to solve. Both p and x are vectors. Suppose we possess a system of equations $q(x) = 0$ whose solutions are already known. Then the homotopy function

$$H(x,t) = q(x)(1-t) + tp(x) = q(x) + t(p(x) - q(x)) , \tag{3.1}$$

is such that $H(x,0) = q(x) = 0$ is a vector system with known solutions, and the system we want to solve is $H(x,1) = p(x) = 0$. We deform from the system at $t = 0$ to the one at $t = 1$ in variable steps Δt. The homotopy parameter t is often called "time."

At a given time t, our problem is to solve $H(x,t) = 0$, and this is done using a iterative scheme. Usually this is a second-order (Newton) scheme, but here a third-order scheme will be used. Since the solution is iterative, a starting estimate is needed. This can be obtained either from the known solutions to $q(x) = 0$, or from the solution at an earlier time, t'. Typically this is the previous time step, that is, $t = t' + \Delta t$. Each homotopy step consists of a predictor stage and a corrector stage. During the predictor stage, a starting estimate is generated for the root, then refined in the corrector stage.

An appropriate start system is now described for the homotopy. Let $p(x) = (p_1(x), \ldots, p_n(x)) = 0$ denote the system of n polynomial equations in n unknowns that we wish to solve. Let d_j denote the total degree of the jth equation (that is, the degree of the highest order monomial in the equation). Then such a start system is

$$q_j(x) = e^{i\phi_j}\left[x_j^{d_j} - \left(e^{i\theta_j}\right)^{d_j}\right] = 0 \qquad (j = 1, 2, \ldots, n), \tag{3.2}$$

where ϕ_j, θ_j are random real numbers in the interval $[0, 2\pi]$. The equation above has the obvious particular solution $x_j = e^{i\theta_j}$ and the complete set of starting solutions for $j = 1, 2, \ldots, n$, is given by:

$$\{\exp(i\theta_j + 2\pi ik/d_j) : k = 0, 1, \ldots, d_j - 1\}. \tag{3.3}$$

Bézout's Theorem states that the number of isolated roots of such a system is bounded above by $d_1 d_2 \cdots d_n$. In particular, with the above start system, with probability 1, each isolated root will lie at the end of an analytic homotopy path originating from one of the starting roots [1]. Bézout's Theorem can be proved by the "method of degeneration," which uses arguments very similar to homotopies [29,27]. More efficient methods use mixed volume to get a smaller upper bound on the number of isolated roots [12], and indeed Bernstein's proof [2] provides a constructive homotopy method.

3.1. *Predictor*

First note that solving $H(x,t) = 0$, where $x(0) = \alpha$ is an exact known root of $q(x) = 0$, is equivalent to solving $\frac{d}{dt}H = 0, x(0) = \alpha$ and so to solving

$$H_x \frac{dx}{dt} + H_t = 0, \; x(0) = \alpha, \;\; 0 \le t \le 1 \tag{3.4}$$

where H_x is the Jacobian of H with respect to $x_1, \ldots, x_n$. Naturally this has led to the use of ODE software for higher order predictors in continuation methods [1, Section 6.3] and [13]. One extreme is not to use any corrector steps.

For convergence the path should stay in the basin of attraction of the sought after isolated root at $t = 1$. To ensure this, most approaches use a combination of corrector and predictor steps. In particular the residuals are monitored to see that they have the characteristic pattern of reduction consistent with the convergence of Newton's method when in the basin of attraction of a root [26]. More recent results on the rigorous identification of the onset of convergence are given in Shub and Smale [17]. If only predictor steps are used, then at the end of a step H will typically have a small nonzero value, and the next step will involve solving a slightly perturbed problem $H = \delta, x(\Delta t) = \widetilde{\alpha}$. This can lead to an accumulation of error, unless residuals are carefully monitored.

ODE integrators usually adapt their step length according to relative error, whereas corrector methods for Newton's method usually use a combination of residual and relative errors. Specifically the residual error $\|H(x,t)\|$ at time t and the relative error Δx, where Δx is the difference between successive values of x, are compared with a working tolerance ε. A challenge then is to reflect the additional residual control in ODE integrators.

Another view of homotopy solving is that of solving a differential equation on a manifold (that is, solving a differential-algebraic equation). Both Visconti [28] and Arponen [24] have implemented DAE solving methods in *Maple*, and it would be of interest to use these in homotopy solving.

Differentiating the ODE (3.4) yields expressions for the higher order derivatives:

$$H_x \frac{d^2x}{dt^2} + \frac{d}{dt} H_x \frac{dx}{dt} + \frac{d}{dt} H_t = 0, \quad H_x \frac{d^3x}{dt^3} + \quad \cdots \quad = 0, \quad \cdots \quad \text{etc}, \tag{3.5}$$

where $\frac{d}{dt} = \frac{\partial}{\partial t} + \Sigma_j \frac{dx_j}{dt} \frac{\partial}{\partial x_j}$. As is well-known, this leads to predictor methods of any order. For example, a higher order predictor method is easily given which is analogous to the higher order corrector method given in the next section.

3.2. *Corrector*

In path-following methods, corrector methods refine the solutions at a fixed time t. The appropriate inclusion of such a corrector phase after each predictor step, can eliminate the accumulation of error involved in pure predictor ODE based methods. In this section we describe a multivariate corrector method (also see a related multivariate Halley method given in [5]). Suppose we have an estimate x, which we wish to improve by computing $x + \Delta x$ at a fixed t. Expanding about (x, t) to second order gives:

$$H(x + \Delta x, t) = H(x, t) + H_x \Delta x + O(\Delta x^2) \; . \tag{3.6}$$

Here H_x is the Jacobian matrix of the vector system H. We use the language of linear algebra here, because it is more convenient for implementation in *Maple.* Then to second order $H(x, t) + H_x \Delta x \approx 0$ and we define this second order approximation of the solution $x + \Delta x$ as $x + \widetilde{\Delta x}$ where $\widetilde{\Delta x}$ satisfies the (vector) system

$$H_x \widetilde{\Delta x} = -H, \tag{3.7}$$

which as usual will be directed to linear solvers, instead of the more expensive computational approach of inverting H_x. Now expanding to third order gives:

$$H(x + \Delta x, t) = H(x, t) + H_x \Delta x + (\Delta x)^T H_{xx}(\Delta x)/2 + O(\Delta x^3)\,, \quad (3.8)$$

where $(\Delta x)^T$ is the transpose of the column vector Δx. If the i-th component of H is denoted by H^i, then in the equation above, H^i_{xx} is an $n \times n$ matrix with entries $H^i_{x_j x_k}$. Suppose $\Delta x = \widetilde{\Delta x} + \widetilde{\widetilde{\Delta x}}$. Then $\widetilde{\Delta x}$ is of order 2 and $\widetilde{\widetilde{\Delta x}}$ satisfies (3.8) to order 3. Substitution of this expression into (3.8), using (3.7), ignoring terms in $\Delta x \widetilde{\Delta x}$, and the above, then shows that $\widetilde{\widetilde{\Delta x}}$ satisfies to third order the following:

$$H_x \widetilde{\widetilde{\Delta x}} = -(\widetilde{\Delta x})^T H_{xx}(\widetilde{\Delta x})/2. \quad (3.9)$$

From a computational point of view, notice the difference between the above expression (3.9) and (2.6). Here the number of computations has been reduced. Also notice that two linear systems must be solved: (3.7) and (3.9). The coefficient matrix is H_x in both cases, so naturally a gain in performance can be realized by computation of an LU factorization, which can then be used twice.

3.3. *Implementation of the Homotopy Algorithm*

We were guided in part by Verschelde's excellent program PHCPack [26]. Another highly developed program is the one by T.Y. Li and his team [12] which enables highly efficient computation of mixed volumes. An efficient and parallel implementation of polyhedral continuation in *MATLAB* is the work of Kim and Kojima [10].

We have made use of the LinearAlgebra Package in *Maple*, and in particular its interface to NAG Library routines.

Using the predictor and corrector methods given above, we now wish to step from $t = 0$ to $t = 1$. The key to efficiency lies in the use of a variable step size. The strategy used in our code is based on counting the number of iterations used at each step. If the corrector step succeeds in only one iteration, then the step size is increased by the factor 1.25 before the next step is taken. If the number of iterations needed at any step is too large, then the step size is decreased. The number of iterations that we count as too large is a parameter that can be tuned. More elaborate stepping schemes have been described elsewhere.

Again for efficiency, the problem $H(x,t) = 0$ is solved to a lower accuracy along the path, and then when a possible solution is obtained at $t = 1$, the "end game" is entered and the solutions are obtained to a greater accuracy. Typically the maximum number of iterative improvements is increased.

Having an environment such as *Maple* opens up the possibilities to use automatic differentiation to calculate the higher order derivatives. Specifically the derivatives are encoded as programs for which good complexity estimates are known [4].

4. Application to Some Polynomial Systems

The algorithms above were coded in *Maple* with a parameter that allowed us to turn the third-order terms on and off. In comparing performance of the codes, we can select between many different metrics. In complicated systems such as *Maple*, there is a particular difficulty of separating the efficiency of the mathematical method from the details of the programming. For this reason, we have selected to test the average size of a step and the number of iterative loops used by the corrector code.

We apply the code to the following simple problems: the intersection of 2 curves given by

$$x^2 + y^2 = 1, \quad x + 2y - 6 = 0\,; \tag{4.1}$$

a univariate problem similar to the well-known Wilkinson polynomial

$$(x-1)(x-2)(x-3)(x-4)(x-5) = 0\,; \tag{4.2}$$

and a cyclic 3 roots problem

$$a + b + c = 0\ ,\ ab + bc + ca = 0,\ abc - 1 = 0 \tag{4.3}$$

Results for these three problems are denoted by the rows labelled *Wilkonson5*, *curves2* and *cyclic3* in the Table below. Also included in that table are some relatively small problems taken from the selection of demo problems given at the Demos link on Jan Verschelde's web site [26]. These problems have rows labelled by their designation at Verschelde's web site (*noon3, lorentz, eco5d*).

The six problems listed were solved using a second order homotopy code based on the algorithms described above. Then the same problems were solved using basically the same code, but with a third-order iterative scheme. For each case, the number of steps taken was recorded and the average step size computed. Also counted was the total number of times the iterative solver was called (abbreviated as Iters below). In all cases, the third-order method used fewer steps and fewer iterations. These statistics are presented in the table below, using N for second order and H for third order.

Problem	N-Time	N-Steps	N-Iters	H-Time	H-Steps	H-Iters
Wilkinson5	2.83	161	285	2.59	119	163
curves2	0.97	60	93	1.05	51	71
cyclic3	4.20	207	360	3.84	149	218
noon3	26.90	1202	2381	24.06	722	1242
lorentz	9.64	391	904	8.80	236	422
eco5d	83.93	2481	5479	78.71	1528	2723

5. Pure Predictor Method in *Maple*

This section contains a description, illustrated by an example, of the use of existing tools in *Maple* to apply homotopy methods to the computation of the roots of a polynomial system. Consider a system of three quadratic equations in three variables,

$$\begin{aligned} p_1 &= 26x^2-55xy+37xz-94x-65y^2+90yz-38y-46z^2+28z+88=0,\\ p_2 &= 64x^2-22xy-37xz+68x-84y^2+80yz+23y-20z^2-7z+4=0,\\ p_3 &= -77x^2+40xy+21xz+55x+61y^2+5yz+66y-83z^2-26z+27=0. \end{aligned} \tag{5.1}$$

We solve this system by deforming the roots of the simpler exactly solvable system

$$q_1=(x-s_1)^2-1=0,\ q_2=(y-s_2)^2-1=0,\ q_3=(z-s_3)^2-1=0. \tag{5.2}$$

The homotopy equations are

$$H_j = tp_j + \alpha_j(1-t)q_j = 0 \qquad (\alpha_j \neq 0,\ j=1,2,3). \tag{5.3}$$

Following the method of Section 3.1, we differentiate each equation in (5.3) with respect to t and obtain a differential system for $x(t), y(t), z(t)$. The system, with initial conditions corresponding to the known solutions of (5.2), is numerically integrated from $t=0$ to $t=1$.

The complex parameters s_j in (5.2) are used in the heuristic method of this section to avoid path problems. For example these problems include path crossing when the starting points are real. Values of $s_1 = 2+\mathrm{i}$, $s_2 = 1-\mathrm{i}$, $s_3 = -1+2\mathrm{i}$ were chosen for the experiment, though a more rigorous choice would have been the starting system (3.2). In addition, the free constants α_j, used in construction of the homotopy equation (5.3), were chosen to be the leading coefficients of the input equations (5.1) with respect to the same variable as the corresponding known solution equation from (5.2).

Integration of the system should be done with care. Direct techniques are of little use, as they require bringing the system into a symbolic solved form for its derivatives before application of the integration technique. The coefficients of the system can be arbitrarily large, depending on the order and density of the system. Performing Gaussian elimination on such a system to bring the system into a solved form for its derivatives is very expensive and often results in numerical instability. Here this problem is addressed by retaining the matrix form of the system, and numerically solving for the derivatives at each predictor step after the coefficients are evaluated at the corresponding time t. This is done in the numerical ODE solvers in *Maple* 8 using the new option *implicit=true* in the call to the numerical ODE solution procedure.

Many existing numerical ODE solvers are restricted to the use of real data, though the path of the system solutions must clearly be complex (as the solutions themselves are complex valued in general). *Maple*'s numerical solvers can handle complex data, but in general the use of the real solver in *Maple* applied to the real and imaginary parts of the system is more efficient for low degree polynomial systems.

The following *Maple* 8 script applies the process described above.

```
# The input system
pols1:=[26*x^2-55*x*y+37*x*z-94*x-65*y^2+90*y*z-38*y-46=z^2+28*z+88,
        64*x^2-22*x*y-37*x*z+68*x-84*y^2+80*y*z+23*y-20=z^2-7*z+4,
      -77*x^2+40*x*y+21*x*z+55*x+61*y^2+5*y*z+66*y-83*z^2-26*z+27]:
# Polynomials with known roots
pols0 := [coeff(pols1[1],x,2)*(x^2-1), coeff(pols1[1],y 2)*(y^2-1),
   coeff(pols1[1],z,2)*(z^2-1)        ]:
# A complex shift - for evaluation of known solution polynomials
sx := 2+I:  sy := 1-I:  sz := -1+2*I:
pols0 := eval(pols0,{x=x-sx,y=y-sy,z=z-sz}):
# Now construct the homotopy system
hsys := eval({seq((1-t)*pols0[i]+t*pols1[i],i=1..3)},
        {x=x(t),y=y(t),z=z(t)}):
dsys := diff(hsys,t):
# Split the system into real and imaginary parts
split:={x(t)=xr(t)+I*xi(t),y(t)=yr(t)+I*yi(t),z(t)=zr(t)+I*zi(t)}:
tmp := collect(eval(expand(eval(dsys,split)),I=II),II):
dsys:={seq(coeff(tmp[i],II,0),i=1..3),
        seq(coeff(tmp[i],II,1),i=1..3)}:
```

```
# The initial conditions corresponding to the known solutions
idata :=[seq(seq(seq({xr(0)=2*i-1+Re(sx),xi(0)=Im(sx),
 yr(0)=2*j-1+Re(sy),yi(0)=Im(sy), zr(0)=2*k-1+Re(sz),zi(0)=Im(sz)},
         k=0..1),j=0..1),i=0..1)]:
# Loop through the data, and obtain the solution for each
for id in idata do
   # Construct the dsolve/numeric procedure
   dsn := dsolve(dsys union id, numeric, implicit=true):
   # Obtain the solution at t=1 and print it.
   sol := dsn(1);
   print(evalf[7](
        eval([x=xr(t)+I*xi(t),y=yr(t)+I*yi(t),z=zr(t)+I*zi(t)],sol)
   ));
end do:
```

The output from running this script is

```
[x = 1.384601 + 0.5873485 I,
 y = -2.516459 - 0.1233784 I,
 z = -1.181569 + 0.7662005 I]
[x = 1.041660 + 2.196067 I,
 y = -0.1078828 - 3.715860 I,
 z = 1.630095 - 1.988129 I]
[x = 0.1122859 + 0.3201893*10^(-6) I,
 y = 0.1227375 + 0.1602351*10^(-6)I,
 z = -0.8616124 - 0.3224805*10^(-7)I]
[x = 0.4493258 + 0.8280515*10^(-6)I,
 y = 1.316899 - 0.7538583*10^(-6)I,
 z = 1.685232 - 0.1398326*10^(-6)I]
[x = 1.384601 - 0.5873477 I,
 y = -2.516459 + 0.1233783 I,
 z = -1.181569 - 0.7661998 I]
[x = 1.041658 - 2.196065 I,
 y = -0.1078817 + 3.715856 I,
 z = 1.630094 + 1.988127 I]
[x = -2.656328 + 5.385065 I,
 y = -1.330291 + 4.515593 I,
 z = 1.681109 + 4.152297 I]
[x = -2.656328 - 5.385073 I,
 y = -1.330290 - 4.515598 I,
 z = 1.681113 - 4.152299 I]
```

The computation took around 1 second on a 1.5GHz machine. Of course a Newton improvement could be done at the end to increase accuracy. Evaluation of the original quadratic equations Eqn. (5.1) yields residuals of magnitude less than 10^{-4}. This reflects the working tolerances of the default method (10^{-7}). Tightening of these tolerances to 10^{-10} provides solutions having residuals of less than 10^{-6} (in 2.5 sec.) and a further reduction of the tolerances to 10^{-12} provides solutions having residuals of less than 10^{-8} (in 5.5 sec.).

6. Application to a Nonlinear System of Partial Differential Equations

In this section we give a new application of homotopy methods to systems of partial differential equations. Specifically we apply our generalization [15] of the methods of Sommese, Verschelde and Wampler [20] to the following nonlinear system of PDE for $u = u(x, y)$:

$$\frac{\partial^2 u}{\partial y^2} - \frac{\partial^2 u}{\partial x \partial y} = 0, \quad \left(\frac{\partial u}{\partial x}\right)^p + \frac{\partial u}{\partial x} - u = 0. \tag{6.1}$$

The aim is to complete (6.1) to an involutive system as defined by the geometric theory of PDE (see [14] and the references therein). This system first appeared in an article to illustrate a new exact elimination algorithm for simplifying systems of PDE [16]. We present, for the first time, a PDE example of the interpolation-free homotopy method described in [15, §6.4], made possible by the works [21,22].

We confine ourselves to the case $p = 2$. In terms of jet coordinates (which are formal indeterminates corresponding to derivatives of the dependent variables, etc.) this is a differential polynomial system in the jet space of second order $J^2 \approx \mathbb{C}^6$. The zero set of the maps defining the PDE, its so-called jet variety, is:

$$\{(u, u_x, u_y, u_{xx}, u_{xy}, u_{yy}) \in J^2 : u_{yy} - u_{xy} = 0,\ u_x^2 + u_x - u = 0\}. \tag{6.2}$$

Here we have suppressed the independent variables x, y since they don't appear explicitly in the PDE. In [15] some homotopy tools for the new area of Numerical Jet Geometry were described, which are now applied to identify the missing constraints of the system (6.2). This completion process can be viewed in J^∞ as generating a descending sequence of manifolds, until that sequence stabilizes.

Setting

$$\phi^1 = u_{yy} - u_{xy},\ \phi^2 = u_x^2 + u_x - u\ , \tag{6.3}$$

the system (6.3) is differentiated (prolonged) up to and including order 2 yielding a system denoted by R, with the associated jet variety:

$$\begin{aligned} V(R) := \{&(u, u_x, u_y, u_{xx}, u_{xy}, u_{yy}) \\ &\in J^2 : \phi^1 = 0, \phi^2 = 0, D_x\phi^2 = 0, D_y\phi^2 = 0\}. \end{aligned} \tag{6.4}$$

Here D_x and D_y are the usual formal total derivatives so that $D_x\phi^2 = (2u_x + 1)u_{xx} - u_x = 0$, etc. Thus we have 4 equations in the 6 unknowns $(u, u_x, u_y, u_{xx}, u_{xy}, u_{yy})$. Now regarded as a submanifold of J^2, the dimension of $V(R)$ satisfies $\dim V(R) \leq 4$ since we already have 2 obviously independent PDEs $\phi^1 = 0$ and $\phi^2 = 0$, and $\dim J^2 = 6$. To check if $V(R)$ has components of dimension 4 in J^2, we intersect it with a random 2 dimensional linear subspace of $\mathbb{C}^6$. This linear space is the solution set of 4 random linear equations of the form:

$$\psi^j := a_{j0}+a_{j1}u+a_{j2}u_x+a_{j3}u_y+a_{j4}u_{xx}+a_{j5}u_{xy}+a_{j6}u_{yy} = 0, \qquad (6.5)$$

where $j = 1, 2, 3, 4$ and the a_{jk} are random complex floating point numbers. The equations (6.5) together with those in (6.4) form a system of 8 equations for 6 variables for the intersection of $V(R)$ with this subspace. Following the procedure in [20] we square the system by incorporating 2 slack variables z_1, z_2. The resulting square system now has 8 equations in 8 unknowns:

$$\begin{aligned} \phi^1 &+ \nu_{11}z_1 + \nu_{12}z_2 = 0, \\ \phi^2 &+ \nu_{21}z_1 + \nu_{22}z_2 = 0, \\ D_x\phi^2 &+ \nu_{31}z_1 + \nu_{32}z_2 = 0, \\ D_y\phi^2 &+ \nu_{41}z_1 + \nu_{42}z_2 = 0, \\ \psi^j &+ \gamma_{j1}z_1 + \gamma_{j2}z_2 = 0 \qquad (j = 1, 2, 3, 4), \end{aligned} \qquad (6.6)$$

where the ν's and γ's are random floating point complex numbers.

Applying our *Maple* program Homotopy to this system yields no solutions with $z_1 = 0, z_2 = 0$. Thus we conclude, numerically, that there are no 4 dimensional components in $V(R)$. Next we check for 3 dimensional components by removing one of the random linear equations, and one of the slack variables (e.g. z_2 by setting $\nu_{j2} = \gamma_{j2} = 0$ in (6.6)) so that the system remains square. Again no solutions are found with $z_1 = 0$ so we conclude that there are no 3 dimensional components. Removing z_1 and one of the remaining random linear equations, and running Homotopy, we find that there are solutions, so we conclude that there exist 2 dimensional components in $V(R)$. A more thorough analysis using the methods of [20] shows that there is only one irreducible component. We conclude that $\dim V(R) = 2$.

We prolong (differentiate) the system R to order 3 in $J^3 \approx \mathbb{C}^{10}$ resulting in the system of equations whose variety we will denote by DR:

$$\begin{aligned} &\phi^1 = 0, \phi^2 = 0, D_x\phi^2 = 0, D_y\phi^2 = 0, D_x\phi^1 = 0, \\ &D_y\phi^1 = 0, D_{xx}\phi^2 = 0, D_{xy}\phi^2 = 0, D_{yy}\phi^2 = 0. \end{aligned} \qquad (6.7)$$

In addition to prolongation, yet another fundamental operation in the jet geometry of differential equations is the projection $\pi : J^q \longrightarrow J^{q-1}$. As described in [15] we implement projection (geometric elimination) onto J^{q-1} by adjoining random linear equations in the J^{q-1} variables alone.

For example to compute $\dim \pi(DR)$ we adjoin the random linear equations in the form (6.5). Slack variables are incorporated to square the system, and similarly it is determined that $\dim \pi(DR) = 1$. Next we calculate D^2R, its dimension, and the dimensions $\dim \pi(D^2R)$, $\dim \pi^2(D^2R)$ of its projections. We summarize the dimensions of these systems below:

$\dim R = 2$		
$\dim \pi(DR) = 1$	$\dim DR = 1$	
$\dim \pi^2(D^2R) = 1$	$\dim \pi(D^2R) = 1$	$\dim D^2R = 1$

For a projection of the output system $\pi^\ell(D^kR)$ to be involutive, it should satisfy a projected dimension test and have involutive symbol [30]. Verifying the projected dimension test involves checking for the maximum $\ell \leq k$, if it exists, such that its dimension satisfies $\dim \pi^\ell(D^kR) = \dim \pi^{\ell+1}(D^{k+1}R)$. This is first satisfied when $k = \ell = 1$, since $\dim \pi(DR) = \dim \pi^2(D^2R)$. Without going into technical details, the involutive symbol test is achieved by computing dimensions. In particular, the dimensions of the symbols of the systems above can be determined from their dimensions above. We find that the dimension of the symbol of $\pi(DR)$ is zero, and hence is involutive. This is in agreement with the results found in [16]. A finer analysis, using homotopy methods, shows that $V(R)$ in J^1 factors into 2 irreducible 1 dimensional components, in accordance with the exact results found in [16]. In particular the exact constraints found by [16] in J^1 are $u_y(u_y - u_x) = 0, u_x^2 + u_x - u = 0$.

The involutive system $\pi(DR)$ can be used to state existence and uniqueness theorems for solutions, and give also the conditions for initializing consistently numerical solvers [24], thus improving their stability.

Finally we mention that if $p > 2$ in (6.1) then as shown in [16] components with higher multiplicity than one can occur (we note that it is always numerically possible to determine when we are in the multiplicity one case, using the methods of Sommese, Verschelde and Wampler, and to bound the multiplicity in the other cases). In the exact case this means that formal derivatives of PDE may not yield the same results as geometric derivatives, and our interpolation-free method may terminate prematurely, before all constraints are found. In the case that the given ideals are radical, then this problem does not occur (this is a generalization of the algebra-geometry correspondence to PDE), and is achieved in the exact case for our example by constructing representations for radicals of algebraic ideals occurring in the computation. In the approximate case the interpolation dependent methods play the same role. However constructing an interpolation-free method in the higher multiplicity case remains an open problem, which is important because of the higher complexity of the interpolation dependent methods.

7. Acknowledgments

Two of the authors (GR and KH) thank Jan Verschelde for helpful discussions. GR thanks Ilias Kotsireas, and Chris Smith for discussions.

References

1. E. L. Allgower, K. Georg. Numerical path following. In P. G. Ciarlet, J. L. Lions, eds. *Scientific Computing* (*Part 2*), 3–203. Volume 5 of *Handbook of Numerical Analysis*, North-Holland, 1997.
2. D. N. Bernstein. The number of roots of a system of equations. (Russian) *Functional Anal. Appl.* **9**(3) (1975), 183–185 (English Translation, 1976).
3. R. M. Corless, A. Galligo, I. S. Kotsireas, S. M. Watt. A geometric-numeric algorithm for absolute factorization of multivariate polynomials. In T. Mora ed., *Proceedings of ISSAC 2002, Lille, France*, 37–45. ACM Press, 2002.
4. G. Corliss, C. Faure, A. Griewank, L. Hascoët, U. Naumann eds., *Automatic Differentiation 2000: From Simulation to Optimization.* Springer, New York, 2001.
5. A. A. M. Cuyt, L. B. Rall. Computational implementation of the multivariate Halley method for solving nonlinear systems of equations. ACM Transactions on Mathematical Software (TOMS), Archive **11**(1) (1985), 20–36.
6. G. Fee. Computing Roots of Truncated Zeta Functions. Poster. MITACs Annual Meeting, Pacific Institute of the Mathematical Sciences, June, 2002.
7. E. Halley. Methodus nova, accurata & facilis inveniendi radices aequationum quarumcunque generaliter, sine praevia reductione. *Philos. Trans. Roy. Soc. London* **18** (1694), 139–148.
8. D. J. Jeffrey, M. W. Giesbrecht, R. M. Corless. Integer roots for integer-power-content calculations. In X.-S. Gao, D. Wang, eds., *Computer mathematics, Proceedings of the Fourth Asian Symposium* (*ASCM 2000*), 195–203. *Lecture Notes Series on Computing* **8**, World Scientific, Singapore, 2000.
9. H. Jeffreys. *Cartesian tensors.* Cambridge University Press, 1965.
10. S. Kim, M. Kojima. CMPSm: A continuation method for polynomial systems (*MATLAB* Version). In A. M. Cohen, X. Gao, N. Takayama eds., *Mathematical Software* (*ICMS2002 Beijing, China, Aug 17–19, 2002*). World Scientific, Singapore, 2002.

11. I. S. Kotsireas. Homotopies and polynomial system solving I. Basic Principles. *SIGSAM Bulletin* **5**(1) (2001), 19–32.
12. T. Y. Li. Numerical solution of multivariate polynomial systems by homotopy continuation methods. *Acta Numerica* **6** (1997), 399–436.
13. B. N. Lundberg, A. B. Poore. Variable order Adams-Bashforth predictors with error-stepsize control for continuation methods. *SIAM J. Sci. Statist. Comp.* **12**(3) (1991), 695–723.
14. G. J. Reid, P. Lin, A. D. Wittkopf. Differential elimination-completion algorithms for DAE and PDAE. *Studies in Applied Mathematics* **106**(1) (2001), 1–45.
15. G. J. Reid, C. Smith, J. Verschelde Geometric completion of differential systems using numeric-symbolic continuation. *SIGSAM Bulletin* **36**(2) (2002), 1–17.
16. G. J. Reid, A.D. Wittkopf, A. Boulton. Reduction of systems of nonlinear partial differential equations to simplified involutive forms. *Eur. J. of Appl. Math.* **7**, 604–635.
17. M. Shub, S. Smale. Complexity of Bezout's theorem V: Polynomial time. *Theoretical Computer Science* **133**(1) (1994), 141–164.
18. C. Smith. Further Development in HomotopySolve for *Maple* 7. Undergraduate Thesis, Department of Applied Mathematics, University of Western Ontario. 2002.
19. A. J. Sommese, J. Verschelde. Numerical homotopies to compute generic points on positive dimensional algebraic sets. *J. Complexity* **16**(3) (2000), 572–602.
20. A. J. Sommese, J. Verschelde, C. W. Wampler. Numerical decomposition of the solution sets of polynomial systems into irreducible components. *SIAM J. Numer. Anal.* **38**(6) (2001), 2022–2046.
21. A. J. Sommese, J. Verschelde, C. W. Wampler. Using monodromy to decompose solution sets of polynomial systems into irreducible components. In C. Ciliberto, F. Hirzebruch, R. Miranda, M. Teicher eds., *Application of Algebraic Geometry to Coding Theory, Physics, and Computation*, 297–315. *Proceedings of a NATO Conference (February 25–March 1, 2001, Eilat, Israel)*. Kluwer Academic Publishers, 2001.
22. A. J. Sommese, J. Verschelde, C. W. Wampler. Symmetric functions applied to decomposing solution sets of polynomial systems. *SIAM J. Numer. Anal.* **40**(6) (2002), 2026–2046.
23. A. J. Sommese, C.W. Wampler. Numerical algebraic geometry. In J. Renegar, M. Shub, S. Smale eds., *The Mathematics of Numerical Analysis*, 749–763. *Proceedings of the AMS-SIAM Summer Seminar in Applied Mathematics (July 17–August 11, 1995, Park City, Utah). Lectures in Applied Mathematics* **32**, 1996.
24. J. Tuomela, T. Arponen. On the numerical solution of involutive ordinary differential systems. *IMA J. Numer. Anal.* **20** (2000), 561–599.
25. J. L. Varona. Graphic and numerical comparison between iterative methods. *Mathematical Intelligencer* **24** (2002), 37–46.

26. J. Verschelde. Algorithm 795: PHCpack: A general-purpose solver for polynomial systems by homotopy continuation. *ACM Transactions on Mathematical Software* **25**(2) (1999), 251–276.
Software site: `http://www.math.uic.edu/~jan`.
27. J. Verschelde. Polynomial homotopies for dense, sparse and determinantal systems. *Mathematical Sciences Research Institute Preprint* # 1999-041, 1999. Available online at `http://www.msri.org`.
28. J. Visconti. *Numerical Solution of Differential Algebraic Equations, Global Error Estimation and Symbolic Index Reduction.* Ph.D. Thesis. Laboratoire de Modélisation et Calcul. Grenoble. 1999.
29. A. Weil. *Foundations of Algebraic Geometry. AMS Colloquium Publications.* Volume XXIX, Providence, Rhode Island, 1962.
30. A. D. Wittkopf, G. J. Reid. Fast differential elimination in C: The CDiffElim Environment. *Comp. Phys. Comm.* **139**(2) (2001), 192–217.

DENSITIES AND FLUXES OF DIFFERENTIAL-DIFFERENCE EQUATIONS

MARK S. HICKMAN
Department of Mathematics and Statistics
University of Canterbury
Private Bag 4800, Christchurch, New Zealand
Email address: M.Hickman@math.canterbury.ac.nz

WILLY A. HEREMAN
Department of Mathematical and Computer Sciences
Colorado School of Mines
Golden CO 80401–1887, USA
Email address: whereman@mines.edu

An algorithm is presented that uses direct methods to find conserved densities and fluxes of differential-difference equations. The use of the code is illustrated with the modified Volterra lattice. This algorithm has been implemented in *Maple* in the form of a toolbox.

1. Differential-Difference Equations

Dating back to the work of Fermi, Pasta, and Ulam in the 1950's [2] differential-difference equations (DDEs) have been the focus of many nonlinear studies. A number of physically interesting problems can be modeled with nonlinear DDEs, including particle vibrations in lattices, currents in electrical networks, pulses in biological chains, etc. DDEs play important roles in queuing problems and discretizations in solid state and quantum physics. Last but not least, they are used in numerical simulations of nonlinear PDEs.

Consider a nonlinear (autonomous) DDE of the form

$$\frac{d}{dt}u_n = f(u_{n-\ell}, u_{n-\ell+1}, \ldots, u_n, \ldots, u_{n+m-1}, u_{n+m}) \tag{1.1}$$

with

$$\frac{\partial f}{\partial u_{n-\ell}}\frac{\partial f}{\partial u_{n+m}} \neq 0,$$

where n is an arbitrary integer. In general, f is a vector-valued function of a finite number of dynamical variables and each u_k is a vector-valued function of t.

The index n may lie in $\mathbf{Z}$, or the u_k may be periodic, $u_k = u_{k+N}$. The integers ℓ and m measure the degree of non-locality in (1.1). If $\ell = m = 0$ then the equation is local and reduces to a system of ordinary differential equations.

The (up-)shift operator D is defined by

$$\mathrm{D}\, u_k = u_{k+1}.$$

Its inverse, called the down-shift operator, is given by $\mathrm{D}^{-1}\, u_k = u_{k-1}$. Obviously, $u_k = \mathrm{D}^k\, u_0$. The actions of D and D^{-1} are extended to functions by acting on their arguments. For example,

$$\begin{aligned} &\mathrm{D}\, g(u_p, u_{p+1}, \ldots, u_q) \\ &= g(\mathrm{D}\, u_p, \mathrm{D}\, u_{p+1}, \ldots, \mathrm{D}\, u_q) \\ &= g(u_{p+1}, u_{p+2}, \ldots u_{q+1}). \end{aligned}$$

In particular,

$$\mathrm{D}\left(\frac{\partial}{\partial\, u_k}\, g(u_p, u_{p+1}, \ldots, u_q)\right) = \frac{\partial}{\partial\, u_{k+1}}\, g(u_{p+1}, u_{p+2}, \ldots, u_{q+1}).$$

Moreover, for equations of type (1.1), the shift operator commutes with the time derivative; that is,

$$\mathrm{D}\left(\frac{d}{dt}\, u_n\right) = \frac{d}{dt}\,(\mathrm{D}\, u_n)\,.$$

Thus, with the use of the shift operator, the entire system (1.1) which may be an infinite set of ordinary differential equations is generated from a single equation

$$\frac{d}{dt}\, u_0 = f(u_{-\ell}, u_{-\ell+1}, \ldots, u_0, \ldots, u_{m-1}, u_m) \tag{1.2}$$

with

$$\frac{\partial\, f}{\partial\, u_{-\ell}}\, \frac{\partial\, f}{\partial\, u_m} \neq 0.$$

Next, we define the (forward) difference operator, $\Delta = \mathrm{D} - \mathrm{I}$, by

$$\Delta\, u_k = (\mathrm{D} - \mathrm{I})\, u_k = u_{k+1} - u_k,$$

where I is the identity operator. The difference operator extends to functions by $\Delta\, g = \mathrm{D}\, g - g$. This operator takes the role of a spatial derivative on the shifted variables as many examples of DDEs arise from discretization of a PDE in $(1+1)$ variables [6].

For any function $g = g(u_p, u_{p+1}, \ldots, u_q)$, the total time derivative $\mathrm{D}_t\, g$ is computed as

$$\mathrm{D}_t\, g = \sum_{k=p}^{q} \frac{\partial\, g}{\partial\, u_k} \frac{d}{dt}\, u_k = \left(\sum_{k=p}^{q} \frac{\partial\, g}{\partial\, u_k}\, \mathrm{D}^k \right) f(u_{-\ell}, u_{-\ell+1}, \ldots, u_m)$$

on solutions of (1.1). A simple calculation shows that the shift operator D commutes with D_t, and so does D with Δ.

A function $\rho = \rho(u_p, u_{p+1}, \ldots, u_q)$ is a (conserved) *density* of (1.2) if there exists a function $J = J(u_r, u_{r+1}, \ldots, u_s)$, called the (associated) *flux*, such that

$$\mathrm{D}_t\, \rho + \Delta\, J = 0 \tag{1.3}$$

is satisfied on the solutions of (1.2). Eq. (1.3) is called a *local conservation law.*

Any shift of a density is trivially a density since

$$\mathrm{D}_t\, \mathrm{D}^k\, \rho + \Delta\, \mathrm{D}^k\, J = \mathrm{D}^k\, (\mathrm{D}_t\, \rho + \Delta\, J) = 0,$$

with associated flux $\mathrm{D}^k\, J$.

Constants of motion for (1.2) are easily obtained from a density and its shifts. Indeed, for any density ρ with corresponding flux J, consider

$$\Omega = \sum_{k=p}^{q} \mathrm{D}^k \rho.$$

The total time derivative of Ω is

$$\mathrm{D}_t\, \Omega = -\sum_{k=p}^{q} \Delta \mathrm{D}^k J = -\sum_{k=p}^{q} \left(\mathrm{D}^{k+1} - \mathrm{D}^k\right) J = -\left(\mathrm{D}^{q+1} - \mathrm{D}^p\right) J.$$

Applying appropriate boundary conditions (e.g. all $u_k \to 0$, as $k \to \pm\infty$) one gets the conservation law

$$\mathrm{D}_t \left(\sum_{k=-\infty}^{\infty} \mathrm{D}^k \rho \right) = \lim_{q\to\infty} \mathrm{D}^{q+1} J - \lim_{p\to -\infty} \mathrm{D}^p J = 0.$$

For a periodic chain where $u_k = u_{k+N}$, after summing over a period, one obtains

$$\mathrm{D}_t \left(\sum_{k=0}^{N} \mathrm{D}^k \rho \right) = \mathrm{D}^{N+1} J - \mathrm{D}^0 J = J - J = 0.$$

In either case, Ω is a constant of motion of (1.2) since Ω does not change with time.

A function $g = g(u_p, u_{p+1}, \ldots, u_q)$ is a *total difference* if there exists another function $h = h(u_p, u_{p+1}, \ldots, u_{q-1})$, such that $g = \Delta\, h$. A density which is a total difference,

$$\rho = \Delta\, F \tag{1.4}$$

(so that $\mathrm{D}_t\, \rho = \Delta\, \mathrm{D}_t\, F$ and therefore $J = -\mathrm{D}_t\, F$ is an associated flux), is called *trivial*. These densities lead to trivial conservation laws since

$$\Omega = \sum_{k=p}^{q} \mathrm{D}^k \Delta\, F = \mathrm{D}^{q+1} F - \mathrm{D}^p F$$

holds identically (and not just on solutions of (1.2)).

Two densities $\rho, \tilde{\rho}$ are called *equivalent* if $\rho - \tilde{\rho} = \Delta\, F$ for some F. Equivalent densities, denoted by $\rho \sim \tilde{\rho}$, differ by a trivial density and yield the same conservation law. Also, $\rho \sim D^k \rho$, and (1.4) expresses that $\rho \sim 0$.

The Discrete Euler Operator (Variational Derivative)

A necessary and sufficient condition for a function g to be a total difference is that

$$\mathrm{E}(g) = 0, \tag{1.5}$$

where E is the discrete Euler operator (variational derivative) [1,7] defined by

$$\mathrm{E}(g) = \sum_{k=p}^{q} \mathrm{D}^{-k} \left(\frac{\partial}{\partial\, u_k}\, g \right).$$

Note that we can rewrite the Euler operator as

$$\mathrm{E}(g) = \frac{\partial}{\partial\, u_0} \left(\sum_{k=p}^{q} \mathrm{D}^{-k}\, g \right). \tag{1.6}$$

Also note that (1.5) implies

$$0 = \mathrm{D}^p \left(\frac{\partial}{\partial\, u_{q-p}}\, \mathrm{E}(g) \right) = \mathrm{D}^p \left(\frac{\partial}{\partial\, u_{q-p}} \left(\mathrm{D}^{-p}\, \frac{\partial\, g}{\partial\, u_p} \right) \right) = \frac{\partial^2\, g}{\partial\, u_p \partial\, u_q}. \tag{1.7}$$

2. The Algorithm for Computing Densities and Fluxes

Densities ρ and fluxes J are related by (1.3). In principle, we need to first solve $E(D_t\,\rho) = 0$ to find the density. This integrability condition for ρ is a rather unusual "PDE" since it involves derivatives of both ρ and shifts of ρ (and so involves ρ with *different* arguments). Surprisingly, perhaps, the integrability condition is amenable to analysis. Next, to compute $J = -\Delta^{-1}(\mathrm{D}_t\,\rho)$, we need to invert the operator $\Delta = \mathrm{D} - \mathrm{I}$. Working with the formal inverse,

$$\Delta^{-1} = \mathrm{D}^{-1} + \mathrm{D}^{-2} + \mathrm{D}^{-3} + \ldots,$$

is impractical, perhaps impossible.

We therefore present a simple algorithm which will yield both the density and the flux, and circumvent the above infinite formal series. The algorithm does not require the densities or fluxes to be polynomial (see [3,4,5] for an algorithm to find polynomial densities). The idea is to split expressions into a total derivative term plus a term involving lower order shifts. Using

$$\mathrm{I} = (\mathrm{D} - \mathrm{I} + \mathrm{I})\,\mathrm{D}^{-1} = \Delta\,\mathrm{D}^{-1} + \mathrm{D}^{-1}, \tag{2.1}$$

any expression T can be split as follows,

$$T = \Delta\,\mathrm{D}^{-1}T + \mathrm{D}^{-1}\,T.$$

The first term will contribute to the flux whilst the second term has a strictly lower shift than the original expression. These decompositions are applied to terms that do not involve the lowest-order shifted variables. Once all terms are "reduced" in this manner, left-over terms (all of which involve the lowest-order shifted variable) yield the constraints for the undetermined coefficients or unknown functions in the density.

Without loss of generality, we can assume that $\rho = \rho(u_0, \ldots, u_q)$. For (1.2), we have

$$\begin{aligned}\mathrm{D}_t\,\rho &= \sum_{k=0}^{q} \frac{\partial\,\rho}{\partial\,u_k}\mathrm{D}^k f(u_{-\ell}, \ldots, u_m)\\ &= \frac{\partial\,\rho}{\partial\,u_0} f(u_{-\ell}, \ldots, u_m) + \sum_{k=1}^{q} \frac{\partial\,\rho}{\partial\,u_k}\mathrm{D}^k f(u_{-\ell}\ \ldots, u_m).\end{aligned}$$

Applying (2.1) to the second term, with $f = f(u_{-\ell}, \ldots, u_m)$, we obtain

$$\mathrm{D}_t\,\rho = \frac{\partial\,\rho}{\partial\,u_0} f + (\Delta\,\mathrm{D}^{-1} + \mathrm{D}^{-1}) \left(\sum_{k=1}^{q} \frac{\partial\,\rho}{\partial\,u_k}\mathrm{D}^k f \right)$$

$$= \frac{\partial \rho}{\partial u_0} f + \Delta \sum_{k=0}^{q-1} \frac{\partial \mathrm{D}^{-1} \rho}{\partial u_k} \mathrm{D}^k f + \sum_{k=0}^{q-1} \frac{\partial \mathrm{D}^{-1} \rho}{\partial u_k} \mathrm{D}^k f$$

$$= \left(\frac{\partial}{\partial u_0} \left(\rho + \mathrm{D}^{-1} \rho \right) \right) f + \Delta \sum_{k=0}^{q-1} \frac{\partial \mathrm{D}^{-1} \rho}{\partial u_k} \mathrm{D}^k f + \sum_{k=1}^{q-1} \frac{\partial \mathrm{D}^{-1} \rho}{\partial u_k} \mathrm{D}^k f$$

Next, we repeat this procedure by applying (2.1) to the last term. After a further $q - 2$ applications, we get

$$\begin{aligned} \mathrm{D}_t \rho &= \left(\frac{\partial}{\partial u_0} \left(\rho + \mathrm{D}^{-1} \rho + \mathrm{D}^{-2} \rho \right) \right) f \\ &\quad + \Delta \left(\sum_{k=0}^{q-1} \frac{\partial \mathrm{D}^{-1} \rho}{\partial u_k} \mathrm{D}^k f + \sum_{k=0}^{q-2} \frac{\partial \mathrm{D}^{-2} \rho}{\partial u_k} \mathrm{D}^k f \right) + \sum_{k=1}^{q-2} \frac{\partial \mathrm{D}^{-2} \rho}{\partial u_k} \mathrm{D}^k f \\ &= \cdots \\ &= \mathrm{E}(\rho) f + \Delta \left(\sum_{j=1}^{q} \sum_{k=0}^{q-j} \frac{\partial \mathrm{D}^{-j} \rho}{\partial u_k} \mathrm{D}^k f \right) \end{aligned}$$

by (1.6).

If $\mathrm{E}(\rho) f = 0$ then ρ is a trivial density. For ρ to be a non-trivial density, we require

$$\mathrm{E}(\rho) f = \Delta h \tag{2.2}$$

for some h with $h \neq 0$. In this case the associated flux is

$$J = -h - \sum_{j=1}^{q} \sum_{k=0}^{q-j} \frac{\partial \mathrm{D}^{-j} \rho}{\partial u_k} \mathrm{D}^k f.$$

One could apply the discrete Euler operator to $\mathrm{E}(\rho) f$ to determine conditions such that (2.2) holds. Alternatively, one could repeat the above strategy by splitting this expression into a part, $A^{(0)}$, that does not depend on the lowest shifted variable and the remaining terms. One then applies (2.1) repeatedly to $A^{(0)}$ (removing the total difference terms generated by (2.1) from $A^{(0)}$) until one obtains a term that involves both the lowest and highest shifted variables. From (1.7), one obtains conditions for (2.2) to hold. This procedure must be repeated until all the terms in $\mathrm{E}(\rho) f$ have been recast as a total difference. We now illustrate this algorithm with a simple example.

The Modified Volterra Lattice

Consider the modified Volterra (mV) lattice [1]

$$\frac{d\,v_n}{dt} = {v_n}^2(v_{n+1} - v_{n-1}),$$

or, equivalently,

$$\frac{d\,v_0}{dt} = {v_0}^2(v_1 - v_{-1}). \tag{2.3}$$

To keep the exposition simple, we search for densities of the form

$$\rho = \rho(v_0, v_1, v_2)$$

for the mV lattice (2.3), where $f = {v_0}^2(v_1 - v_{-1})$. The construction of densities involving a greater spread of dynamic variables can be accomplished with the *Maple* code.

We write

$$\mathrm{D}_t\,\rho = \sigma + \Delta\,K$$

with $\sigma = \mathrm{E}(\rho)\,f$ given by

$${v_0}^2\,(v_1 - v_{-1})\left(\frac{\partial\rho\,(v_0, v_1, v_2)}{\partial v_0} + \frac{\partial\rho\,(v_{-1}, v_0, v_1)}{\partial v_0} + \frac{\partial\rho\,(v_{-2}, v_{-1}, v_0)}{\partial v_0}\right)$$

and

$$\begin{aligned} K &= \sum_{j=1}^{q}\sum_{k=0}^{q-j}\frac{\partial\,\mathrm{D}^{-j}\rho}{\partial\,u_k}\mathrm{D}^k\,f \\ &= {v_0}^2\,(v_1 - v_{-1})\,\frac{\partial\rho\,(v_{-1}, v_0, v_1)}{\partial v_0} + {v_1}^2\,(v_2 - v_0)\,\frac{\partial\rho\,(v_{-1}, v_0, v_1)}{\partial v_1} \\ &\quad + {v_0}^2\,(v_1 - v_{-1})\,\frac{\partial\rho\,(v_{-2}, v_{-1}, v_0)}{\partial v_0}. \end{aligned}$$

The application of (1.7) directly to σ yields nothing since no term depends on both the lowest shifted variable (v_{-2}) and the highest shifted variable (v_2). We therefore split σ into a term which depends on v_{-2} and a term that does not. We then apply (2.1) to the latter term.

Thus,

$$\begin{aligned} \sigma &= {v_0}^2\,(v_1 - v_{-1})\,\frac{\partial\rho\,(v_{-2}, v_{-1}, v_0)}{\partial v_0} \\ &\quad + (\Delta + \mathrm{I})\,\mathrm{D}^{-1}\left({v_0}^2\,(v_1 - v_{-1})\left(\frac{\partial\rho\,(v_0, v_1, v_2)}{\partial v_0} + \frac{\partial\rho\,(v_{-1}, v_0, v_1)}{\partial v_0}\right)\right). \end{aligned}$$

Next, we update σ and K by removing the total difference term from σ and adding it to K. Thus,

$$\sigma = {v_0}^2 (v_1 - v_{-1}) \frac{\partial \rho (v_{-2}, v_{-1}, v_0)}{\partial v_0} + {v_{-1}}^2 (v_0 - v_{-2}) \left(\frac{\partial \rho (v_{-1}, v_0, v_1)}{\partial v_{-1}} + \frac{\partial \rho (v_{-2}, v_{-1}, v_0)}{\partial v_{-1}} \right)$$

and

$$K = {v_0}^2 (v_1 - v_{-1}) \left(\frac{\partial \rho (v_{-1}, v_0, v_1)}{\partial v_0} + \frac{\partial \rho (v_{-2}, v_{-1}, v_0)}{\partial v_0} \right) + {v_1}^2 (v_2 - v_0) \frac{\partial \rho (v_{-1}, v_0, v_1)}{\partial v_1} + {v_{-1}}^2 (v_0 - v_{-2}) \left(\frac{\partial \rho (v_{-1}, v_0, v_1)}{\partial v_{-1}} + \frac{\partial \rho (v_{-2}, v_{-1}, v_0)}{\partial v_{-1}} \right).$$

Now we are ready to apply (1.7) to σ. Thus,

$$\mathrm{D}^2 \frac{\partial^2 \sigma}{\partial v_{-2} \partial v_1} = {v_2}^2 \frac{\partial^2 \rho (v_0, v_1, v_2)}{\partial v_0 \partial v_2} - {v_1}^2 \frac{\partial^2 \rho (v_1, v_2, v_3)}{\partial v_1 \partial v_3} = 0. \qquad (2.4)$$

A differentiation with respect to v_0 yields

$${v_2}^2 \frac{\partial^3 \rho (v_0, v_1, v_2)}{\partial {v_0}^2 \partial v_2} = 0.$$

So,

$$\begin{aligned} \rho(v_0, v_1, v_2) &= \rho^{(1)}(v_0, v_1) - \rho^{(2)}(v_0, v_1) + \rho^{(2)}(v_1, v_2) + \rho^{(3)}(v_1, v_2) v_0 \\ &= \rho^{(1)}(v_0, v_1) + \Delta \rho^{(2)}(v_0, v_1) + \rho^{(3)}(v_1, v_2) v_0 \end{aligned}$$

for some unknown functions $\rho^{(1)}$, $\rho^{(2)}$ and $\rho^{(3)}$. The term $\Delta \rho^{(2)}(v_0, v_1)$ leads to a trivial density and can be ignored. Thus

$$\rho = \rho^{(1)}(v_0, v_1) + \rho^{(3)}(v_1, v_2) v_0.$$

The integrability condition (2.4) is now

$${v_2}^2 \frac{\partial}{\partial v_2} \rho^{(3)} (v_1, v_2) - {v_1}^2 \frac{\partial}{\partial v_3} \rho^{(3)} (v_2, v_3) = 0.$$

A subsequent differentiation with respect to v_3 gives

$$- {v_1}^2 \frac{\partial^2}{\partial {v_3}^2} \rho^{(3)} (v_2, v_3) = 0.$$

Consequently, $\rho^{(3)}(v_2, v_3) = \rho^{(4)}(v_2) + \rho^{(5)}(v_2) v_3$. Equation (2.4) has become

$${v_2}^2 \rho^{(5)} (v_1) - {v_1}^2 \rho^{(5)} (v_2) = 0$$

from which we get

$$\rho^{(5)}(v_2) = c^{(1)}{v_2}^2$$

for some constant $c^{(1)}$.

The density now has the form $\rho = \rho^{(1)}(v_0, v_1) + (\rho^{(4)}(v_1) + c^{(1)}{v_1}^2 v_2)v_0$. The term $\rho^{(4)}(v_1)v_0$ can be absorbed into the term $\rho^{(1)}(v_0, v_1)$. Thus

$$\rho(v_0, v_1, v_2) = \rho^{(1)}(v_0, v_1) + c^{(1)} v_0 {v_1}^2 v_2.$$

Hence,

$$\begin{aligned}\sigma = {} & c^{(1)} v_{-2}{v_{-1}}^3 v_0(v_0 - 2v_{-2}) - v_{-2}{v_{-1}}^2 \frac{\partial \rho^{(1)}(v_{-1}, v_0)}{\partial v_{-1}} \\ & + {v_{-1}}^2(v_0 - v_{-2}) \frac{\partial \rho^{(1)}(v_{-2}, v_{-1})}{\partial v_{-1}} \\ & + (\Delta + \mathrm{I}) \left(c^{(1)}{v_{-2}}^2{v_{-1}}^3 v_0 + {v_{-2}}^2 v_{-1} \frac{\partial \rho^{(1)}(v_{-2}, v_{-1})}{\partial v_{-2}} \right)\end{aligned}$$

by (2.1). As before, we update σ by moving the total difference term into K. Therefore,

$$\begin{aligned}\sigma = {} & c^{(1)} v_{-2}{v_{-1}}^3 v_0(v_0 - v_{-2}) - v_{-2}{v_{-1}}^2 \frac{\partial \rho^{(1)}(v_{-1}, v_0)}{\partial v_{-1}} \\ & + {v_{-1}}^2(v_0 - v_{-2}) \frac{\partial \rho^{(1)}(v_{-2}, v_{-1})}{\partial v_{-1}} + {v_{-2}}^2 v_{-1} \frac{\partial \rho^{(1)}(v_{-2}, v_{-1})}{\partial v_{-2}}\end{aligned}$$

and

$$\begin{aligned}K = {} & c^{(1)} v_{-1} v_0 \left(v_{-1}{v_0}^2 v_1 - v_0 {v_1}^2 v_2 - v_{-2}{v_{-1}}^2 v_0 - {v_0}^2{v_1}^2 + 3{v_{-2}}^2{v_{-1}}^2 \right) \\ & + {v_0}^2(v_{-1} - v_1) \frac{\partial}{\partial v_0} \rho^{(1)}(v_{-1}, v_0) + {v_{-2}}^2 v_{-1} \frac{\partial}{\partial v_{-2}} \rho^{(1)}(v_{-2}, v_{-1}) \\ & + {v_{-1}}^2(v_{-2} - v_0) \left(\frac{\partial \rho^{(1)}(v_{-1}, v_0)}{\partial v_{-1}} + \frac{\partial \rho^{(1)}(v_{-2}, v_{-1})}{\partial v_{-1}} \right).\end{aligned}$$

Applying (1.7) to σ yields

$$\begin{aligned}\mathrm{D}^2 \frac{\partial^2 \sigma}{\partial v_{-2} \partial v_0} & = {v_1}^2 \left(2c^{(1)} v_1(v_2 - v_0) - \frac{\partial^2 \rho^{(1)}(v_1, v_2)}{\partial v_1 \partial v_2} + \frac{\partial^2 \rho^{(1)}(v_0, v_1)}{\partial v_0 \partial v_1} \right) \\ & = 0\end{aligned}$$

which readily integrates to

$$\rho^{(1)}(v_0, v_1) = \tfrac{1}{2} c^{(1)}{v_0}^2{v_1}^2 + \rho^{(6)}(v_0) + c^{(2)} v_0 v_1.$$

At this stage

$$\sigma = {v_{-2}}^2 v_{-1} \frac{d\rho^{(6)}\left(v_{-2}\right)}{dv_{-2}} - v_2 {v_{-1}}^2 \frac{d\rho^{(6)}\left(v_{-1}\right)}{dv_{-1}}$$

and

$$K = {v_{-2}}^2 v_{-1} \frac{d\rho^{(6)}\left(v_{-2}\right)}{dv_{-2}} - {v_{-1}}^2 v_{-2} \frac{d\rho^{(6)}\left(v_{-1}\right)}{dv_{-1}} + c^{(1)}\, v_{-1} {v_0}^3 {v_1}^2$$
$$+ {v_{-1}}^2 v_0 \frac{d\rho^{(6)}\left(v_{-1}\right)}{dv_{-1}} + c^{(1)}\, v_{-1} {v_0}^2 {v_1}^2 v_2 + c^{(2)}\, v_{-1} {v_0}^2 v_1.$$

One more application of (1.7) to σ yields

$$\mathrm{D}^2 \frac{\partial^2 \sigma}{\partial v_{-2} \partial v_{-1}} = \frac{d}{dv_0}\left({v_0}^2 \frac{d\rho^{(6)}\left(v_0\right)}{dv_0}\right) - \frac{d}{dv_1}\left({v_1}^2 \frac{d\rho^{(6)}\left(v_1\right)}{dv_1}\right) = 0.$$

This equation integrates to yield

$$\rho^{(6)}(v_0) = c^{(3)} \frac{1}{v_0} + c^{(4)} \log v_0.$$

Thus the solution is given by

$$\rho = \tfrac{1}{2} c^{(1)} ({v_0}^2 {v_1}^2 + 2 v_0 {v_1}^2 v_2) + c^{(2)} v_0 v_1 + c^{(3)} \frac{1}{v_0} + c^{(4)} \log v_0.$$

Consequently,

$$\sigma = c^{(3)} v_{-2} - c^{(3)} v_{-1} = \; - c^{(3)} \Delta\, v_{-2},$$

which can be absorbed into K to yield $\sigma = 0$. Finally, the associated flux is

$$J = \; - K$$
$$= -c^{(1)} v_{-1} {v_0}^2 {v_1}^2 (v_0 + v_2) - c^{(2)} v_{-1} {v_0}^2 v_1 + c^{(3)} (v_0 + v_{-1}) - c^{(4)} v_{-1} v_0.$$

Splitting the density ρ and flux J according to the independent constants $c^{(i)}$ we obtain four non-trivial conversation laws.

3. Implementation

The strategy outlined in the above example has been implemented in *Maple*. This implementation is in the form of a toolbox which provides code to compute the reductions and generate the integrability conditions. The solution of the integrability conditions has not been automated. The code is available from the author.

Acknowledgments

The first author wishes to thank the Department of Mathematical and Computer Sciences of the Colorado School of Mines for its hospitality during his sabbatical visit where this work was completed.

References

1. V. E. Adler, S. I. Svinolupov, R. I. Yamilov. Multi-component Volterra and Toda type integrable equations. *Phys. Lett. A* **254** (1999), 24–36.
2. E. Fermi, J. Pasta, S. Ulam. *Collected papers of Enrico Fermi*, **II**. University of Chicago Press, Chicago, Illinois, 1965, 978.
3. Ü. Göktaş, W. Hereman. Computation of conserved densities for nonlinear lattices. *Physica D* **123** (1998), 425–436.
4. Ü. Göktaş, W. Hereman, G. Erdmann. Computation of conserved densities for systems of nonlinear differential-difference equations. *Phys. Lett. A* **236** (1997), 30–38.
5. W. Hereman, Ü. Göktaş, M. Colagrosso, A. Miller. Algorithmic integrability tests for nonlinear differential and lattice equations. In *Special Issue on Computer Algebra in Physics Research. Comp. Phys. Comm.* **115** (1998), 428–446.
6. A. B. Shabat, R. I. Yamilov, Lattice representations of integrable systems. *Phys. Lett. A* **130** (1988), 271–275.
7. A. B. Shabat, R. I. Yamilov. Symmetries of nonlinear chains. *Leningrad Math. J.* **2** (1991), 377–400.

A COMPLETE MAPLE PACKAGE FOR NONCOMMUTATIVE RATIONAL POWER SERIES

V. HOUSEAUX, G. JACOB, N.E. OUSSOUS, M. PETITOT
LIFL, Bât. M3–Informatique, Université Lille I
59655 Villeneuve d'Ascq Cedex France
Email: {houseaux, jacob, oussous, petitot}@lifl.fr

The noncommutative rational power series constitute a very important class of noncommutative power series. They allow the encoding of the Input/Output behaviour of bilinear dynamical systems which can be used as approximant of nonlinear systems. In the study of multiple zeta values (MZV), they allow elegant proofs of some formulae. In this paper, we give an original way to represent these power series: a rational power series is represented by a noncommutative polynomial and a rewriting system. This new kind of representation allows us to define and implement a unique canonical representation of rational series. The usual operations on rational series : sum, Cauchy product, quasi-inverse, shuffle products, can be implemented in the representation.

1. Introduction

The formal power series in non commutative variables were used in connection with the theory of automata in 1959 by M.P. Schützenberger [24]. Their study benefits from headaways in theoretical data processing and in return brings often a new light on computer algebra and on other domains such as algebra, geometry, combinatorics and control theory [9,8,10,11,12,21]. The noncommutative formal power series lead to well adapted combinatorial manipulations in programming and in computer algebra.

The rational power series constitute the smallest class of noncommutative formal power series which contains polynomials and which is closed under the sum, Cauchy product and star (or quasi-inverse).

On computers it is common to implement the rational power series by a matrix representation [13,22]. Here, the work consists in representing the rational power series by using their definition by "*finite expansion*" (i.e. a noncommutative polynomial) and "*rewriting system*".

For several years, mathematicians have used algebraic rewriting [7] to obtain effective criteria for equality between vector spaces (resp. ideals) of polynomials defined by a finite number of generators.

The computation of standard bases[a] was introduced by Buchberger [6] to study the ideals of commutative polynomials. For the ideals of noncommutative polynomials this motivates the desire for an effective implementation of rational power series in noncommutative variables, and to compute them a unique canonical form. This unified representation is obtained by introducing the concept of a *filtering*, which is a partial order relation on all the words being used.

More generally, the definition of rationality would be described in the formalism of Hopf algebra, and then the algebra of rational series can be viewed as the Sweedler dual of some graded Hopf algebra [1,25,19].

Our specific interest is to derive from this work new demonstrations of identities, which strictly polynomial computation did not yield. This could bring new applications in the study of special functions, the Riemann zeta function [3,17,18] and quantum groups.

We present here first the definitions and the basic operations on the noncommutative formal power series. Then, we present the rational power series, their representation in Maple, and the techniques used to implement the basic operations.

2. Motivation

We call *Euler-Zagier sum*, *polyzeta*, or MZV (*Multiple Zeta Value*) the following sum, which appears as an extension of the Riemann zeta function to multi-indices:

$$\zeta(s) = \zeta(s_1, s_2, \ldots, s_p) \;=\; \sum_{n_1 > n_2 > \cdots > n_p > 0} \frac{1}{n_1^{s_1} n_2^{s_2} \cdots n_p^{s_p}}. \tag{2.1}$$

This sum converges if and only if $s_1 \geqslant 2$. The following equality was conjectured by Zagier [26]:

$$\zeta(\underbrace{3, 1, \ldots, 3, 1}_{2n}) = 4^{-n}\zeta(\underbrace{4, \ldots, 4}_{n})$$

If we put $\langle 3,1\rangle^n = \underbrace{\widehat{3,1}, \ldots, \widehat{3,1}}_{n}$ and $\langle 4\rangle^n = \underbrace{4, \ldots, 4}_{n}$, this equality can be rewritten as:

$$\zeta(\langle 3,1\rangle^n) = 4^{-n}\zeta(\langle 4\rangle^n)$$

[a] Buchberger called them Gröbner bases.

This conjecture was proved by Borwein *et al.* [4,5]. Here we shall prove it by a pure syntactic computation using noncommutative rational power series. Indeed, let any multi-index $\boldsymbol{s} = (s_1, s_2, \ldots, s_p)$ be coded by the word $w = a^{s_1-1}ba^{s_2-1}b \cdots a^{s_p-1}b$ on the alphabet $X = \{a, b\}$. Then, by a very simple computation of automata, we deduce the following equality between rational power series [16]:

$$(t^2ab)^* ⧢ (-t^2ab)^* = (-4t^4a^2b^2)^* \tag{2.2}$$

where t is some formal parameter and ⧢ denotes the shuffle product to be defined in Section 5. This equality implies the following identity between two generating series of polylogarithm functions:

$$\left(\sum_{i=0}^{\infty}(t^2)^i \mathrm{Li}_{\langle 2\rangle^i}(z)\right) \times \left(\sum_{j=0}^{\infty}(-t^2)^j \mathrm{Li}_{\langle 2\rangle^j}(z)\right) = \sum_{n=0}^{\infty}(-4t^4)^n \mathrm{Li}_{\langle 3,1\rangle^n}(z), \tag{2.3}$$

which when evaluated at $z = 1$ becomes:

$$\left(\sum_{i=0}^{\infty}\zeta(\langle 2\rangle^i)(t^2)^i\right) \times \left(\sum_{j=0}^{\infty}\zeta(\langle 2\rangle^j)(-t^2)^j\right) = \sum_{n=0}^{\infty}\zeta(\langle 3,1\rangle^n)(-4t^4)^n. \tag{2.4}$$

On the other hand, let us code each word of the form $a^{s-1}b$ by a letter y_s on a new infinite alphabet $Y = \{y_i,\, i > 0\}$ indexed by positive integers. We prove the following identity on rational series on Y with the quasi–shuffle operation $\uplus$ to be defined in Section 5 [15]:

$$(t^2y_2)^* \uplus (-t^2y_2)^* = (-t^4y_4)^*. \tag{2.5}$$

By interpreting the words on Y as quasi-symmetric functions [14], we obtain the following equality between MZV:

$$\left(\sum_{i=0}^{\infty}\zeta(\langle 2\rangle^i)(t^2)^i\right) \times \left(\sum_{j=0}^{\infty}\zeta(\langle 2\rangle^j)(-t^2)^j\right) = \sum_{n=0}^{\infty}\zeta(\langle 4\rangle^n)(-t^4)^n. \tag{2.6}$$

From (2.4) and (2.6), we deduce:

$$\sum_{n=0}^{\infty}\zeta(\langle 3,1\rangle^n)(-4t^4)^n = \sum_{n=0}^{\infty}\zeta(\langle 4\rangle^n)(-t^4)^n. \tag{2.7}$$

Finally, identifying the coefficients of $(-t^4)$ on both sides yields:

$$4^n\zeta(\langle 3,1\rangle^n) = \zeta(\langle 4\rangle^n). \tag{2.8}$$

This example shows clearly how identities of noncommutative rational series appear as an elegant way to produce identities on MZV. And this justifies the implementation of symbolic tools for handling the noncommutative rational series.

3. The Algebra of Polynomials and Power Series

Let $X = \{x_0, x_1, ..., x_m\}$ be a nonempty finite *alphabet*. The *free monoid* X^* generated by X is the set of all words (finite sequences of letters) over X, including the *empty word*, denoted by ε. We denote by X^+ the set $X^* \setminus \{\varepsilon\}$. The *length* of a word w, denoted by $|w|$, is the number of letters that compose it.

A *noncommutative formal power series* S on the alphabet X with coefficients in the field k is a map from X^* to k which associates to any word $w \in X^*$ a scalar denoted by $\langle S|w\rangle$ and called the *coefficient* of the word w in the power series S. This series will be written as the formal sum:

$$S = \sum_{w\in X^*} \langle S|w\rangle\, w. \tag{3.1}$$

The set of formal power series thus defined will be denoted by $k\langle\langle X\rangle\rangle$.

Let S be a noncommutative power series. The *support* of S is the language $\mathrm{Supp}(S) = \{w \in X^* \mid \langle S|w\rangle \neq 0\}$. If $\langle S|\varepsilon\rangle = 0$ then S is said to be *proper*.

A *noncommutative polynomial* is a noncommutative formal power series with finite support. The set of all noncommutative polynomials will be denoted by $k\langle X\rangle$. It is a subset of $k\langle\langle X\rangle\rangle$. The *degree* of a polynomial $P \in k\langle X\rangle$, denoted by $\deg(P)$, is equal to $\sup\{|w|,\ w \in \mathrm{Supp}(P)\}$ if $P \neq 0$ and $-\infty$ if $P = 0$.

Clearly, $k\langle\langle X\rangle\rangle$ is a k-vector space. Let S and T be two series in $k\langle\langle X\rangle\rangle$. We define the *Cauchy product* as follows:

$$S \cdot T = \sum_{w\in X^*} \Big(\sum_{u\cdot v=w} \langle S|u\rangle \times \langle T|v\rangle \Big)\, w, \tag{3.2}$$

where $u\cdot v$ denotes the concatenation product defined in X^*. Endowed with this product, $k\langle\langle X\rangle\rangle$ has a structure of an *associative and noncommutative algebra*.

Let $S \in k\langle\langle X\rangle\rangle$ be a proper series. We denote by S^n the n-th power of S for the Cauchy product. In that case, the family $(S^n)_{n\geq 0}$ is *locally finite*. That is:

$$\forall\, w \in X^*, \quad n > |w| \Longrightarrow \langle S^n|w\rangle = 0$$

and consequently this family is summable. We denote by S^* the sum of this family and call it the *star* (or *quasi-inverse*) of S:

$$S^* = \sum_{n\geq 0} S^n. \tag{3.3}$$

Also, we denote by S^+ the power series: $S^+ = \sum_{n\geq 1} S^n$.

The sum, the Cauchy product, and the quasi-inverse are called *rational operations.* The algebra of *rational series* [2], denoted by $\mathcal{R}at_k(X)$, is the smallest subset of $k\langle\langle X\rangle\rangle$ containing constants, letters, and closed under the rational operations. We can also define $\mathcal{R}at_k(X)$ as the smallest sub-algebra of $k\langle\langle X\rangle\rangle$ containing $k\langle X\rangle$ and closed under the star operation.

To implement rational series, we need the action called *right remainder.* Let $u \in X^*$ be a word and $S \in k\langle\langle X\rangle\rangle$ be a series. The *right remainder* of S by u, denoted by $S \triangleright u$, is defined as follows:

$$S \triangleright u = \sum_{w\in X^*} \langle S|w\rangle \; w \triangleright u \quad \text{or by duality} \quad \langle S \triangleright u|w\rangle = \langle S|uw\rangle, \tag{3.4}$$

where $w \triangleright u$ is equal to v if $w = uv$ and zero otherwise. This defines a *right action* of X^* on $k\langle\langle X\rangle\rangle$: $S\triangleright(uv) = (S\triangleright u)\triangleright v$. We define in a symmetric way the *left remainder.* These two actions commute: $(u \triangleleft S) \triangleright v = u \triangleleft (S \triangleright v)$. Let S and T be power series, w a word and x a letter. The right remainder verifies the following rules[b]:

$$(S+T) \triangleright w = S \triangleright w + T \triangleright w, \tag{3.5}$$

$$(S \cdot T) \triangleright w = (S \triangleright w) \cdot T + \sum_{\substack{uv = w \\ v \neq \varepsilon}} \langle S|u\rangle \cdot (T \triangleright v), \tag{3.6}$$

$$S^* \triangleright w = (S \triangleright w) \cdot S^* + \sum_{\substack{uv = w \\ u, v \neq \varepsilon}} \langle S|u\rangle \cdot (S^* \triangleright v). \tag{3.7}$$

A series $S \in k\langle\langle X\rangle\rangle$ is said to be *recognizable* if and only if the set $\{\, S \triangleright P \mid P \in k\langle X\rangle \,\}$ is a k-vector space of finite dimension.

Theorem 3.8. (Kleene-Schützenberger 1961) *A formal power series is rational if and only if it is recognizable.*

4. Representation of Rational Power Series

The various notions presented in this part allow us to introduce a canonical representation of the rational power series. This representation constitutes an alternative to the matrix representation, which is not unique, even when it is minimal. We show that it is possible to represent a rational series by a given noncommutative polynomial (finite series) and a set of rewriting rules [2,23].

---

[b] (3.7) holds for $w \neq \varepsilon$.

Let X be a totally ordered alphabet. We consider the *lexicographic-by-length order* on the words. The support of a rational series is always defined on a *finite sub-alphabet.* So we shall restrict their representations to some finite alphabet X, thus ensuring that Theorem 3.8 holds true.

Definition 4.1. (prefix code) We call any finite set C of words of X^+ a *prefix code* if for all $u \in C,\ v \in X^*$, we have: $uv \in C \Longrightarrow v = \varepsilon$.

With any prefix code C, we can associate its *prefixial part* $\mathbb{P}_C$ formed by the empty word and the proper left factors of the words of C:

$$\mathbb{P}_C = \{u \in X^* \mid \exists\, v \neq \varepsilon, uv \in C\}.$$

A prefix code C will be called *complete* if any word of X^* except its prefixial part begins with a word of C. In other words,

$$C \text{ complete} \iff X^* - \mathbb{P}_C \;=\; CX^* \quad \iff \quad X^* = C^*\mathbb{P}_C.$$

The monoid X^* can be represented by an *infinite n-ary tree*, with nodes labelled by the words on X, expanded in lexicographic-by-length order (see Figure 1).

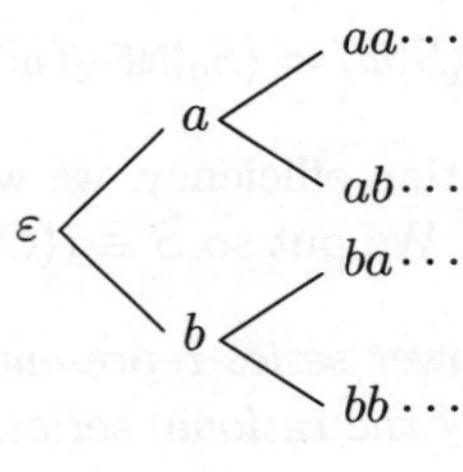

Figure 1. Representation of X^*

With any finite complete code C, we can associate a finite tree T_C by removing from the complete tree of X^* any branch issued from a node labelled by a word in the code C. The words of $\mathbb{P}_C$ are exactly the labels of the internal nodes of T_C.

Definition 4.2. (Rewriting System) A *rewriting system* is a pair $(C, \mathcal{R})$, where C is a finite complete prefix code, and $\mathcal{R} : C \to k\langle X\rangle$ is a map such that:

$$\forall\, u \in C \quad \mathrm{Supp}(\mathcal{R}(u)) \subset \{v \in \mathbb{P}_C \mid v < u\}.$$

The pair $(u, \mathcal{R}(u))$, denoted also by $u \to \mathcal{R}(u)$, is called a *rewriting rule.*

Given a rewriting system $(C, \mathcal{R})$, we can define a linear endomorphism $\widetilde{\mathcal{R}}$ of $k\langle X\rangle$ in the following way:

- if $u \in \mathbb{P}_C$, then no rule is applicable: $\widetilde{\mathcal{R}}(u) = u$,
- if $u \notin \mathbb{P}_C$, there is a unique $c \in C$ which is a left factor of u. A single rule applies then: $\widetilde{\mathcal{R}}(u) = \mathcal{R}(c) \cdot (u \triangleright c)$.

The condition on $\mathcal{R}$ in Definition 4.2 and the fact that the lexicographic-by-length order is a *well-order* warrant us that for any $P \in k\langle X\rangle$, the sequence[c] $(\widetilde{\mathcal{R}}^k(P))_{k\geqslant 0}$ is stationary and that $\widetilde{\mathcal{R}}(P) = P$ if and only if the support of P is included in $\mathbb{P}_C$.

In other words, after a finite number of rewriting steps, the resulting polynomial is called a *normal form of P* and is denoted by $\mathsf{NF}_{\mathcal{R}}(P)$. Clearly, its support is included in $\mathbb{P}_C$. Let us note that for any words u and v, we have $\mathsf{NF}_{\mathcal{R}}(uv) = \mathsf{NF}_{\mathcal{R}}(\mathsf{NF}_{\mathcal{R}}(u)v)$.

Definition 4.3. By a *rewriting representation* we mean a rewriting system $(C, \mathcal{R})$ together with a polynomial S_0 with support included in $\mathbb{P}_C$, called the *finite expansion*. The series S represented by $((C, \mathcal{R}), S_0)$ is then defined by:

$$\langle S|w\rangle = \langle S_0|\mathsf{NF}_{\mathcal{R}}(w)\rangle. \tag{4.4}$$

For reasons of implementation efficiency, we will include the prefixial part in the representation of S. We put so $S \equiv ((C, \mathcal{R}), S_0, \mathbb{P}_C)$.

Proposition 4.5. *The power series represented by rewriting system and finite expansion are exactly the rational series.*

Proof. The series defined by (4.4) is rational. Indeed, we show easily that for all $w \in X^*$ we have $\langle S \triangleright u|w\rangle = \langle S \triangleright \mathsf{NF}_{\mathcal{R}}(u)|w\rangle$. Then $S \triangleright u = S \triangleright \mathsf{NF}_{\mathcal{R}}(u)$.

Conversely, if S is a rational series, we can easily build a rewriting system and a polynomial that represent S. □

Example 4.6. We consider the prefix code $C = \{aa, aba, abb, ba, bb\}$, the finite expansion $S_0 = 1 + 2ab + b$ and the following rules: $(aa, 0)$, $(aba, 2a)$, $(abb, 0)$, $(ba, a+ab)$, and $(bb, 2b)$. It is shown in Figure 2 how the automaton is obtained from the tree of C.

---

[c] $\widetilde{\mathcal{R}}^0(P) = P$ and $\widetilde{\mathcal{R}}^{k+1}(P) = \widetilde{\mathcal{R}}^k(\widetilde{\mathcal{R}}(P))$.

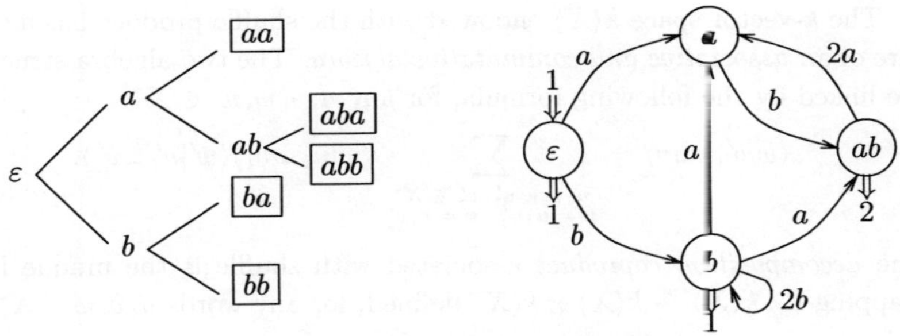

Figure 2. The tree and the associated automaton

A rewriting representation of any rational series S (given for example by a rational expression) can be obtained by computing the right remainders of S, following the lexicographic-by-length order.

Example 4.7. Let $S = (ab)^* = 1 + ab + abab + \cdots$ be a rational series on the alphabet $X = \{a, b\}$. We deduce the rewriting rules of the series S by computing the right remainders by the words of X^* taken in lexicographic-by-length order.

$$S \triangleright aa = 0 \quad aa \to 0$$

$$S \triangleright a = bS$$

$$S \triangleright \varepsilon = S \qquad S \triangleright ab = S \quad ab \to \varepsilon$$

$$S \triangleright b = 0 \quad b \to 0$$

The prefixial part consists of the internal nodes: $\mathbb{P}(S) = \{\varepsilon, a\}$. The finite expansion of S is $S_0 = \langle S|\varepsilon\rangle + \langle S|a\rangle a = 1 + 0 \cdot a = 1$. The leaves of the tree yield the heads of the rewriting rules. This leads to the following rewriting representation:

$$S = (\ \{b \to 0,\ aa \to 0,\ ab \to \varepsilon\},\ 1,\ \{\varepsilon,\ a\}\ ).$$

5. Shuffle Products

The *shuffle (or Hurwitz) product* is a commutative and associative product defined, for the series S and T, as follows:

$$S ⧢ T = \sum_{u,v \in X^*} (\langle S|u\rangle \times \langle T|v\rangle)\ u ⧢ v \tag{5.1}$$

where $u ⧢ v$ is defined recursively by $u ⧢ \varepsilon = \varepsilon ⧢ u = u$ and $(xu) ⧢ (yv) = x(u ⧢ (yv)) + y((xu) ⧢ v)$ if $x, y \in X$ and $u, v \in X^*$.

The k-vector space $k\langle X\rangle$ endowed with the shuffle product has a structure of an *associative and commutative algebra.* The two algebra structures are linked by the following formula, for any $u, v, w, w' \in X^*$:

$$\langle ww'|u \sqcup\!\sqcup v\rangle = \sum_{\substack{u_1, v_1, u_1', v_1' \in X^* \\ u = u_1 u_1',\ v = v_1 v_1'}} \langle w|u_1 \sqcup\!\sqcup v_1\rangle\langle w'|u_1' \sqcup\!\sqcup v_1'\rangle. \tag{5.2}$$

The *decomposition coproduct* associated with shuffle is the unique linear mapping $\Gamma : k\langle X\rangle \to k\langle X\rangle \otimes k\langle X\rangle$ defined, for any words $u, v, w \in X^*$, by:

$$\langle \Gamma(w)|u \otimes v\rangle = \langle w|u \sqcup\!\sqcup v\rangle. \tag{5.3}$$

It is easily verified that any letter $x \in X$ is a *primitive element* for Γ, *i.e.* $\Gamma(x) = x \otimes 1 + 1 \otimes x$, and that Γ is a morphism for the Cauchy product.

Let us consider now the *infinite* alphabet $Y = \{y_i,\ i \geqslant 1\}$ indexed by positive integers. We denote again by ε or y_0 the empty word of Y^*. The Cauchy product on $k\langle Y\rangle$ is defined as above. On the other hand, we define a *second shuffle product*, called *quasi-shuffle* [19,20] as follows: for any $y_i, y_j \in Y$ and $u, v \in Y^*$, we set

$$\begin{cases} u \uplus \varepsilon = \varepsilon \uplus u = u, \\ (y_i u) \uplus (y_j v) = y_i(u \uplus y_j v) + y_j(y_i u \uplus v) + y_{i+j}(u \uplus v). \end{cases} \tag{5.4}$$

This product is extended to polynomials and series, giving $k\langle Y\rangle$ a structure of an *associative and commutative algebra.* The *decomposition coproduct* associated with the quasi-shuffle is denoted here by Δ and is defined by duality:

$$\langle \Delta(R)|P \otimes Q\rangle = \langle R|P \uplus Q\rangle, \qquad \forall P, Q, R \in k\langle Y\rangle. \tag{5.5}$$

Its value on the letters is:

$$\Delta(y_n) = \sum_{i+j=n} y_i \otimes y_j \tag{5.6}$$

and it is also a morphism for the Cauchy product.

Proposition 5.7. *The shuffle (resp. quasi-shuffle) product of two rational series is a rational series.*

Proof. The computation in Section 6.3 proves that the resulting series is recognizable, and Theorem 3.8 ends the proof[d]. □

---

[d] In the case of the quasi-shuffle, the theorem can be used because if S and T are rational, then $S \uplus T$ still uses only a finite number of letters.

6. The Package rationalSeries

The algorithms presented in this part for the computation of a series S all have a similar global structure: they explore the ordered tree of the words. For every word w, a search is made for a linear dependence of $S \triangleright w$ from the remainders of S by the words previously examined and kept in the prefixial part. If such a dependence is found, it is used to establish a new rewriting rule and the branches of the tree leaving from w will not be explored (w will be in the prefix code). Otherwise, we keep w for the prefixial part and add its successors $(wx)_{x \in X^*}$ in the list `fifo`[e] so that they will be examined later. The algorithm ends when `fifo` is found empty.

Obviously, the main difficulty lies in the methods used to search for linear dependences. But it is not necessary to find them all, since a specific algorithm is devoted to minimize the obtained rewriting representation.

To search for these linear dependences, every algorithm computes for the current word of the list `fifo` a certain expression varying in a vector space of finite dimension[f] which is represented by one *row-vector*. The row-vectors corresponding to words kept for the prefixial part are stacked as lines in a matrix which is used for each search of linear dependences.

For each of these algorithms, it is then necessary to clarify the expression that the row-vectors will represent, and to check that any linear dependence between these expressions implies the same dependence among the remainders of S by the corresponding words.

In the package `rationalSeries`, the rational power series are represented by a table named `Series` with four entries such that:

- `Series[Alph]` is the ordered alphabet (a list),
- `Series[Prefix]` is the prefixial part (ordered list of words),
- `Series[Rules]` is a table of rewriting rules,
- `Series[DL]` is a finite expansion (polynomial).

The words are represented by unaffected variables of type `indexed`, *i.e.* a word $a_1 a_2 \cdots a_n$ is represented by $\texttt{X}[\texttt{a}_1, \texttt{a}_2, \ldots, \texttt{a}_n]$. For the table of rules, the indices are the words w of the prefix code and the associated values are $\mathcal{R}(w)$ (`Series[Rules][X[w]]:=`$\mathcal{R}(w)$).

---

[e] Initially containing only ε.

[f] This is what ascertains that all these algorithms do stop: as soon as there are as many words in the prefixial part as this dimension, linear dependences will be found for all remaining words in the list `fifo`, which will then be emptied.

In the following sections, we will describe the explicit algorithms for addition, the Cauchy product, and the shuffle product. Let S_1 and S_2 be two rational series represented respectively by variables S1 and S2.

6.1. *Addition*

If $S = S_1 + S_2$ is being represented by the variable S, then the alphabet S[Alph] will be the union of S1[Alph] and S2[Alph]. It is the tree of the words of this new alphabet that is explored.

For any word w, the couple $(\mathsf{NF}_{S_1}(w), \mathsf{NF}_{S_2}(w))$ is represented by the row L_w consisting of the coefficients of each normal forms on the prefixial parts of their respective series. These row-vectors are indeed in a space of finite dimension. Furthermore, if $L_w = \sum_i \lambda_i L_{w_i}$ then

$$\begin{aligned} S \triangleright w = S_1 \triangleright w + S_2 \triangleright w &= S_1 \triangleright \mathsf{NF}_{S_1}(w) + S_2 \triangleright \mathsf{NF}_{S_2}(w) \\ = \sum_i \lambda_i \big(S_1 \triangleright \mathsf{NF}_{S_1}(w_i) + S_2 \triangleright \mathsf{NF}_{S_2}(w_i)\big) &= \sum_i \lambda_i S \triangleright w_i. \end{aligned}$$

Thus the search for linear dependences can be made with these L_w, standing for $(\mathsf{NF}_{S_1}(w), \mathsf{NF}_{S_2}(w))$.

To compute the finite expansion of S, it suffices to compute the coefficient $\langle S_1|u\rangle + \langle S_2|u\rangle$ for each word in the prefixial part.

6.2. *Cauchy Product*

By using the relation (3.6) and by setting

$$Q^w_{S_1} = \sum_{uv=w,\ v\neq 1} \langle S_1|u\rangle v$$

we can write

$$S_1 \cdot S_2 \triangleright w = (S_1 \triangleright w) S_2 + S_2 \triangleright Q^w_{S_1}.$$

We associate with $S_1 \cdot S_2 \triangleright w$ the pair of polynomials $(\mathsf{NF}_{S_1}(w), \mathsf{NF}_{S_2}(Q^w_{S_1}))$. We compute both normal forms and build the row-vector of their coordinates on the prefixial parts of S_1 and S_2. It is easy to see that as for the sum, since $Q^w_{S_1}$ is linear in S_1, any linear dependence between these row-vectors also holds between the corresponding remainders of the product series.

In the loop controlled by the list fifo, each time a word is kept the prefixial part, its coefficient in $S_1 \cdot S_2$ is computed, being equal to $\langle S_1|w\rangle\langle S_2|\varepsilon\rangle + \langle S_2|Q^w_{S_1}\rangle$.

6.3. *Shuffle Product*

The easy computation of the shuffle product of S_1 and S_2 go through the coproduct Γ as follows:

$$\begin{aligned}\langle (S_1 \sqcup\!\sqcup S_2) \triangleright w | u\rangle = \langle S_1 \sqcup\!\sqcup S_2 | wu\rangle &= \langle S_1 \otimes S_2 | \Gamma(w)\Gamma(u)\rangle \\ &= \big\langle (S_1 \otimes S_2) \triangleright \Gamma(w) | \Gamma(u)\big\rangle \\ &= \big\langle (S_1 \otimes S_2) \triangleright \mathsf{NF}_{S_1\otimes S_2}(\Gamma(w)) | \Gamma(u)\big\rangle,\end{aligned}$$

where $\mathsf{NF}_{S_1\otimes S_2}$ is defined by:

$$\mathsf{NF}_{S_1\otimes S_2}(v\otimes v') = (\mathsf{NF}_{S_1} \otimes \mathsf{NF}_{S_2})(v\otimes v') = \mathsf{NF}_{S_1}(v) \otimes \mathsf{NF}_{S_2}(v').$$

The shuffle algorithm then searches for linear dependences between elements $\mathsf{NF}_{S_1\otimes S_2}(\Gamma(w))$, dependences that will hold for the remainders of $S_1 \sqcup\!\sqcup S_2$.

So it is necessary to implement tensor products of words and polynomials. This is done by representing the tensor product $w_1 \otimes w_2$ of two words by the indexed variable `Theta[X[w1],X[w2]]`. Some procedures are then defined for computing algebraic operations on these tensor products, as well as a procedure `Tnormalf` computing the joint normal form with respect to both series.

The row-vector which represents $\mathsf{NF}_{S_1\otimes S_2}(\Gamma(w))$ is that of its coordinates on the obvious basis consisting of $u \otimes u'$ for $u \in$ `S1[CP]` and $u' \in$ `S2[CP]`.

When a word w is kept for the prefixial part (if no dependence was found), the expression $\langle S_1 \otimes S_2 | \mathsf{NF}_{S_1\otimes S_2}(\Gamma(w))\rangle \cdot w$ is added to the finite expansion S_0 of the resulting series.

7. Example: R_add

```
> with(rationalSeries):
> S:=R_star(X[a,b]):
> T:=R_star(-X[a,b]):
> R_print(S):
      The series alphabet:           [a, b]
      The series prefix part:        [X[], X[a], X[a,b]]
      The truncated series expansion: X[]+X[a,b]
      The set of rewriting rules:
        Rule number 1 :   X[b] -> 0
        Rule number 2 :   X[a,a] -> 0
        Rule number 3 :   X[a,b,a] -> X[a]
        Rule number 4 :   X[a,b,b] -> 0
```

```
> R_print(T):
      The series alphabet:          [a, b]
      The series prefix part:       [X[], X[a], X[a,b]]
      The truncated series expansion: X[]-X[a,b]
      The set of rewriting rules:
        Rule number 1 :   X[b] -> 0
        Rule number 2 :   X[a,a] -> 0
        Rule number 3 :   X[a,b,a] -> -X[a]
        Rule number 4 :   X[a,b,b] -> 0
> t:=time():
> R:=R_add(S,T):
> time()-t;
                                  .640

> R_print(R):
      The series alphabet:          [a, b]
      The series prefix part:       [X[], X[a], X[a,b],
                                     X[a,b,a], X[a,b,a,b]]
      The truncated series expansion: 2*X[]+2*X[a,b,a,b]
      The set of rewriting rules:
        Rule number 1 :   X[b] -> 0
        Rule number 2 :   X[a,a] -> 0
        Rule number 3 :   X[a,b,b] -> 0
        Rule number 4 :   X[a,b,a,a] -> 0
        Rule number 5 :   X[a,b,a,b,a] -> X[a]
        Rule number 6 :   X[a,b,a,b,b] -> 0

> R_print(R_minimize(R)):
      The series alphabet:          [a, b]
      The series prefix part:       [X[], X[a], X[a,b], X[a,b,a]]
      The truncated series expansion: 2*X[]
      The set of rewriting rules:
        Rule number 1 :   X[b] -> 0
        Rule number 2 :   X[a,a] -> 0
        Rule number 3 :   X[a,b,b] -> 0
        Rule number 4 :   X[a,b,a,a] -> 0
        Rule number 5 :   X[a,b,a,b] -> X[]
```

8. Conclusion

The work presented in this paper describes a package using an original representation of rational series (with regard to the pure matrix methods). In this package, all the operations were implemented to allow computation with the infinite alphabet Y and computation of the quasi-shuffle. Beyond the basic operations presented here, the package provides various tools for further computation on rational series[g] (Hadamard product, rational approximation of a formal series, etc.).

This package allowed us to verify the motivation we presented in the Section 2. It will be used extensively to produce new shuffle relations between rational series and potentially new MZV identities.

References

1. E. Abe. *Hopf algebras.* Cambridge University Press, 1980.
2. J. Berstel, C. Reutenauer. *Les séries rationnelles et leurs langages.* (French) *Études et Recherches en Informatique*, Masson, Paris, 1984
3. J. M. Borwein, D. M. Bradley, D. J. Broadhurst. Evaluation of k-fold Euler/Zagier sums : a compendium of results for arbitrary k. *Electronic J. Combinatorics*, **4** (2) (1997), #R5.
4. J. M. Borwein, D. M. Bradley, D. J. Broadhurst, P. Lisoněk. Combinatorial aspects of multiple zeta values. *Electronic J. Combinatorics* **5**(1) (1998), #R38.
5. J. M. Borwein, D. M. Bradley, D. J. Broadhurst, P. Lisoněk. Special values of multidimensional polylogarithms. *Trans. Amer. Math. Soc.* **353**(3) (2001), 907–941.
6. B. Buchberger. An algorithm for finding a basis for the residue class ring of zero-dimensional polynomial ideal. Ph. d. thesis, Univ. of Innsbruck, Austria, Math. Inst., 25 Juin 1965.
7. N. Dershowitz, J.-P. Jouannaud. *Rewrite Systems.* In *Handbook of Theoretical Computer Science, Vol. B*, 243–320. Elsevier, Amsterdam, 1990.
8. M. Fliess. Matrices de Hankel. *J. Math. Pure Appl.*, **53** (1974), 197–222.
9. M. Fliess. Sur divers produits de séries formelles. *Bull. Soc. Math. Fr.* **102** (1974), 181–191.
10. M. Fliess. Séries formelles non commutatives et automatique non linéaire. In Berstel, J., editor, *Séries Formelles en Variables Non Commutatives et Applications. Proc. 5e École de Printemps d'Informatique Théorique.* Vieux-Boucau les Bains, France (1977), 69–118.
11. M. Fliess. Réalisation locale des systèmes non linéaires, algèbres de Lie filtrées transitives et séries génératrices. *Invent. Math.* **71** (1983), 521–537.
12. M. Fliess, M. Lamnabhi, F. Lamnabhi-Lagarrigue. An algebraic approach to nonlinear functional expansions. *IEEE Trans. Circ. Syst.* **30**(8) (1983), 554–570.
13. M. Flouret. Contribution à l'algorithmique non commutative. Thèse de Doctorat, Université de Rouen, 19 janvier 1999.
14. I. Gessel. Multipartite P-partitions and inner product of skew Schur functions. In C. Greene, ed., *Combinatorics and Algebra, Contemporary Mathematics* **34** (1984) 289–301.

[g] The current version of the package can be asked for by contacting the authors.

15. N. M. Hoang. Poly-Bernoulli numbers, identities of MZV's and non commutative rational power series. *Manuscrit*, 2000.
16. N. M. Hoang, M. Petitot. Contribution à l'étude des MZV. *Manuscrit*, 1999.
17. N. M. Hoang, M. Petitot. Lyndon words, polylogarithms and Riemann ζ function. *Discrete Mathematics* **217** (2000), 273–292.
18. N. M. Hoang, M. Petitot, J. Van Der Hoeven. Shuffle algebra and polylogarithms. In *Proc. of FPSAC'98, 10-th International Conference on Formal Power Series and Algebraic Combinatorics, (Toronto, 1998)*. Also *Discrete Math.* **225** (2000), 217–230.
19. M. E. Hoffman. The algebra of multiple harmonic series. *Journal of Algebra* **194**(2) (1997), 477–495.
20. M. E. Hoffman. Quasi-shuffle products. *J. Algebraic Combin.* **11**(1) (2000), 49–68.
21. G. Jacob. Réalisation des systèmes réguliers (ou bilinéaires) et séries génératrices non commutatives. In I.D. Landau, ed., *Outils et modèles mathématiques pour l'automatique, l'analyse des systèmes et le traitement du signal* **1** (1981), 325–357. CNRS.
22. J. G. Luque. Monoïdes et automates admettant un produit de mélange. Thèse de Doctorat, Université de Rouen, 2000.
23. M. Petitot. Algèbre non commutative en Scratchpad: Application au problème de la réalisation minimale analytique. Thèse de Doctorat, Université Lille I, Janvier 1992.
24. M. P. Schützenberger. On the definition of a family of automata. *Information and Control* **4** (1961), 245–270.
25. M. Sweedler. *Hopf algebras.* Benjamin, 1969.
26. D. Zagier. Values of zeta functions and their applications. In *First European Congress of Mathematics, Vol. 2 (Paris, 1992)*, 497–512. *Progr. Math.* **120**, Birkhäuser, Basel, 1994.

GLOBAL SUPERCONVERGENCE OF BIQUADRATIC LAGRANGE ELEMENTS FOR POISSON'S EQUATION

HUNG-TSAI HUANG, ZI-CAI LI

Department of Applied Mathematics
National Sun Yat-Sen University,
Kaohsiung, Taiwan 80424
E-mail: {huanght, zcli}@math.nsysu.edu.tw

AIHUI ZHOU

Institute of Computational Mathematics and
Scientific/Engineering Computing
Academy of Mathematics and System Sciences
Chinese Academy of Sciences,
P.O. Box 2719, Beijing 100080, China
E-mail: azhou@lsec.cc.ac.cn

Biquadratic Lagrange elements are important in application, because they are used most often among high order finite element methods (FEMs). In this paper, we report some new discoveries of biquadratic Lagrange elements for the Dirichlet problem of Poisson's equation in error estimates and global superconvergences. It is well known in Ciarlet [3] that the optimal convergence rate $\|u - u_h\|_1 = O(h^2|u|_3)$ is obtained, where u_h and u are the biquadratic Lagrange element solution and the true solution, respectively. In Lin, Yan and Zhou [8], the superclose $\|u_I - u_h\|_1 = O(h^4\|u\|_5)$ can be obtained for uniform rectangles $\square_{ij}$, where u_I is the biquadratic Lagrange interpolant of the true solutions u. Hence, the global superconvergence $\|u - \prod_{2h}^4 u_h\|_1 = O(h^4\|u\|_5)$ can be gained, where $\Pi_{2h}^4 u_h$ is an *a posteriori* interpolant of polynomials with order four, based on the obtained solution u_h of biquadratic Lagrange elements. In the paper, we report the new results: For solving $- \triangle u = f$ where $f_{xxyy} = 0$, and using the uniform squares, the higher superclose estimates: $\|u_I - u_h\|_\ell = O(h^{6-\ell}\|u\|_6)$, $\ell = 0, 1$, can be achieved. The global superclose $\|u_I - u_h\|_1 = O(h^5)$ is three order higher than the optimal convergence rate in [3], and the superclose, $\|u_I - u_h\|_0 = O(h^6)$, is two orders higher than that in [8]. Numerical experiments are provided to verify the theoretical analysis obtained. To our best knowledge, this is the first time to report the numerical verification for supercloses $O(h^4)$–$O(h^5)$ and global superconvergence $O(h^4)$ in H^1.

1. Introduction

It is well known that there are three types of superconvergence: *locally pointwise*, *average* and *global*. There exist many reports on superconvergence at special points (that is, *locally pointwise*), see MacKinnon and Carey [9], Nakao [10], Pekhlivanov *et al.* [11], Wheeler and Whiteman [13], and in particular in the monograph of Wahlbin [12]. In Krizek and Neittaanmaki [6], superconvergence on average (or majority) of the nodal derivative solutions is introduced for Poisson's equation. The *global* superconvergence over the entire solution domain is studied in Krizek and Neittaanmaki [5] and Lin, Yan and Zhou [8]. In this paper we report some new discoveries of global superconvergences of biquadratic Lagrange elements for Poisson's equation.

The Lagrange elements using the *point-line-area* variables are different from the traditional Lagrange interpolations using the nodal variables only. The *biquadratic Lagrange elements* in this paper are defined on rectangle $\square_{ij}$, by means of solutions at the corners Z_i (i=1,2,3,4), on the line integrals along four edges of $\partial\square_{ij}$, and on the area integral on $\square_{ij}$. The biquadratic Lagrange interpolant functions on $\square_{ij}$ can be expressed by the polynomials, $Q_2(x,y) = span\{1, x, y, xy, x^2, y^2, x^2y, xy^2, x^2y^2\}$. On the other hand, the traditional Lagrange interpolants with order two are defined on rectangle $\square_{ij}$, by means of nodal solutions at the corners Z_i (i=1,2,3,4), at the midpoints of the edges $\partial\square_{ij}$, and at the centroid of $\square_{ij}$. To link the two interpolant methods on $\square_{ij}$, the point-line-area variables can be viewed as the corner values, the mean values along edges on $\partial\square_{ij}$, and the mean value on $\square_{ij}$.

The aim of this paper is to achieve *high* global superconvergence for Poisson's equation, $-\triangle u = f$, in the rectangular domain S with the Dirichlet boundary condition. When $u \in H^3(S)$ and the biquadratic Lagrange elements are used, the optimal convergence rate $\|u-u_h\|_{1,S} = O(h^2)$ is obtained for quasiuniform $\square_{ij}$ in Ciarlet [3], where u_h and u are the biquadratic Lagrange element solution and the true solution, respectively. In this paper, we assume that the solution has high smoothness: $u \in H^6(S)$. Based on careful integration estimates we can obtain the superclose $\|u_I - u_h\|_{1,S} = O(h^4)$ for uniform $\square_{ij}$, where u_I is the biquadratic Lagrange interpolant by the point-line-area variables of the true solution u. Hence the global superconvergence $\|u - \prod_{2h}^4 u_h\|_{1,S} = O(h^4)$ can be gained, where $\prod_{2h}^4 u_h$ is an *a posteriori* interpolant of polynomials with order four, based on the obtained solution u_h of the biquadratic Lagrange elements.

Moreover for $f_{xxyy} = 0$ and the uniform squares, the higher supercloses, $\|u_I - u_h\|_{\ell,S} = O(h^{6-\ell})$, $\ell = 0, 1$, can be achieved. Note that the superclose $\|u_I - u_h\|_{0,S} = O(h^6)$ is two orders higher than that in [8].

This paper is organized as follows. In the next section the biquadratic Lagrange elements are used for the Dirichlet problem of Poisson's equation. In Section 3 an outline is presented for new error estimates in supercovergence, and in Section 4 numerical experiments are provided to verify perfectly the new theoretical results made.

2. Biquadratic Lagrange Elements

Consider Poisson's equation with the Dirichlet boundary condition (Fig. 1):

$$-\triangle u = -\left(\frac{\partial^2 u}{\partial x^2} + \frac{\partial^2 u}{\partial y^2}\right) = f(x, y), \quad \text{in } S, \tag{2.1}$$

$$u = g, \quad \text{on } \partial S, \tag{2.2}$$

where S is a rectangle and ∂S is its boundary, and f and g are functions smooth enough.

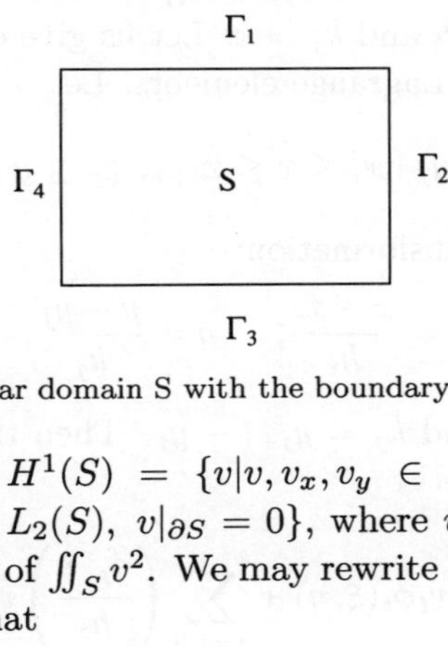

Figure 1. The rectangular domain S with the boundary $\partial S = \Gamma_1 \cup \Gamma_2 \cup \Gamma_3 \cup \Gamma 4$.

Denote the spaces $H^1(S) = \{v | v, v_x, v_y \in L_2(S),\ v|_{\partial S} = g\}$ and $H^1_0(S) = \{v | v, v_x, v_y \in L_2(S),\ v|_{\partial S} = 0\}$, where $v_x = \frac{\partial v}{\partial x}$, and $v \in L^2(S)$ implies bounded values of $\iint_S v^2$. We may rewrite (2.1) as a weak form: To seek $u \in H^1(S)$ such that

$$a(u, v) = f(v), \qquad \forall\, v \in H^1_0(S),$$

where

$$a(u, v) = \iint_S \nabla u \nabla v, \qquad f(v) = \iint_S fv, \tag{2.3}$$

where $\nabla u = u_x \vec{i} + u_y \vec{j}$, and $\vec{i}$ and $\vec{j}$ are the unit vectors along x and y, respectively. For simplicity, we always omit the integration variables, e.g., $\iint_S fv = \iint_S fv ds$.

In this paper we consider the Lagrange elements with order two on rectangles, called the *biquadratic Lagrange elements.* The interpolant functions u_I on $\Box_{ij}$ are designed by means of solutions at the corner values at Z_i (i=1,2,3,4), on the integrals along the four edges of $\partial\Box_{ij}$, and on the integral on $\Box_{ij}$, see Lin, Yan and Zhou [8]. It is called the *point-line-area interpolant* which was first introduced in Girault and Raviart [4]. The piecewise interpolant functions $u_I \in Q_2(x,y)$ are formed as follows:

$$u_I(Z_t) = u(Z_t), \quad t = 1,2,3,4, \tag{2.4}$$

$$\int_{\ell_r \cap \Box_{ij}} (u - u_I) d\ell = 0, \quad r = 1,2,3,4,$$

$$\iint_{\Box_{ij}} (u - u_I) = 0.$$

Let S be split into small rectangles $\Box_{ij}$, e.g., $S = \bigcup_{ij} \Box_{ij}$. Denote by h_i and k_j the boundary lengths of $\Box_{ij}$, and $h = \max_{ij}\{h_i, k_j\}$. The rectangles $\Box_{ij}$ are said to be *quasiuniform* if $\frac{h}{\min_{ij}\{h_i,k_j\}} \le C$, C is a constant independent of h. Also the rectangles $\Box_{ij}$ are said to be *uniform* if $\Box_{ij}$ are quasiuniform, and $h_i = h$ and $k_j = k$. Let us give explicitly the nine basis functions of biquadratic Lagrange elements. Let

$$\Box_{ij} = \{(x,y) \mid x_i \le x \le x_{i+1},\ y_j \le y \le y_{j+1}\},$$

and choose the affine transformation:

$$\xi = \frac{x - x_i}{h_i}, \qquad \eta = \frac{y - y_j}{k_j},$$

where $h_i = x_{i+1} - x_i$ and $k_j = y_{j+1} - y_j$. Then the admissible functions on $\Box_{ij}$ are expressed as

$$v(x,y) = \sum_{t=1}^{4} v_t \phi_t(\xi,\eta) + \sum_{t=1,3} \left(\frac{\int_{\ell_t} v}{h_i} \right) \psi_t(\xi,\eta) \tag{2.5}$$

$$+ \sum_{t=2,4} \left(\frac{\int_{\ell_t} v}{k_j} \right) \psi_t(\xi,\eta) + \left(\frac{\iint_{\Box_{ij}} v}{h_i k_j} \right) \varphi(\xi,\eta),$$

where the nodal points 1,2,3,4 denote (i,j), $(i+1,j)$, $(i,j+1)$, $(i+1,j+1)$, respectively, see Fig. 2. The integrals, $\frac{1}{h_i}\int_{\ell_1} v$, $\frac{1}{k_j}\int_{\ell_2} v$, $\frac{1}{h_i}\int_{\ell_3} v$ and $\frac{1}{k_j}\int_{\ell_4} v$, also illustrate the mean values of v along the edges ℓ_i, and $\frac{1}{h_i k_j}\iint_{\Box_{ij}} v$ illustrates the mean value of v on $\Box_{ij}$. The nine basis functions on $[0,1]^2$

are given explicitly by

$$
\begin{aligned}
\phi_1(x,y) &= (1-3x)(1-x)(1-y)(1-3y),\\
\phi_2(x,y) &= x(3x-2)(1-3y)(1-y),\\
\phi_3(x,y) &= (1-3x)(1-x)y(3y-2),\\
\phi_4(x,y) &= x(2-3x)y(2-3y),\\
\phi_5(x,y) &= \psi_1(x,y) = 6x(x-1)y(2-3y),\\
\phi_6(x,y) &= \psi_2(x,y) = 6x(3x-2)y(1-y),\\
\phi_7(x,y) &= \psi_3(x,y) = 6x(1-x)(1-3y)(1-y),\\
\phi_8(x,y) &= \psi_4(x,y) = 6(1-3x)(1-x)y(1-y),\\
\phi_9(x,y) &= \varphi(x,y) = 36x(1-x)y(1-y).
\end{aligned}
$$

Also denote by $V_h^* \subset H^1(S)$ and $V_h^0 \subset H_0^1(S)$ the finite dimensional collections of v in (2.5) satisfying $v|_{\partial S} = g_I$ and $v|_{\partial S} = 0$, respectively. Then the biquadratic Lagrange elements are designed to seek $u_h \in V_h^*$ such that

$$a(u_h, v) = f(v), \qquad \forall\, v \in V_h^0.$$

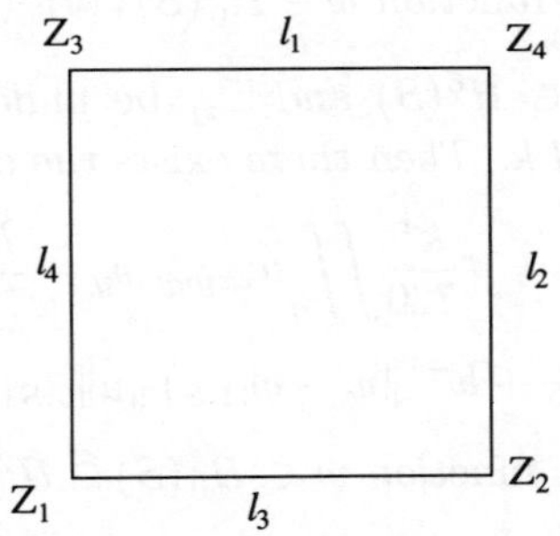

Figure 2. The small rectangular element $\Box_{ij}$ with corners Z_i and edges ℓ_i.

3. Global Superconvergence

We will provide Theorem 3.4 with an outlined proof. Let S be divided into small rectangles $e = \Box_{ij}$ with the boundary lengths h_e and k_e, and $h = \max\{h_e, k_e\}$. Then we have

$$e = \Box_{ij} = [x_e - \frac{h_e}{2}, x_e + \frac{h_e}{2}] \times [y_e - \frac{k_e}{2}, y_e + \frac{k_e}{2}],$$

where $x_e = \frac{x_{i+1}+x_i}{2}$ and $y_e = \frac{y_{j+1}+y_j}{2}$. First, we give the following lemmas without proofs which are tedious, mainly by means of integration by parts.

Lemma 3.1. *Let u_I be the biquadratic Lagrange interpolants of the true solutions u. Then for $v \in V_h^0$ there exists the following equality,*

$$\iint_e (u - u_I)_x \; v_x(x, y_e) = 0.$$

Lemma 3.2. *Let $u \in H^6(S)$ and $\square_{ij}$ be uniform with $k = k_e$. Then for $v \in V_h^0$,*

$$\begin{aligned}&\sum_e \iint_e (y - y_e)\,(u - u_I)_x \; v_{xy}(x, y_e)\\ &= -\frac{k^4}{720} \sum_e \iint_e u_{xxyyy} \; v_y + O(h^6)\|u\|_{6,S}\|v\|_{2,S}.\end{aligned}$$

Lemma 3.3. *Let $u \in H^6(S)$ and $\square_{ij}$ be uniform. Then for $v \in V_h^0$,*

$$\begin{aligned}&\sum_e \iint_e \frac{(y - y_e)^2}{2}(u - u_I)_x \; v_{xyy}\\ &= O(h^6)(h^{-1}\|v - w\|_{1,S} + \|w\|_{2,S})\|u\|_{6,S},\end{aligned}$$

where w is an auxiliary function $w \in H_0^1(S) \cap H^2(S)$.

Theorem 3.4. *Let $u \in H^6(S)$ and $\square_{ij}$ be uniform rectangles with the boundary lengths h and k. Then there exists the equality*

$$\begin{aligned}\iint_S \nabla\,(u_h - u_I)\,\nabla\, v = -\frac{k^4}{720}\iint_S u_{xxyyy}\; v_y - \frac{h^4}{720}\iint_S u_{xxxyy}\; v_x \quad (3.5)\\ +O(h^6)\left(\|w\|_{2,S} + h^{-1}\|w - v\|_{1,S}\right)\|u\|_{6,S}, \quad \forall v \in V_h^0,\end{aligned}$$

where w is an auxiliary function $w \in H_0^1(S) \cap H^2(S)$. Also if $h = k$ and $f \in H^4(S)$, then

$$\begin{aligned}\iint_S \nabla\,(u_h - u_I)\,\nabla\, v = -\frac{h^4}{720}\iint_S f_{xxyy}\; v \quad (3.6)\\ +O(h^6)(h^{-1}\|w - v\|_{1,S} + \|w\|_{2,S})\|u\|_{6,S}.\end{aligned}$$

Moreover, if $h = k$ and $f_{xxyy} = 0$ then

$$\iint_S \nabla\,(u_h - u_I)\,\nabla\, v = O(h^6)(h^{-1}\|w - v\|_{1,S} + \|w\|_{2,S})\|u\|_{6,S}. \quad (3.7)$$

Proof. Since for $v \in V_h^0$,

$$\iint_S \nabla\,(u - u_h)\,\nabla\, v = \int_{\partial S} \frac{\partial(u_h - u)}{\partial n} v - \iint_S \triangle\,(u_h - u)v = 0,$$

we have

$$\iint_S \nabla(u_h - u_I)\nabla v = \iint_S \nabla(u_h - u)\nabla v + \iint_S \nabla(u - u_I)\nabla v$$
$$= \iint_S \nabla(u - u_I)\nabla v. \tag{3.8}$$

Below, we focus only on estimations of bounds for $\iint_S \nabla(u - u_I)\nabla v$.

Using Taylor's expansion at y_e we have

$$v_x(x,y) = v_x(x,y_e) + (y - y_e)v_{xy}(x,y_e) + \frac{(y-y_e)^2}{2}v_{xyy}. \tag{3.9}$$

Based on Lemmas 3.1–3.3, we obtain from (3.9)

$$\iint_S (u - u_I)_x\ v_x = -\frac{k^4}{720}\iint_S u_{xxyyy}\ v_y \tag{3.10}$$
$$+O(h^6)\left(\|w\|_{2,S} + h^{-1}\|w - v\|_{1,S}\right)\|u\|_{6,S}.$$

Similarly,

$$\iint_S (u - u_I)_y\ v_y = -\frac{h^4}{720}\iint_S u_{xxxyy}\ v_x \tag{3.11}$$
$$+O(h^6)\left(\|w\|_{2,S} + h^{-1}\|w - v|_{1,S}\right)\|u\|_{6,S}.$$

Combining (3.10) and (3.11) gives

$$\iint_S \nabla(u - u_I)\nabla v = \iint_S \left((u - u_I)_x\ v_x + (u - u_I)_y\ v_y\right)$$
$$= -\frac{k^4}{720}\iint_S u_{xxyyy}\ v_y - \frac{h^4}{720}\iint_S u_{xxxyy}\ v_x$$
$$+O(h^6)\left(\|w\|_{2,S} + h^{-1}\|w - v\|_{1,S}\right)\|u\|_{6,S}.$$

This is the first result (3.5) in Theorem 3.4.

Next, when $\square_{ij}$ are squares, i.e., $h = k$, we have from integration by parts

$$\frac{k^4}{720}\iint_S u_{xxyyy}\ v_y + \frac{h^4}{720}\iint_S u_{xxxyy}\ v_x$$
$$= \frac{h^4}{720}\iint_S (u_{xxyyy}\ v_y + u_{xxxyy}\ v_x)$$
$$= -\frac{h^4}{720}\iint_S (u_{xxyyyy} + u_{xxxxyy})\,v$$
$$= -\frac{h^4}{720}\iint_S (\triangle u)_{xxyy}\cdot v = \frac{h^4}{720}\iint_S f_{xxyy}\ v.$$

Eq. (3.5) leads to the second result (3.6) and the last result (3.7) follows from (3.6) when $f_{xxyy} = 0$. This completes the proof of Theorem 3.4. $\square$

From Theorem 3.4, we can obtain the following corollaries easily.

Corollary 3.12. *Let all conditions in Theorem 3.4 hold. Then*

$$\|u_h - u_I\|_{1,S} = O(h^4).$$

Corollary 3.13. *Let all conditions in Theorem 3.4 hold with $h = k$ and $f_{xxyy} = 0$. Then*

$$\|u_h - u_I\|_{1,S} = O(h^5), \quad \|u_h - u_I\|_{0,S} = O(h^6).$$

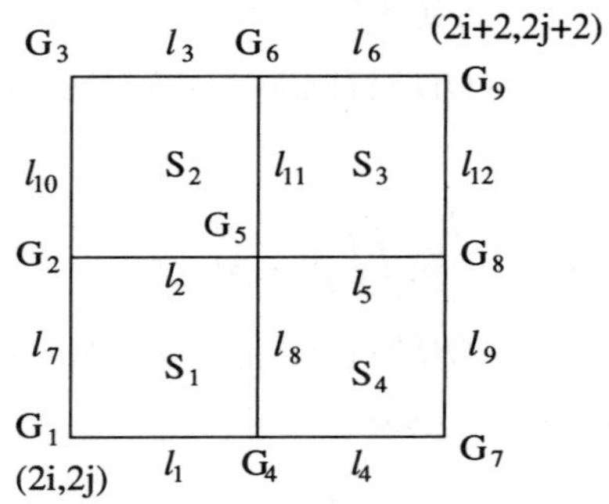

Figure 3. $\square^{2\times 2}_{2i+1,2j+1}$ in the 2×2 fashion of partition.

Below, we may employ an *a posteriori* interpolant to gain global superconvergence. The *a posteriori* interpolant may be formulated based on the obtained solution u_h as follows: $\Pi^4_{2h} v \in Q_4(x,y)$ on $\square^{2\times 2}_{2i+1,2j+1}$, where $Q_4(x,y) = \sum^4_{i,j=0} a_{ij} x^i y^j$ with coefficients a_{ij}. Therefore, there are 25 coefficients in $\prod^4_{2h} v$, which can be determined uniquely by the following 25 equations (Fig. 3):

$$v(G_t) = (\prod_{2h}^4 v)(G_t), \quad t = 1, 2, ..., 9, \tag{3.14}$$

$$\int_{\ell_r} v = \int_{\ell_r} \prod_{2h}^4 v, \quad r = 1, 2, ..., 12, \tag{3.15}$$

$$\iint_{S_t} v = \iint_{S_t} \prod_{2h}^4 v, \quad t = 1, 2, 3, 4. \tag{3.16}$$

The computational formulas are expressed as

$$\prod_{2h}^{4} v(x,y) = \sum_{i=1}^{9} v(G_i)\phi_i^*(\xi,\eta) + \sum_{i=1}^{6}\left(\frac{\int_{\ell_i} v}{h}\right)\psi_i^*(\xi\ \eta) \qquad (3.17)$$

$$+\sum_{i=7}^{12}\left(\frac{\int_{\ell_i} v}{k}\right)\psi_i^*(\xi,\eta) + \sum_{i=1}^{4}\left(\frac{\iint_{S_i} v}{hk}\right)\varphi_i^*(\xi,\eta),$$

where $\xi = \frac{x-x_{2i}}{h}$ and $\eta = \frac{y-y_{2j}}{k}$. In (3.17), G_i, ℓ_i, and S_i denote the vertices, edges and areas of $\Box_{ij}$, respectively, shown in Fig. 3. By *Mathematica*, the 25 basis functions ϕ_i^*, ψ_i^* and φ_i^* on $[0,2]^2$ can be obtained explicitly from (3.14)–(3.16); detailed functions are omitted here.

For $h = k$ and $f_{xxyy} = 0$, we may similarly construct an *a posteriori* interpolant polynomial $\prod_{3h}^{6} u_h$ of order six on 3×3 rectangles $\Box_{ij}$. Then we have the following theorem.

Theorem 3.18. *Let all conditions in Theorem 3.4 hold. Then*

$$\|u - \prod_{2h}^{4} u_h\|_{1,S} = O(h^4).$$

Moreover, for $h = k$ and $f_{xxyy} = 0$,

$$\|u - \prod_{3h}^{6} u_h\|_{\ell,S} = O(h^{6-\ell}), \qquad \ell = 0, 1.$$

Remark 3.19. Let all conditions in Theorem 3.4 hold. Suppose that $f_{xxyy} \neq 0$ and $h = k$. We have the following relations,

$$\|u_h - u_I + \frac{h^4}{720} w\|_{1,S} = O(h^5),$$

$$\|u_h - u_I + \frac{h^4}{720} w\|_{0,S} = O(h^6),$$

where $w \in H_0^1(S) \bigcap H^2(S)$ satisfy

$$\iint_S \nabla w \nabla v = \iint_S f_{xxyy}\, v, \qquad \forall v \in V_h^0. \qquad (3.20)$$

Hence, we may seek the solution w_h to (3.20) again by the same biquadratic Lagrange elements, and choose

$$\bar{u}_h = u_h + \frac{h^4}{720} w_h \qquad (3.21)$$

to reach the same high supercloses and superconvergences:

$$\|\bar{u}_h - u_I\|_{\ell,S} = O(h^{6-\ell}), \quad \|u - \prod_{3h}^{6} \bar{u}_h\|_{\ell,S} = O(h^{6-\ell}), \quad \ell = 0,1. \quad (3.22)$$

The alternative is to adopt the extrapolation techniques in Blum, Lin and Rannacher [2] and Lin [7], to also retain high superconvergences

$$\|\frac{16}{15}\hat{u}_{\frac{h}{2}} - \frac{1}{15}\hat{u}_h - u\|_{\ell,S} = O(h^{6-\ell}), \qquad \ell = 0,1,$$

where $\hat{u}_h = \prod_{3h}^{6} u_h$.

4. Numerical Experiments

In this section, we provide three numerical experiments of biquadratic Lagrange elements for solving (2.1) and (2.2).

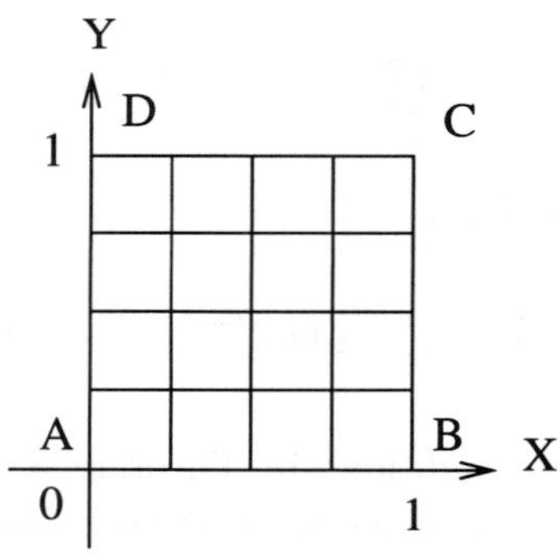

Figure 4. Partition of square elements with $N = 4$.

4.1. *Global Superconvergence*

Consider an example with $f_{xxyy} \neq 0$. Let $S = \{(x,y)|0 < x,y < 1\}$, and choose the true solution $u = \sin(\pi x)\sin(\pi y)$ which satisfies Poisson's equation in S and the Dirichlet condition on boundary ∂S (Fig. 4):

$$-\triangle u = f = 2\pi^2 \sin(\pi x)\sin(\pi y) \quad \text{in S},$$
$$u = 0 \quad \text{on } \partial S = \overline{AB} \cup \overline{BC} \cup \overline{CD} \cup \overline{AD},$$

where $f_{xxyy} = 2\pi^6 \sin(\pi x)\sin(\pi y) \neq 0$. Let $\square_{ij}$ be uniform squares with the boundary length h. Denote by N the division number of $\square_{ij}$ along the boundary $\overline{AB}$. Hence $h = 1/N$. The errors and condition numbers are listed in Table 1. We can see the following asymptotic rates:

$$\|u - u_h\|_{\ell,S} = O(h^{3-\ell}), \quad \ell = 0, 1,$$
$$\|u_I - u_h\|_{\ell,S} = O(h^4), \quad \ell = 0, 1, \tag{4.1}$$
$$\|u - \prod_{2h}^{4} u_h\|_{\ell,S} = O(h^4), \ \ell = 0, 1, \tag{4.2}$$
$$\text{Cond}(\mathbf{A}) = O(h^{-2}). \tag{4.3}$$

In (4.3), the associated matrix $\mathbf{A}$ results from (2.3), and the condition numbers are defined by $\text{Cond}(\mathbf{A}) = \lambda_{\max}(\mathbf{A})/\lambda_{\min}(\mathbf{A})$, where $\lambda_{\max}(\mathbf{A})$ and $\lambda_{\min}(\mathbf{A})$ are the maximal and minimal eigenvalues of $\mathbf{A}$, respectively. Eq.(4.2) coincides very well with Theorem 3.18. Since $f_{xxyy} = \pi^4 f$ in this example, we have $\bar{u}_h = (1 + \frac{h^4}{720}\pi^4)u_h$ from (3.21), and then obtain the high superconvergences (3.22).

Table 1: The error norms and condition numbers for Poisson's equation with $f_{xxyy} \neq 0$ by the biquadratic Lagrange elements.

N	2	4	8	16	32
$\|u - u_h\|_0$	0.183(−1)	0.207(−2)	0.250(−3)	0.309(−4)	0.385(−5)
$\|u - u_h\|_1$	0.206	0.511(−1)	0.128(−1)	0.319(−2)	0.798(−3)
$\|u_I - u_h\|_0$	0.372(−2)	0.256(−3)	0.164(−4)	0.103(−5)	0.645(−7)
$\|u_I - u_h\|_1$	0.181(−1)	0.119(−2)	0.750(−4)	0.470(−5)	0.294(−6)
$\|u - \prod_{2h}^4 u_h\|_0$	0.349(−2)	0.266(−3)	0.165(−4)	0.103(−5)	0.645(−7)
$\|u - \prod_{2h}^4 u_h\|_1$	0.219(−1)	0.139(−2)	0.847(−4)	0.526(−5)	0.328(−6)
$\lambda_{\max}(\mathbf{A})$	42.8	48.0	49.3	49.3	49.3
$\lambda_{\min}(\mathbf{A})$	1.15	0.304	0.768(−1)	0.193(−1)	0.482(−2)
$\text{Cond}(\mathbf{A})$	0.371(2)	0.158(3)	0.642(3)	0.256(4)	0.102(5)

4.2. *Special Case of $h = k$ and $f_{xxyy} = 0$*

To test the case $f_{xxyy} = 0$. We choose the true solutions $u = \sin(\pi y)\sinh(\pi x)$ which satisfies Laplace's equation (i.e., $f = 0$) in S and the Dirichlet condition on ∂S:

$$\Delta u = 0, \tag{4.4}$$
$$u|_{\overline{AB}} = u|_{\overline{CD}} = u|_{\overline{AD}} = 0, \quad u|_{\overline{BC}} = g = \sinh(\pi)\sin(\pi y). \tag{4.5}$$

The computed errors and condition numbers are listed in Table 2. We can see that

$$\begin{aligned} &\|u-u_h\|_{\ell,S} = O(h^{3-\ell}), \quad \ell = 0,1, \\ &\|u_I-u_h\|_{\ell,S} = O(h^{6-\ell}), \quad \ell = 0,1, \\ &\text{Cond}(\mathbf{A}) = O(h^{-2}). \end{aligned} \tag{4.6}$$

Eqs. (4.6) verify perfectly Corollary 3.13.

It is new and important that the error norm $\|u_I - u_h\|_{1,S} = O(h^5)$ be three order higher than the optimal $O(h^2)$ in H^1 in Ciarlet [3], and one order higher than $O(h^4)$ in Lin, Yan and Zhou [8].

Table 2: The error norms and condition numbers for Laplace's equation by the biquadratic Lagrange elements.

N	2	4	8	16	32
$\|u-u_h\|_0$	0.892(−1)	0.123(−1)	0.159(−2)	0.200(−3)	0.251(−4)
$\|u-u_h\|_1$	1.27	0.329	0.830(−1)	0.208(−1)	0.521(−2)
$\|u_I-u_h\|_0$	0.343(−2)	0.697(−4)	0.118(−5)	0.189(−7)	0.296(−9)
$\|u_I-u_h\|_1$	0.345(−1)	0.138(−2)	0.467(−4)	0.150(−5)	0.475(−7)
$\lambda_{\max}(\mathbf{A})$	42.8	48.0	49.3	49.3	49.3
$\lambda_{\min}(\mathbf{A})$	1.15	0.304	0.768(−1)	0.193(−1)	0.482(−2)
Cond($\mathbf{A}$)	0.371(2)	0.158(3)	0.642(3)	0.256(4)	0.102(5)

4.3. *Comparisons*

In this subsection, we compare the biquadratic Lagrange elements using the point-line-area variables (2.4), with those using the nine-nodal variables in the traditional Lagrange elements in [3]. First, we use the traditional biquadratic Lagrange elements for solving (4.4) and (4.5), and obtain the solution u_h^*. The errors and condition numbers are listed in Table 3, where u_I^* is the nine nodal interpolant, and $\mathbf{A}^*$ is the associated matrix. We are more interested in errors by *a posteriori* interpolants of u_h^*. Below, we propose two methods to formulate the *a posteriori* interpolants of order four on $\square_{2i+1,2j+1}^{2\times 2}$ in Fig. 3.

Method I: The traditional Lagrange interpolation with 25 nodal variables, see Atkinson [1],

$$\pi_{2h}^4 u_h^* = \sum_{i=1}^{25} a_i \Psi_i(x,y),$$

where a_i are coefficients, and the basis functions $\Psi_i(x,y) \in Q_4(x,y)$. There are 25 coefficients to be determined uniquely by 25 equations at vertices, G_1 ,..., G_9, the midpoints of lines, l_i, $i = 1, ..., 12$, and the centroids of S_i, $i = 1, ..., 4$ (Fig. 3). The errors of the *a posteriori* interpolant solutions are listed in Table 3. It is easy to see the asymptotic rates

$$\|u - u_h^*\|_{\ell,S} = O(h^{3-\ell}), \quad \ell = 0, 1,$$

$$\|u_I^* - u_h^*\|_{\ell,S} = O(h^{4-\ell}), \quad \ell = 0, 1, \tag{4.7}$$

$$\|u - \pi_{2h}^4 u_h^*\|_{1,S} = O(h^3), \quad \|u - \pi_{2h}^4 u_h^*\|_{0,S} = O(h^4), \tag{4.8}$$

$$\text{Cond}(\mathbf{A}) = O(h^{-2}).$$

Obviously, the global superconvergence $\|u - \pi_{2h}^4 u_h^*\|_{1,S} = O(h^3)$ is only one order higher than the optimal convergence rate $O(h^2)$.

Table 3: The error norms and condition numbers for Laplace's equation by using traditional nine-nodal Lagrange elements, where u_I^* and u_h^* are the Lagrange interpolant of the true solution and the finite element solution, respectively.

N	2	4	8	16	32
$\|u - u_h^*\|_0$	0.900(−1)	0.124(−1)	0.159(−2)	0.200(−3)	0.251(−4)
$\|u - u_h^*\|_1$	1.27	0.329	0.830(−1)	0.208(−1)	0.521(−2)
$\|u_I^* - u_h^*\|_0$	0.902(−2)	0.522(−3)	0.317(−4)	0.197(−5)	0.123(−6)
$\|u_I^* - u_h^*\|_1$	0.698(−1)	0.790(−2)	0.101(−2)	0.128(−3)	0.162(−4)
$\|u - \pi_{2h}^4 u_h^*\|_0$	0.148(−1)	0.912(−3)	0.411(−4)	0.219(−5)	0.130(−6)
$\|u - \pi_{2h}^4 u_h^*\|_1$	0.936(−1)	0.119(−1)	0.150(−2)	0.191(−3)	0.242(−4)
$\|u - \prod_{2h}^4 u_h^*\|_0$	0.147(−1)	0.917(−3)	0.383(−4)	0.189(−5)	0.108(−6)
$\|u - \prod_{2h}^4 u_h^*\|_1$	0.961(−1)	0.671(−2)	0.425(−3)	0.311(−4)	0.248(−5)
$\lambda_{\max}(\mathbf{A}^*)$	7.02	7.40	7.44	7.43	7.41
$\lambda_{\min}(\mathbf{A}^*)$	1.15	0.304	0.768(−1)	0.193(−1)	0.482(−2)
$\text{Cond}(\mathbf{A}^*)$	0.611(1)	0.243(2)	0.968(2)	0.386(3)	0.154(4)

Method II: Based on u_h^* with the nine-nodal solutions on $\Box_{ij}$, we may compute the point-line-area values of the solution of u_h^* as follows: (1) The corner values are the same. (2) The line and area values in (2.5) can be evaluated from the following integrations:

$$\frac{1}{h_i}\int_{\ell_1} u_h^*, \quad \frac{1}{k_j}\int_{\ell_2} u_h^*, \quad \frac{1}{h_i}\int_{\ell_3} u_h^*, \quad \frac{1}{k_j}\int_{\ell_4} u_h^*, \quad \frac{1}{h_i k_j}\iint_{\Box_{ij}} u_h^*.$$

Then, we formulate the a *posteriori* interpolant polynomial of order four again from (3.17). The errors of the *a posteriori* interpolant solutions are also listed in Table 3. We can see that

$$\|u - \prod_{2h}^{4} u_h^*\|_{1,S} = O(h^{3.5}), \quad \|u - \prod_{2h}^{4} u_h^*\|_{0,S} = O(h^4).$$

Evidently, using the *a posteriori* interpolant of the point-line-area variables is more advantageous, because the error order in H^1 is a half order higher than that in (4.8).

At last, let us compare (4.7) with (4.6). The error orders of $(u_I^* - u_h^*)$ in (4.7) are two order lower than $O(h^{6-\ell})$ in (4.6)! This fact clearly displays that the methods given in this paper are remarkably advantageous. It is due to the very high superconvergence, see Theorem 3.4, as well as (4.1), (4.2) and (4.6), that the biquadratic Lagrange elements with the point-line-area variables are strongly recommended for solving Poisson's equation.

In summary, we address a few remarks on the novelties of this paper.

1. The new error estimates of supercloses are provided in Theorem 3.4, which under the special cases $f_{xxyy} = 0$ and $h = k$ leads to the superclose in H^1 three orders higher than the optimal one in [3], and one order higher than that in [8]. The numerical experiments given in this paper have verified perfectly the global supercloses and superconvergences made. Moreover, by using an *a posteriori* interpolation of at least order six, the superconvergence of the biquadratic Lagrange elements may achieve $O(h^5)$ in H^1 and $O(h^6)$ in L^2 high, under the case of $f_{xxyy} = 0$ and $h = k$. When $f_{xxyy} \neq 0$, the same high superconvergences may be retained by the techniques in Remark 3.19.

2. Theorem 3.4 is new and important, where a new auxiliary function $w \in H_0^1(S) \cap H^2(S)$ is employed to express the error bounds in (3.5)–(3.7). When $w = v$ and $w = 0$, the error bounds in L^2 and H^1 are obtained easily. The function w here plays a role in combining the error bounds in H^1 and L^2 together into concise mathematics.

3. From the numerical experiments in Section 4.3, the pure point-line-area variables used throughout from finite element methods (FEMs) to an *a posteriori* interpolation are superior to the nodal variables used in the traditional Lagrange elements in [3], which cause some loss in accuracy even using an *a posteriori* interpolation. The reason is that the real high supercloses and superconvergences can be made only by using the point-line-area variables. This fact indicates that we may raise the point-line-area

FEMs to be *new kinds* of FEMs, departing from the traditional Lagrange FEMs discussed in most papers and textbooks (see [3]).

Acknowledgments

We are very grateful to the editors and reviewers for their valuable comments and suggestions.

References

1. K. E. Atkinson. *An Introduction to Numerical Analysis*. John Wiley and Sons, New York, 1989.
2. H. Blum, Q. Lin, R. Rannacher. Asymptotic error expansion and Richardson extrapolation for linear finite elements. *Numer. Math.* **49** (1986), 11–37.
3. P. G. Ciarlet. Basic error estimates for elliptic problems. In P. G. Ciarlet, J. L. Lions eds., *Finite Element Methods* (Part I), 17–351. North-Holland, Amsterdam, 1991.
4. V. Girault, P. A. Raviart. *Finite Element Methods for Navier-Stokes Equations, Theory and Algorithms. Springer Series in Computational Mathematics* **5**, Springer-Verlag, 1986.
5. M. Krizek, P. Neittaanmaki. On a global-superconvergence of the gradient of linear triangular elements. *J. Comput. Appl. Math.* **18** (1987), 221–233.
6. M. Krizek, P. Neittaanmaki. Superconvergence phenomenon in the finite element method arising from averaging gradients. *Numer. Math.* **45** (1984), 105–116.
7. Q. Lin. High performance FEMs. *Inter. Symposium Computational and Applied PDEs*, China, July 1–7, 2001.
8. Q. Lin, N. Yan, A. Zhou. A rectangle test for interpolated finite elements. *Proc. Sys. Sci. and Sys. Engrg.*, 217–229. Great Wall Culture Publ. Co., Hong Kong, 1991.
9. R. J. MacKinnon, G. F. Carey. Nodal superconvergence and solution enhancement for a class of finite-element and finite difference methods. *SIAM. J. Sci. Stat. Comput.* **11** (1990), 343–353.
10. M. T. Nakao. Superconvergence of the gradient of Galerkin approximations for elliptic problems. *J. Comput. Appl. Math.* **20** (1987), 341–348.
11. A. I. Pekhlivanov, R. D. Lazarov, G. F. Carey, S. S. Chow. Superconvergence analysis of approximate boundary-flux calculations. *Numer. Math.* **63** (1992), 483–501.
12. L. L. Wahlbin. Local Behaviour in Finite Element Method, Chap. VII., Superconvergence. In P. G. Ciarlet, J. L. Lions, eds., *Finite Element Methods* (Part I), 501–522. North-Holland, Amsterdam, 1991.
13. M. F. Wheeler, J. R. Whiteman. Superconvergent recovery of gradients of on subdomains from piecewise linear finite-element approximation. *Numer. Methods for PDE* **3** (1987), 65–82 and 357–374.

AUTOMATION OF PERTURBATION ANALYSIS IN COMPUTER ALGEBRA: PROGRESS AND DIFFICULTIES

RAYA KHANIN

Department of Mechanical Engineering
J. Watt Building, University of Glasgow
Glasgow, G12 8QQ, UK
Email: R.Khanin@mech.gla.ac.uk

The goal of this paper is to give an overview of the present state of automation of perturbation analysis in Computer Algebra. The paper reviews existing computer algebra packages thatimplement various perturbation methods as well as applications to specific problems. The main directions of a more systematic development in this area are underlined. A summary of the author's work on solving singularly perturbed boundary value problems is also presented.

1. Introduction

Major advances have been made in the last two decades in finding solutions of differential equations in an automatic way based on research and methods from Computer Algebra. There have been much work done on ordinary differential equations, particularly for linear ones (see [1,31]). As a result of these on-going and extensive research, modern computer algebra systems (CASs) now provide good and fast supports for solving ordinary differential equations (see [26] for a review of the capabilities of different systems). They are designed to find, where possible, exact or closed form solutions for differential and algebraic equations. At the same time, with a few exceptions, they provide little assistance for qualitative analysis, which includes finding approximate and asymptotic symbolic solutions to differential equations.

This paper is about perturbation analysis, a powerful analytical tool which can be applied to a wide class of weakly nonlinear problems. The basic idea of all perturbation methods is the expansion of the unknown variable in series wherein each term is calculated by induction, given all the previous terms. Perturbation methods are conceptually simple but their actual implementation by hand involves lengthy and complex algebraic calculations.

Despite a large number of specific applications of perturbation methods to particular problems, the major CASs are still missing the robust implementation of such methods from perturbation analysis. The goal of this paper is to give an overview of the present state of implementation of perturbation methods in CASs and to underline the main directions of a more systematic development in this area.

2. General Solution Procedures

There is a vast literature on perturbation methods both from the point of view of their applications as well as their theoretical foundations. For excellent surveys of perturbation techniques, see [14,25,22].

There are several well-established perturbation methods for solving weakly non-linear differential equations. Examples include the methods of *straightforward expansion*, *matched asymptotic expansions*, *multiple scales*, *harmonic balance*, and *averaging*. Let us briefly discuss features common to several perturbation methods by considering an initial value problem for $x = x(t) \in R^n$ with a small parameter ε $(0 < \varepsilon \ll 1)$:

$$\begin{aligned} &F(t, x, x', x'', \ldots, x^{(n)}; \varepsilon) = 0, \\ &x(0) = x_0, \; x'(0) = x'_0, \; \ldots, \; x^{(n-1)}(0) = x_0^{(n-1)} \end{aligned} \tag{2.1}$$

If one cannot solve this differential equation exactly, it is useful to determine whether one can calculate the asymptotic expansion of the solution by considering a sequence of simpler differential equations governing each term in the expansion.

The perturbation procedure is based on an expansion of the unknown function $x(t; \varepsilon)$ in an asymptotic series of the small parameter ε:

$$x(t; \varepsilon) = g_0(\varepsilon)x_0(\tau) + g_1(\varepsilon)x_1(\tau) + \cdots = \sum_{j=0}^{N} g_j(\varepsilon)x_j(\tau), \tag{2.2}$$

where $g_j(\varepsilon)$ is an asymptotic sequence and $\tau = \tau(t; \varepsilon)$ is a new *time-scaling* (that is, a scaled time variable). The simplest and most commonly used expansion is the power series in ε

$$x(t; \varepsilon) = \sum_{j=0}^{N} \varepsilon^{\alpha_j} x_j(\tau) \qquad (\alpha_j \geq 0), \tag{2.3}$$

where the scaling $\tau = t$ is used for *regular perturbation problems*, in which the solution $x(t;\varepsilon)$ of the perturbed problem (2.1) tends to the solution $x(t)$ of the unperturbed problem

$$F(t, x, x', x'', \ldots, x^{(n)}; 0) = 0$$

as $\varepsilon \to 0$. When the regular perturbation limit property fails to hold, the perturbation is called a *singular perturbation.* Solving singlular perturbation problems involves defining fast ($\tau = O(t)$) and slow time-scales ($\tau(t;\varepsilon) = o(t)$) with the corresponding *outer* and *inner* solutions which are matched on the boundaries [25].

The asymptotic series expansion of the solution $x(t;\varepsilon)$ sought can also use several scalings, as, for example, in the multiple scales method:

$$x(t;\varepsilon) = \sum_{j=0}^{M} \varepsilon^{\alpha_j} x_j(T_0, T_1, T_2, \ldots, T_M) + O(\varepsilon T_M). \tag{2.4}$$

Here the single independent variable t is split up into several variables $T_0, T_1, ..., T_M$ related to each other by $T_j = \varepsilon^j t$ but considered independent [22]. The newly constructed series (2.2) in the form (2.3) or (2.4) is substituted into the original problem (2.1), and the perturbation equations are found by collecting terms of the same order of the small parameter ε. The study of the lowest order equations is aimed at checking the validity of the asymptotic expansion. If secular terms exist in the lowest order equations, then the original asymptotic expansion has to be adjusted accordingly. In the case of the multiple scales method, secular terms are equated to zero, thus providing conditions for determining certain unknown constants.

The solutions to the lowest order ε-equation are then substituted into the next order ε-equations and so on. In this way, perturbation methods have a hierarchical structure (the k-th equation has to be solved before the $(k+1)$-st step is attempted) for potentially well-defined algorithms, and are all based on using asymptotic expansions. These features make the methods of perturbation analysis ideal for computerization.

3. A Brief Review of Existing Packages

CASs have been used for generating perturbation expansions or expansions in the series of a (usually) small parameter for some time already. Rand and Armbruster [27] developed *MACSYMA* programs that implemented a number of popular perturbation techniques about 20 years ago. Nowadays almost every textbook on perturbation analysis mentions the advantages of using CASs for automating the solution procedure or dealing with cumbersome algebra (e.g. [5]). A number of packages which efficiently implement certain perturbation methods in different computer algebra systems have been developed. A *Mathematica* package developed by Kaufmann [13] covers asymptotic expansions of the solutions of ordinary differential equations with respect to polynomial or general gauge functions. The methods of straightforward expansions, strained coordinates and matched and composite solutions are implemented. The package has a very good range of functionalities but its main purpose is to demonstrate a possible implementation of a number of perturbation methods in *Mathematica* and to generate educational examples of perturbation expansions. Solution methods for singularly perturbed large control systems using non-commutative algebra packages have been successfully implemented in *Mathematica* by Helton *et al.* [9]. The powerful multiple scales method has been implemented in *MACSYMA* [27], *Maple* [28] and *Mathematica* [18]. An excellent series of *Mathematica* and *Maple* notebooks have been developed by Nayfeh to accompany his books on perturbation (`www.esm.vt.edu/~anayfeh/`).

But the main focus in this area has been to automate and speed up a particular computation for certain applications of perturbation analysis to specific problems. Examples of these problems include those in hydrodynamics [4], mechanics [20], structural mechanics [7], celestial mechanics [32], quantum physics [6,8], and molecular chemistry [10], to mention just a few papers.

Still, however, the progress in automation of perturbation methods has been somewhat limited. There is no general computer algebra package yet for finding symbolic asymptotic solutions.

4. Choosing Scalings for Asymptotic Sequences

4.1. *Examples of Asymptotic Expansions*

Actual generation of a perturbation expansion using a computer algebra system is a trivial task provided all parameters and their properties are defined. Suppose the solution of the following initial value problem

$$x''(t) + \varepsilon x'(t)^3 + x(t) = 0, \qquad x(0) = a,\ x'(0) = b \tag{4.1}$$

is to be found by the expansion $x(t) = \sum_{j=0}^{N} x_j(t)\varepsilon^j$. Using the function `PolyOrderList` from the `PerturbationODE.m` *Mathematica* package developed by Kaufmann [13] yields the following perturbation equations up to order 1 of ε:

```
In[1]:=PolyOrderList[{x''[t]+ε x'[t]^3+x[t]==0,
                       x[0]==a,x'[0]==b},x[t],ε,1]
Out[1]:={{x[0][t]+x[0]''[t]==0, x[0][0]==a,x[0]'[0]==b,
          x[1][t]+x[0]'[t]^3+x[1]''[t]==0, x[1][0]==0,x[1]'[0]==0}}
```

For regular perturbation problems in the polynomial form the series expansion is easy to find. For example, for the following problem

$$x''(t) + \varepsilon^{1/2}x'(t)^2 + x(t) = 0, \qquad x(0) = a,\ x'(0) = b, \tag{4.2}$$

substitution of the expansion in the form (2.3) results in $\alpha = 1/2$, which is found by omitting the nonlinearity in the zero-th order equation and equating terms of the similar power at the next order.

Finding other types of asymptotic expansions is less trivial. In the case of the following problem, for example,

$$x''(t) + \sin(\varepsilon x(t)) = 0, \qquad x(0) = a, x'(0) = b, \tag{4.3}$$

a solution is sought in the form $x(t) = x_0(t)+\sin(\varepsilon)x_1(t)+\sin(\varepsilon)^2x_2(t)+\cdots$, where $1, \sin(\varepsilon), \sin(\varepsilon^2), \cdots$ are part of the asymptotic sequence. The perturbation equations based on this expansion can be generated using the function `GeneralOrderList` from Kaufmann's package. The question that remains, however, is how those asymptotic sequences used in the perturbation expansions are to be found. There is no constructive recipe.

The following example of using asymptotic expansions in logarithmic form proves this case even further. Consider two model problems of heat conduction outside a sphere and a cylinder with a small nonlinear heat source [11]. The problems differ by only a numerical factor. Namely, the first problem is represented by

$$y''(x) + 2/xy'(x) + \varepsilon y(x)y'(x) = 0,\ y(1) = 0,\quad y(x) \to 1 \text{ as } x \to \infty, \tag{4.4}$$

and the second problem is given by

$$y''(x) + 1/xy'(x) + \varepsilon y(x)y'(x) = 0,\ y(1) = 0,\quad y(x) \to 1 \text{ as } x \to \infty. \tag{4.5}$$

The expansions used in the two *outer* problems are different [11]: the first one uses the Poincare expansion in the form

$$y(x;\varepsilon) = (1 - \frac{1}{x}) + \varepsilon \log \frac{1}{\varepsilon} y_1(x) + \varepsilon y_2(x), \tag{4.6}$$

while the cylinder problem requires expansion, in less obvious form

$$y(x;\varepsilon) = \frac{y_1(x)}{\frac{1}{\log\frac{1}{\varepsilon}}} + \frac{y_1(x)}{\frac{1}{\log\frac{1}{\varepsilon}}^{2}}. \tag{4.7}$$

4.2. *Scalings for Singular Perturbation Problems*

For singular perturbation problems, the search for scalings remains rather heuristic with one or possibly two available algorithms for determining the right order of local expansion [2,23]. Consider the first-order nonlinear initial value problem

$$\varepsilon \dot{x} = (1-t)x - x^2, \quad x(0) = x_0 > 0 \tag{4.8}$$

Provided $x_0 \neq 1$ or $x_0 \neq 0$, there is non-uniform convergence in an initial layer [25]. Introduction of the commonly used scalings for the initial layer $t = \varepsilon\tau$, $x = \widetilde{x}$ yields the zero-th order of ε solution $1/\widetilde{x}_0(\tau) = 1 + C_1 e^{-\tau}$, $\lim_{\tau\to\infty} \widetilde{x}_0(\tau) = 1$. Seeking an outer solution yields

$$X(t;\varepsilon) = (1-t) + \frac{\varepsilon}{1-t} + \frac{2}{(1-t)^3}\varepsilon^2 + \cdots, \tag{4.9}$$

which breaks down at $t = 1$ and another local approximation is to be sought. The automated solution procedure should seek for singularities and this is where most CASs fail. Shifting into the singularity by transformation $t = 1 - s$ in the original system yields

$$\varepsilon \frac{dx}{ds} = sx - x^2.$$

A scaling transformation is then sought in the form $x = \varepsilon^\alpha \widehat{x}$, $s = \varepsilon^\beta \widehat{t}$, $\alpha, \beta \geq 0$ resulting in the following equation in terms of new variables

$$\varepsilon^{1+\alpha-\beta} \frac{d\widehat{x}}{d\widehat{t}} = \varepsilon^{\alpha+\beta} \widehat{t}\widehat{x} - \varepsilon^{2\alpha} \widehat{x}^2. \tag{4.10}$$

The question remains on how to determine α and β. Following the considerations from [25], notice first that the term $\varepsilon^{1+\alpha-\beta} d\widehat{x}/d\widehat{t}$ must be bounded as $\varepsilon \to 0$, that is, $1+\alpha-\beta > 0$, so that, for example, $\alpha = \beta$ will work. Secondly, the two nonlinear terms must be of the same order as the differential term, that is, $2\alpha = 1$. Therefore, $\alpha = \beta = 1/2$ results in an equation

$$\frac{d\widehat{x}}{d\widehat{t}} = -\widehat{x}(\widehat{x} + \widehat{t}), \tag{4.11}$$

whose solution $\widehat{x}(\widehat{t})$ allows a transition (or corner) layer to occur between the outer limit $X_0(t)$ $(0 < t < 1)$ and the trivial limit $X_0(t) = 1$ (for $t > 1$).

For the above "toy" problem (4.8), it is easy to follow the considerations leading to the selection of appropriate scalings α and β for the new dependent and independent variables. But can such considerations actually be coded? Once the problem gets more complicated, the procedure for the selection of scalings would greatly benefit from a constructive algorithm.

4.3. *Constructive Approaches*

One constructive approach for finding local approximations to singular perturbation problems is based on the Newton polyhedron algorithm [2], which is implemented in [30]. Using edges and unit vectors for the Newton polyhedron, differential equation can be regularised and its solutions can be computed. By introducing the Newton ε-polygon, Macutan [21] developed a computational algorithm for constructing the formal solutions of scalar singularly perturbed linear differential equation of order n. Unlike the common method of matched asymptotic expansions, it has so far been applied to linear problems only.

A different type of polyhedron algorithm for finding solutions of the singularly perturbed problems was developed by Nipp [23]. The Nipp polyhedron algorithm can be applied to solve a class of singularly perturbed problems with polynomial right-hand sides. As with the Newton polyhedron algorithm, the correspondence between a system of ordinary differential equations containing a small parameter ε and a convex polyhedron is established. Only one high dimensional polyhedron is required per system of first order differential equations. According to this algorithm, adjacent vertices of the polyhedron correspond to adjacent systems, which are systems with the maximum number of common terms. Therefore, calculating vertices adjacent to the given vertex yields a set of several possible scalings for the approximating system at the singular point corresponding to this vertex. In addition, the Nipp polyhedron algorithm provides simple selection rules for choosing the appropriate approximating system from a set of adjacent systems.

Computationally the procedure for finding the appropriate scalings for the local approximating systems in singular perturbation problem reduces to the linear programming procedure for finding polyhedron vertices which are adjacent to the zero vertex. This algorithm has been implemented in *Maple* [16] and as a part of larger package in *Mathematica* [15] by Khanin.

5. Computer Algebra Package for Perturbation Analysis

The examples in the previous sections shed light on some of the difficulties in automating perturbation analysis procedures in CA. Actual application of perturbation methods on paper is still somewhat of an art rather than a methodological usage of the recipe. A recent review on applications of the multiple scales method to problems in engineering dynamics demonstrates how important the ordering schemes and the form of the series expansions are when attempting to solve anything but the most simplistic problems [3]. However, given a large theoretical background in perturbation analysis, its numerous applications as well as some implementation work which has already been done, it is plausible to develop an efficient CA package which would provide general implementation of perturbation methods. The package should include various tools for finding approximate solutions including regular perturbation methods, multiple scales, singular perturbation methods, averaging methods, methods based on differential inequalities, and others. The challenge here is to design software to deal with various nonlinear equations as generally as possible. The package must also contain tools for efficient implementation of new perturbation algorithms.

The package should be able to choose the best perturbation technique (among those available) to tackle a given problem. Is it feasible to have a database with equations and corresponding solution methods or may be even an artificial intelligence technique to train such a program? An interesting question is whether a perturbation analysis package can be a stand-alone self-contained package or part of a large system? Probably, the latter should be the case as the perturbation solution procedure requires tools for dealing with differential equations.

There remain substantial difficulties, however, mainly due to limited capabilities of CASs in performing qualitative analysis. Perturbation analysis includes several qualitative steps: identifying singularities, studying asymptotics, and matching two local approximations. With a few exceptions (for example, `gdev` [29]), CASs are not very helpful in finding asymptotic behaviour of functions near singularities and locating singularities (in particular those determined from heuristic considerations). It does not seem possible to automate qualitative considerations (like identifying attractive limit cycles) and there are various subtleties which might be encountered in specific applications.

The rather powerful `gdev` package [29] requires non-trivial changes of variables together with a trial-and-error approach for obtaining the result in an explicit form. Solving approximate systems is usually done by existing DESolvers which do not necessarily find solutions in closed-form. Even when the solutions are found, they are not necessarily in a form suitable for further analysis. In the former case, the solution can be attempted using numerical methods which is computationally easier than the original problems. The matching procedure required for singular perturbation problems is yet to be implemented in CA. A good starting point here is the function `MatchedSolutionList` from `PerturbationODE.m` package [13]. Computer automation of the perturbation analysis procedure looks plausible, but human interaction will be needed. A set of symbolic and symbolic-numerical tools (rather than a completely automated package) for solving perturbation problems is the scope for future work.

6. Example: A *Mathematica* Solver for Singularly Perturbed Boundary Value Problems

In this section, a *Mathematica* solver `BVPSolver` that is developed by the author for boundary value problems [15] is discussed briefly. This solver constructs approximate symbolic solutions of second-order singularly-perturbed boundary value problems. The employed symbolic technique is based on the method of matched asymptotic expansion [25]. The Nipp polyhedron algorithm [23] is used for finding appropriate scalings in local approximating systems. Linear non-self-adjoint and self-adjoint problems are solved, including those with turning points. Quasilinear problems are also attempted. The package also includes routines for solving semilinear second order problems using the symbolic-numerical methodology which employs the Newton's quasi-linearisation method. The main command of the solver `BVPSolve` has the same syntax as the built-in *Mathematica* function `DSolve` and enhances its functionality.

Example 6.1. Consider an example of the linear boundary value problem with a turning point [5]:

$$\varepsilon y''(x) + (x - 1/2)y'(x) - y(x) = 0, \qquad y(0) = 2, \;\; y(0) = 3. \tag{6.2}$$

The following *Mathematica* commands

```
example1={ε y''[x]+(x-1/2) y'[x]-y[x]==0, y[0]==2, y[0]==3};
BVPSolve[example1, y, x, EpsOrder->1]
```

yields the output:

```
this is non self-adjoint problem
there is one turning point:  1/2
inner scalings:
```

$$\texttt{s==(x-1/2)/}\varepsilon^{1/2}\texttt{, yy[0]==}\varepsilon^{1/2}\texttt{y}$$

$$\Big\{\texttt{yy[0][s]} \;\rightarrow \Big(\tfrac{1}{2}\texttt{-e}^{\texttt{s}^2/2}\Big(\texttt{-2 CI[0,2] + 2 e}^{\texttt{s}^2/2}\texttt{CI[0,1]s} \\ \texttt{- e}^{\texttt{s}^2/2}\sqrt{2\pi}\,\texttt{CI[0,2] Erf}\Big[\tfrac{\texttt{s}}{\sqrt{2}}\Big]\texttt{ s}\Big)\Big)\Big\},$$

$$\Big\{\texttt{Yl[0][x]}\rightarrow\texttt{(-2 (-1+2 x)),} \quad \texttt{Yr[0][x]}\rightarrow\texttt{(3 (-1+2 x))}\Big\}$$

```
Matching to the zero-th order...
```

$$\{\mathtt{CI[0,1]}\to 1,\ \mathtt{CI[0,2]}\to -5\sqrt{\tfrac{2}{\pi}}\}$$

$$\{\mathtt{yy[0][s]}\to 5-e^{s^2/2}\sqrt{\tfrac{2}{\pi}}+s+5\,s\,\mathrm{Erf}[\tfrac{s}{\sqrt{2}}]\}$$

$$-\tfrac{1}{2}+x+5\,e^{-\frac{\left(-\frac{1}{2}+x\right)^2}{2\varepsilon}}\sqrt{\tfrac{2}{\pi}}\sqrt{\varepsilon}+\left(-\tfrac{5}{2}+5x\right)\mathrm{Erf}\left[\tfrac{-\frac{1}{2}+x}{\sqrt{2}\sqrt{\varepsilon}}\right]$$

An algorithm for tackling higher-order boundary value problems

$$\begin{aligned}&\varepsilon y^{(n)}(t) = f(t, y, y', y'', \ldots, y^{(n-1)}) \qquad (a < t < b);\\ &y^{(j)}(a) = A_j, \quad 0 \le j \le (n-2); \qquad y^{(n-2)}(b) = B_{n-2},\end{aligned} \tag{6.3}$$

based on the classical theory for second-order problems and differential inequalities of higher order [12] has recently been added to the BVPSolver package [17]. This algorithm determines the leading asymptotic solution which can either be found in a fully analytical form or a symbolic-numerical one. In both cases, the task of finding a solution to a higher order boundary value problem is simplified to that of solving an initial value problem of an order lower than the one for the original problem and the second-order boundary value problem.

Example 6.4. Consider an example of a third-order problem with a shock layer behaviour:

$$\varepsilon y'''(t) = y'(t) - y'(t)y''(t),\ y(0) = 1, y'(0) = 1/2, y'(1) = 1 \tag{6.5}$$

The following *Mathematica* commands

```
example2={ε y'''[t]==y'[t]-y'[t]y''[t], y[0]==1, y'[0]==1/2, y'[1]==1};
BVPSolve[example2, y[t], {t,0,1}, EqType->HigherOrder]
```

yields the output:

```
Third order quasilinear equation
Solving Reduced Equation
raw solutions:
```

$$\{\{y[t]\to C[1],\ y[t]\to -\tfrac{t^2}{2}+C[1]+t\,C[2]\}\}$$

stable left solutions: $\{\{1-\frac{t}{2}+\frac{t^2}{2}\}\}$ on $0, \frac{1}{2}]$

stable right solutions: $\{\{C[1]+\frac{t^2}{2}\}\}$ on $[0,1]$

shock at

$t=\frac{1}{4}$, $C[1]=\frac{7}{8}$

```
asymptotic solutions:  shock layer behaviour
```

$$\text{uL}=1-\frac{t}{2}+\frac{t^2}{2}+\frac{\varepsilon}{8}-\frac{1}{8}\,e^{-\frac{1}{4\varepsilon}}\,\varepsilon,\ 0\le t<\frac{1}{4}$$

$$\text{uR}=\frac{7}{8}+\frac{t^2}{2}-\frac{\varepsilon}{8}+\frac{1}{8}e^{-\frac{1}{4\varepsilon}}\,\varepsilon,\ \ \frac{1}{4}<t\le 1.$$

`BVPSolver` can also deal with semilinear systems of reaction-diffusion type represented by

$$\begin{aligned}&\varepsilon y''(t) = \textstyle\prod_{j=1}^{n} f_{ij}(t, y_i) \qquad (i = 1, \ldots, n),\\ &y(a) = \mathbf{A},\ y(b) = \mathbf{B} \qquad (a \le t \le b,\ y \in R^n).\end{aligned} \tag{6.6}$$

Such a system can have multiple solutions which can be constructed asymptotically [24] provided the system satisfies certain stability assumptions in subdomains. Stable piecewise solutions of the reduced systems are found and then smoothed near boundaries and internal singularities by asymptotic corrections. This algorithm has also been implemented as a part of the `BVPSolver`.

7. Conclusion

In the author's view, finding approximate asymptotic solutions to differential equations based on existing theoretical results can in a few years time reach the same level of automation and wide use as the packages for finding closed-form solutions.

References

1. M. Bronstein. Computer algebra algorithms for linear ordinary differential and difference equations. *Proceedings of the Third European Congress of Mathematics, II*, 105–119. *Progress in Mathematics* **202** (2001), Birkhauser.
2. A. D. Bruno. Newton polyhedra and power transformations, *Mathematics and Computers in Simulation* **45** (1998), 429–443.
3. M. P. Cartmell, S.W. Ziegler, R. Khanin, D. I. M. Forehand. Mechanics of systems with weak nonlinearities. *Applied Mechanics Review* (in press).
4. R. M. Corless, D. J. Jeffrey, M. B. Monagan, Pratibha. Two perturbation calculations in fluid mechanics using large-expression management. *J. Symbolic Computation* **23**(4) (1997), 427–443.
5. M. Holmes. *Introduction to Perturbation Methods*, Springer, New York, 1995.
6. F. M. Fernandez, R. Guardiola, J. Rose. Computer algebra and large scale perturbation theory. *Computer Physics Communications* **115** (1998) 170–182.
7. C. Franciosi, S. Tomasiello. The use of *Mathematica* for the analysis of strongly nonlinear two-degree-of-freedom systems by means of the modified Lindstedt-Poincare method. *J. Sound and Vibration* **211** (1998), 145–156.

8. A. Gusev, V. Samoilov, V. Rostovtsev, S. Vinitsky. Symbolic algorithms of algebraic perturbation theory: hydrogen atom in the field of distant charge. In V. G. Ganzha *et al.* eds. *Computer Algebra in Scientific Computing* (*CASC'01*). Springer-Verlag, Berlin, 2001.
9. J. W. Helton, F.D. Kronewitter, W. M. McEneaney, M. Stankus. Singularly perturbed control systems using non-commutative computer algebra. *International J. Robust and Nonlinear Control* **10**(11–12) (2000), 983–1003.
10. J. M. Herbert, W. C. Ermler. Symbolic implementation of arbitrary-order perturbation theory using computer algebra: Application to vibrational-rotational analysis of diatomic molecules. *Computers and Chemistry* **22**(2–3) (1998), 169–184.
11. E. J. Hinch. *Perturbation Methods.* Cambridge University Press, 1991.
12. F. A. Howes. Differential inequalities of higher order and the asymptotic solution of nonlinear boundary value problems. *SIAM J. Math. Anal.* **13** (1982), 61–80.
13. S. Kaufmann. `http://www.ifm.ethz.ch/~kaufmann/news.html`.
14. J. Kevorkian, J. D. Cole. *Perturbation Methods in Applied Mathematics.* Springer-Verlag, New York, 1980.
15. R. Khanin. A *Mathematica* solver for two-point singularly-perturbed Boundary Value Problems. *Proceedings of the 4-th Workshop on Computer Algebra in Scientific Computing* (*CASC'01*), 2001.
16. R. Khanin. On the Nipp polyhedron algorithm for solving singular perturbation problems. *Mathematics and Computers in Simulation* **58** (2002), 255–272.
17. R. Khanin. On asymptotic solutions of higher-order boundary value problems. *Proceedings of the 5-th Workshop on Computer Algebra in Scientific Computing* (*CASC-02*), 2002.
18. R. Khanin, M. P. Cartmell. A computerised implementation of the multiple scales perturbation method using *Mathematica. Computers and Structures* **76** (2000), 565–575.
19. R. Khanin, M. Cartmell. Parallelisation of perturbation analysis: application to large-scale engineering problems. *Journal of Symbolic Computation* **31** (2001), 461–473.
20. D.M. Klimov, V.V. Leonov, V.M. Rudenko. The study of motion of a gyroscope with gimbal suspension obtaining the highest approximations for a drift of magnus *J. Symbolic Computation* **15** (1993), 73–78.
21. Y.O. Macutan. Formal solutions of scalar singularly-perturbed linear differential equations. *Proceedings of ISAAC-99*, 113–120. ACM Press, 1999.
22. A.H. Nayfeh. *Perturbation Methods.* John Wiley & Sons, New York, 1973.
23. K. Nipp. An algorithmic approach for solving singularly perturbed initial value problems. *Dynamics Reported* **1** (1980), 173–263.
24. M. O'Donnell. Boundary and corner behaviour in singularly perturbed semilinear systems of boundary value problems. *SIAM J.Math.Anal* **15** (1984), 317–332.
25. Jr. R. O'Malley. *Singular Perturbation Methods for Ordinary Differential Equations.* Springer-Verlag, New York, 1991.

26. F. Postel, P. Zimmerman. A Review of the ODE Solvers of *Axiom, Derive, Maple, Mathematica, MACSYMA* and *Reduce. Proceedings of the 5th Rhine Workshop on Computer Algebra.* 1996.
27. R. H. Rand, D. Armbruster. *Perturbation methods, Bifurcation Theory and Computer Algebra.* Springer, New York, 1987.
28. N. E. Sanchez. The method of multiple scales: asymptotic solutions and normal forms for nonlinear oscillatory problems. *J. Symbolic Computation* **21**(2) (1996), 245–252.
29. B. Salvy, J. Shackell. Symbolic asymptotics: Multiseries of inverse functions. *J. Symbolic Computation* **27**(6) (1999), 543–564.
30. A. Soleev, A. B. Aranson, *Computation of a polyhedron and normal cones of its faces.* (in Russian). *Inst. Appl. Math.* Preprint #36, Moscow, 1994.
31. E. Tournier. *CATHODE*: Computer Algebra Tools for Handling Ordinary Differential Equations. `http://www-lmc.imag.fr/CATHODE/`.
32. A. A. Vakhidov, N. N. Vasiliev. A new approach for analytical computation of Hamiltonian of a satellite perturbed motion. *J. Symbolic Computation* **24**(6) (1997), 705–710.

IMPLICITIZATION OF POLYNOMIAL CURVES*

ILIAS S. KOTSIREAS, EDMOND S. C. LAU

Wilfrid Laurier University
Computer Algebra Research Group
Department of Computing
Waterloo N2L 3C5, ON, Canada
http://www.cargo.wlu.ca

In this paper we present an implementation of an implicitization algorithm for curves given by polynomial parametric equations. We also establish some new structural properties of the implicitization matrices used in this algorithm. From a suitable viewpoint the implicitization matrices reveal a Hankel-like structure. Several examples illustrate in detail the implementation and some appealing perspectives for further work are briefly touched upon.

1. Introduction

In [1] the authors introduced a new algorithm for implicitization of parametric curves, surfaces and hypersurfaces. The algorithm uses essentially Linear Algebra, works in both symbolic and numeric contexts, and is applicable to a wide variety of types of parametric equations as well as indexed families of parametric equations. This algorithm has been implemented in *Maple* in the algcurves package.

In [2] the authors used various tools from algebraic geometry, in particular sparse elimination theory, in order to predict the support of the implicit equation of a rational parametric hypersurface. These ideas reduce dramatically the size of the implicitization matrices and can also be applied to other implicitization methods based on resultants. The resulting IPSOS algorithm gives optimal results in all of the examples tested. The implementation of the algorithm that we developed, requires interfacing several freely available C/C++ programs and is written in *Maple*.

In this paper we study more closely the case of curves given by polynomial parametric equations. An efficient implementation of the algorithm using exclusively C and GMP[a] arithmetic allows us to treat relatively big

*This work is supported by a grant from the Natural Sciences and Engineering Research Council of Canada.

[a]For more information about the GMP library, please visit http://www.swox.com/gmp/

examples. We also establish some structural properties of the implicitization matrices that could potentially lead to more efficient strategies to compute their nullspaces as well as other optimizations. We show that the implicitization matrices exhibit a Hankel-like structure when we consider blocks with respect to the degrees of the monomials.

2. Curves Given by Polynomial Parametric Equations

In this section we give an overview of the implicitization algorithm in [1] emphasizing the case of algebraic curves given by polynomial parametric equations.

Suppose that a curve is given by polynomial parametric equations

$$\begin{aligned} x &= a_p t^p + a_{p-1} t^{p-1} + \cdots + a_1 t + a_0, \\ y &= b_q t^q + b_{q-1} t^{q-1} + \cdots + b_1 t + b_0 \end{aligned}$$

where the coefficients $a_p, \ldots, a_0, b_q, \ldots, b_0$ are rational numbers. The variable t is called the **parameter** of the parameterization of the curve.

Denote by m the (total) degree of the sought for implicit equation. Generate all the monomials in the two variables x, y up to total degree $2m$. These are:

$$\ell_{2m} = [\underbrace{1}_{\deg 0}, \underbrace{x, y}_{\deg 1}, \underbrace{x^2, xy, y^2}_{\deg 2}, \ldots, \underbrace{x^{2m}, x^{2m-1}y, \ldots, xy^{2m-1}, y^{2m}}_{\deg 2m}] \qquad (2.1)$$

The number of these monomials is equal to $\sum_{i=1}^{2m+1} i = (2m+1)(m+1)$. A combinatorial argument shows that this number is equal to the binomial coefficient $\binom{2m+2}{2}$. The advantage of the combinatorial argument is that it is more easily generalizable in the case of surfaces.

For each of the $(2m+1)(m+1)$ monomials of the form $x^i y^j$ of the list ℓ_{2m} we need to compute the integral

$$\int_a^b x^i y^j \, dt = \int_a^b (a_p t^p + \cdots + a_0)^i (b_q t^q + \cdots + b_0)^j \, dt. \qquad (2.2)$$

We can choose for example $a = 0, b = 1$, since we are dealing with polynomial functions only. The computation of integral (2.2) requires two steps:

- **Expand** the polynomial to be integrated:

  $$(a_p t^p + a_{p-1} t^{p-1} + \cdots + a_1 t + a_0)^i (b_q t^q + b_{q-1} t^{q-1} + \cdots + b_1 t + b_0)^j.$$

  The degree of this expanded univariate polynomial in t will be equal to $pi + qj$.

- **Integrate** the expanded polynomial by using the simple rule to integrate monomials with coefficients:
$$\int_a^b ct^k dt = \frac{c\left(b^{k+1} - a^{k+1}\right)}{k+1},$$
where c is a constant and k is a positive integer.

We construct a matrix M, starting with the list of monomials in x, y up to total degree m,

$$\ell_m = [1, x, y, x^2, xy, y^2, \ldots, x^m, x^{m-1}y, \ldots, y^m]. \tag{2.3}$$

Then we form the product $M = \ell_m^t \ell_m$. In detail we have:

$$\begin{pmatrix} 1 \\ x \\ y \\ \vdots \\ x^m \\ \vdots \\ y^m \end{pmatrix} (1\ x\ y \cdots x^m \cdots y^m) = \begin{pmatrix} 1 & x & y & \cdots & x^m & \cdots & y^m \\ x & x^2 & xy & \cdots & x^{m+1} & \cdots & xy^m \\ y & xy & y^2 & \cdots & x^m y & \cdots & y^{m+1} \\ \vdots & \vdots & \vdots & \vdots & \vdots & \vdots & \vdots \\ x^m & x^{m+1} & x^m y & \cdots & x^{2m} & \cdots & x^m y^m \\ \vdots & \vdots & \vdots & \vdots & \vdots & \vdots & \vdots \\ y^m & xy^m & y^{m+1} & \cdots & x^m y^m & \cdots & y^{2m} \end{pmatrix} \tag{2.4}$$

We call matrices of the form (2.4), *implicitization matrices*. Implicitization matrices are symmetric (by construction) of dimension $d = \binom{m+2}{2}$. When we are looking for an implicit equation of degree m, it can be seen that the implicitization matrix G contains only $(2m+1)(m+1) = \binom{2m+2}{2}$ different elements, which are exactly the elements of the list ℓ_{2m}. Moreover, as m increases, $\binom{2m+2}{2}$ becomes much less that d^2.

A matrix G is constructed by placing the results of the integrations of the elements of the list ℓ_{2m} into the matrix M. This raises some interesting combinatorial and programming problems. If an implicit equation of degree m exists, its coefficients will be given by a nullvector of the matrix G. For a more detailed presentation of the implicitization algorithm as well as many fully worked out examples, the reader can consult [1].

3. Implementation

The algorithm outlined above was implemented in C in a program called IPCurves. Because for most examples the implicitization matrices contain big rational entries with more than 50 digits in both numerators and denominators, we decided to use the GMP library to handle the base data type.

The program IPCurves is divided into two sections:

(1) Functions for handling polynomials operations:

- Initialization
- Memory clean up
- Univariate polynomials multiplications
- Two versions of expander for parametric equations
- Integration of a polynomial over an interval $[a, b]$

(2) Functions for handling computations:

- Main integration function for ℓ_{2m}
- Find common denominator of the elements of matrix G
- Construct the symmetric matrix (2.4) from ℓ_{2m}

The program IPCurves provides some extra functionality that makes it easy to interface with the Computer Algebra System *Maple*. In particular, IPCurves generates implicitization matrices in *Maple* format and saves them in a file. This functionality is important for testing purposes. The main difference of IPCurves with the *Maple* implementation of [1] is the future perspective of interfacing IPCurves with IPSOS and other C/C++ programs in order to produce a software that is suitable for Computer Aided Geometric Design (CAGD) applications.

One important point of efficiency improvement in IPCurves is the generation of the implicitization matrices not by performing the matrix multiplication on the left hand side of (2.4), but by using the vector ℓ_{2m} and the Hankel-like structural properties of the implicitization matrices. Thus IPCurves integrates each monomial only once and places it in the right position. Hankel-like structural properties of the implicitization matrices are studied in the last section of this paper.

4. Examples

In this section we present three examples of implicitization of algebraic curves given by polynomial parametric equations using IPCurves. The results of these examples have been verified in different computer algebra systems, using other available methods for implicitization. For surveys of available implicitization methods of algebraic curves and surfaces see [4] and [3]. For illustration purposes we give the code and timing for these examples using Gröbner bases in the computer algebra system *Magma*. All computations were done on a Pentium III, 1GHz.

Example 4.1. Consider the polynomial parametric equations:

$$X = -59 + 57t + 63t^2 + 49t^3 + 56t^4 + 79t^5$$
$$Y = -5t + 54t^3 + 66t^4 + 77t^5 - 62t^6 + 43t^8$$

Degree arguments as detailed for instance in [2] can be used to show that the total degree of the sought for implicit equation is 8. We choose $m = 9$. We generate the 190 monomials in the two variables x, y up to total degree $2m$ and after performing the integrations and the substitutions we construct a 55×55 symmetric implicitization matrix. The generation of the matrix took 3 seconds. Computing the nullspace of this matrix we find the nullvector as in Figure 1. This corresponds to the (irreducible) implicit equation of total degree 8:

$$\begin{aligned}
&147008443\,x^8 + 320543044931\,x^7 - 302581857082\,x^6y + 27723620938\,2588\,x^6 \\
&-453225281500757\,x^5y + 152219541784174\,x^4y^2 - 32264377496512\,x^3y^3 \\
&+11412679399592433\,7x^5 - 2336750022890924\,34x^4y + 12436003119568190\,x^3y^2 \\
&-4538136278814081\,9x^2y^3 + 10645351560962279\,xy^4 - 1517108809906561\,y^5 \\
&+2298911955568760\,4740\,x^4 - 4857752127772062\,5557\,x^3y \\
&+2308232688715543\,7807\,x^2y^2 - 5736255878318665\,221\,xy^3 \\
&+597166299869832806\,y^4 + 2468570729523692256470\,x^3 \\
&-4962030858026428002201\,x^2y + 1727101617790134679420\,xy^2 \\
&-21020989982296156\,1295\,y^3 + 1447100889963386739\,33265\,x^2 \\
&-253765899902391517525373\,xy + 482483447485666876\,63567\,y^2 \\
&+4357097887893983938081311\,x - 5250056506846982581584327\,y \\
&+5243226426889696968100\,5015 = 0.
\end{aligned}$$

Using Gröbner bases implicitization in *Magma* one can obtain the same implicit equation with the following code, which executed in 1 second.

```
Q:=RationalField();
P<t,x,y>:=PolynomialRing(Q,3);
p1:=-59+57*t+63*t^2+49*t^3+56*t^4+79*t^5-x;
p2:=-5*t+54*t^3+66*t^4+77*t^5-62*t^6+43*t^8-y;
L:=[p1,p2];
time H1:=GroebnerBasis(L);
H1[4];
```

Example 4.2. Consider the polynomial parametric equations:

$$X = 92 - 93t - 8t^2 + 45t^3 - 59t^4 + 57t^5 + 63t^6 + 49t^7$$
$$Y = -12 - 50t - 61t^2 + 99t^3 - 5t^4 + 54t^5 + 66t^6 + 77t^7 - 62t^8 + 43t^9$$

We choose $m = 10$ and generate the 231 monomials in x, y up to degree $2m$. After the integrations and substitutions we construct a 66×66 symmetric

$$\begin{pmatrix} 0 \\ 5243226426889696968 1005015 \\ 0 \\ 4357097887893983938081311 \\ -5250056506846982581584327 \\ 0 \\ 1447100889963386739 33265 \\ -2537658999023915175 25373 \\ 4824834474856668766 3567 \\ 0 \\ 2468570729523692256470 \\ -4962030858026428002201 \\ 1727101617790134679420 \\ -2102098998229615 61295 \\ 0 \\ 229891195556876047 40 \\ -4857752127772062 5557 \\ 230823268871554378 07 \\ -5736255878318665221 \\ 597166299869832806 \\ 0 \\ 114126793995924337 \\ -233675002289092434 \\ 124360032119568190 \\ -45381362788140819 \\ 10645351560962279 \\ -1517108809906561 \\ 0 \\ 277236209382588 \\ -453225281500757 \\ 152219541784174 \\ -32264377496512 \\ 0 \\ 0 \\ 0 \\ 0 \\ 320543044931 \\ -302581857082 \\ 0 \\ 0 \\ 0 \\ 0 \\ 0 \\ 0 \\ 0 \\ 147008443 \\ 0 \\ 0 \\ 0 \\ 0 \\ 0 \\ 0 \\ 0 \\ 0 \\ 0 \end{pmatrix}$$

Figure 1. Nullvector for Example 4.1

$$\begin{pmatrix} 0 \\ 40038653707122814323293870 3229756 \\ 0 \\ -2114884373185911591645136 1330882 \\ 2825262401200651738516822 4708119 \\ 0 \\ 6869023085770192758209220 20869 \\ -1596903663043401074130708 004945 \\ 8672577907537295628548410 38447 \\ 0 \\ -1532164220083711236396860 0462 \\ 4141491944156656237871945 4912 \\ -3865056924450948141768107 7373 \\ 1162264487712974220110513 4155 \\ 0 \\ 1948183803015257911985766 63 \\ -5666663508469973459058524 56 \\ 6639925368392660899377510 79 \\ -3492606436799527520969929 23 \\ 6822828374868504456494766 5 \\ 0 \\ -1317144886837516806016519 \\ 395940732102937709833960 1 \\ -512309184256401992389637 1 \\ 335390220538154411694547 6 \\ -104955226035926679548211 7 \\ 117247234802145672603023 \\ 0 \\ 4444337261554459507433 \\ -1325629068794305187893 4 \\ 1723784119527454159466 1 \\ -1180684339326345215713 9 \\ 4026877697719624702346 \\ -524385527862798988932 \\ 155216526541692112 28 \\ 0 \\ -6331914982244916011 \\ 1802014330908877651 8 \\ -2125181166055922041 0 \\ 1345076512069271295 3 \\ -453893408487822548 7 \\ 664529056096392893 \\ -45405000058022003 \\ 1628413597910449 \\ 0 \\ 1884921382721797 \\ -468497810343426 2 \\ 191560464806099 2 \\ 426978342555821 \\ 0 \\ 0 \\ 0 \\ 0 \\ 0 \\ 0 \\ -271818611107 \\ 0 \\ 0 \\ 0 \\ 0 \\ 0 \\ 0 \\ 0 \\ 0 \\ 0 \\ 0 \end{pmatrix}$$

Figure 2. Nullvector for Example 4.2

implicitization matrix. The generation of the matrix took 9 seconds. The nullspace is generated by the vector in Figure 2. This corresponds to the (irreducible) implicit equation of total degree 9:

$$\begin{aligned} &-271818611107\, x^9 + 1884921382721797\, x^8 + \cdots \\ &\cdots + 4003865370712281432329387032 29756 = 0. \end{aligned}$$

Using Gröbner bases implicitization in *Magma* one can obtain the same implicit equation with the following code, which executed in 16 seconds.

```
Q:=RationalField();
P<t,x,y>:=PolynomialRing(Q,3);
p1:=92-93*t-8*t^2+45*t^3-59*t^4+57*t^5+63*t^6+49*t^7-x;
p2:=-12-50*t-61*t^2+99*t^3-5*t^4+54*t^5+66*t^6
    +77*t^7-62*t^8+43*t^9-y;
L:=[p1,p2];
time H1:=GroebnerBasis(L);
H1[4];
```

Example 4.3. Consider the following polynomial parametric equations:

$$\begin{aligned} X &= -8-85\,t^{13}-55\,t^{12}-37\,t^{11}-35\,t^{10}+97\,t^{9}+50\,t^{8} \\ &\quad +79\,t^{7}+56\,t^{6}+49\,t^{5}+63\,t^{4}+57\,t^{3}-59\,t^{2}+45\,t \\ Y &= 1-93\,t^{16}+92\,t^{15}+43\,t^{14}-62\,t^{13}+77\,t^{12}+66\,t^{11}+54\,t^{10} \\ &\quad -5\,t^{9}+99\,t^{8}-61\,t^{7}-50\,t^{6}-12\,t^{5}-18\,t^{4}+31\,t^{3}-26\,t^{2}-62\,t \end{aligned}$$

Suppose $m = 17$. We generate the 630 monomials in the two variables x, y up to total degree $2m$ and construct a 171×171 symmetric implicitization matrix with rational number entries. To compute the symmetric matrix, our program required 3 minutes. Computing a basis for the nullspace of this matrix is an interesting challenge for computer algebra systems. Using Gröbner bases implicitization in *Magma* the computation does not finish after an hour.

We note that in the examples of this section, we have chosen m to be one more than the known degree of the implicit equation, just to illustrate that the coefficients of the degree m monomials will indeed be zero. This can be observed in the two nullvectors shown in the first two examples.

5. Hankel-like Structural Properties of Implicitization Matrices

In this section we establish some interesting properties pertaining to the structure of the implicitization matrices. In particular we show that if one uses the degree ordering to write the vector of monomials ℓ_m as defined in (2.3), then the associated implicitization matrix is revealed to have a type of

Hankel-like[b] structure. It is interesting to note that the Hankel structure is of a different type if we use the lexicographical ordering to write the vector of monomials ℓ_m. In general, the Hankel structure for the degree ordering will be maintained if we group together the monomials of same degree in the vector ℓ_m. In the sections below, we illustrate the Hankel structure by examining in detail the case of the degree ordering. Similar results hold for the case of the lexicographical ordering.

Hankel-like Structure for the Degree Ordering

We illustrate the Hankel-like structural properties in degree 3. The corresponding general result is easy to state and prove. Define the vector $u = [1\ x\ y\ x^2\ xy\ y^2\ x^3\ x^2y\ xy^2\ y^3]$ and compute $p = u^t \cdot u$:

$$p = \begin{pmatrix} 1 & x & y & x^2 & xy & y^2 & x^3 & x^2y & xy^2 & y^3 \\ x & x^2 & xy & x^3 & x^2y & xy^2 & x^4 & x^3y & x^2y^2 & xy^3 \\ y & xy & y^2 & x^2y & xy^2 & y^3 & x^3y & x^2y^2 & xy^3 & y^4 \\ x^2 & x^3 & x^2y & x^4 & x^3y & x^2y^2 & x^5 & x^4y & x^3y^2 & x^2y^3 \\ xy & x^2y & xy^2 & x^3y & x^2y^2 & xy^3 & x^4y & x^3y^2 & x^2y^3 & xy^4 \\ y^2 & xy^2 & y^3 & x^2y^2 & xy^3 & y^4 & x^3y^2 & x^2y^3 & xy^4 & y^5 \\ x^3 & x^4 & x^3y & x^5 & x^4y & x^3y^2 & x^6 & x^5y & x^4y^2 & x^3y^3 \\ x^2y & x^3y & x^2y^2 & x^4y & x^3y^2 & x^2y^3 & x^5y & x^4y^2 & x^3y^3 & x^2y^4 \\ xy^2 & x^2y^2 & xy^3 & x^3y^2 & x^2y^3 & xy^4 & x^4y^2 & x^3y^3 & x^2y^4 & xy^5 \\ y^3 & xy^3 & y^4 & x^2y^3 & xy^4 & y^5 & x^3y^3 & x^2y^4 & xy^5 & y^6 \end{pmatrix}$$

If we group the elements in p by the degrees of each monomial term into submatrices and denote by i a block of monomials of total degree i, then p can be represented as follows:

$$p = \begin{pmatrix} 0 & 1 & 2 & 3 \\ 1 & 2 & 3 & 4 \\ 2 & 3 & 4 & 5 \\ 3 & 4 & 5 & 6 \end{pmatrix}$$

This representation of p shows clearly its Hankel structure with respect to the degrees of blocks of monomials. Moreover, if we examine the structure

[b]The term Hankel-like here is used to describe a Hankel structure with respect to degrees of blocks of monomials.

of each degree block individually, we see that p can be rewritten as follows:

$$p = \begin{pmatrix} H^0 & C_1 & C_2 & C_3 \\ C_1^t & H^2 & C_3 & C_4 \\ C_2^t & C_3^t & H^4 & C_5 \\ C_3^t & C_4^t & C_5^t & H^6 \end{pmatrix}$$

where the t superscript denotes matrix transposition, the C_i are rectangular **banded** blocks formed by monomials of total degree i and the main diagonal contains square blocks H^i which are **Hankel** matrices of monomials of total degree i. Thus the implicitization matrices, aside from being symmetric and usually singular, demonstrate a much richer structure.

Currently, it is not clear to us how to take advantage of the Hankel-like structure exhibited by the implicitization matrices to improve the algorithm. However, since there is a vast literature on algorithms for structured matrices and in particular for Hankel-like matrices, we believe that this issue deserves further investigation.

6. Conclusions and Future Work

We presented an efficient implementation of the implicitization algorithm in [1] for curves given by polynomial parametric equations. We also showed that the implicitization matrices used in this algorithm exhibit different types of Hankel-like structure according to the orderings employed to write the monomials.

Future research directions that will result in significant speed-ups in the algorithm are the application of modulo p techniques as well as interfacing IPCurves with the implementation of the IPSOS algorithm described in [2]. Another direction is to capitalize on the Hankel-like structure of the implicitization matrices. All of these techniques will subsequently be applied in the case of surfaces.

It is clear that numerical techniques can be applied for performing the integrations and computing the nullspace. This is related to the approximate implicitization problem whose study is outside the scope of this paper.

Acknowledgments

The authors thank Mr. Richard Voino for stimulating discussions and his help with early versions of the implementation. The authors thank the anonymous referees for their constructive comments.

References

1. R. M. Corless, M. W. Giesbrecht, I. S. Kotsireas, S. M. Watt. Numerical implicitization of parametric hypersurfaces with linear algebra. In *AISC'2000 Proceedings* (*Madrid*), 174–183. *LNAI* **1930**, Springer, 2000.
2. I. Z. Emiris, I. S. Kotsireas. On the support of the implicit equation of rational parametric hypersurfaces. ORCCA Technical report TR-02-01, 2002. Available on-line from `http://www.orcca.on.ca`.
3. I. S. Kotsireas. Panorama of methods for exact implicitization of algebraic curves and surfaces. Preprint, 2002.
4. T. Sederberg, J. Zheng. Algebraic methods for computer aided geometric design. In *Handbook of Computer Aided Geometric Design*, 363–387. North Holland, Amsterdam, 2002.

A BRACKET METHOD FOR JUDGING THE INTERSECTION OF CONVEX BODIES

HONGBO LI, YING CHEN

Academy of Mathematics and System Sciences,
Chinese Academy of Sciences,
Beijing 100080, China
Email: hli, ychen@mmrc.iss.ac.cn

In this paper we study a basic problem based on simple bracket manipulations in computational geometry: how to judge if two solid convex polygons or polytopes intersect or not. We will establish a sequence of criteria based on boundary intersection searching, hyperplane separation searching and hybrid search method combining the former two systems. Our simulation results show that the latter method is significantly superior to the former criteria.

1. Introduction

Judging if two convex bodies in space intersect or not is a basic task in computational geometry and computer graphics. This task is easy theoretically, but from the application point of view, the efficiency of the criterion used in the judgement still bears much concern, and finding more efficient detection method is still an active research topic nowadays.

For two solid convex bodies, if they intersect at all, either one is completely situated at the interior of the other, or the boundaries of the two convex bodies, which have lower dimensions, must intersect This intersection-searching idea leads to a recursive algorithm of judging if two convex bodies intersect.

Another idea is based on the fact that if two solid compact convex bodies are separate, then there exists a hyperplane in the space between them but not touching any of them. This paper shows that one only needs to check some boundaries of the convex bodies, and in the 3D case, some extra planes spanned by the edges of the polygons or polytopes. This idea leads to a separation-searching algorithm, which can also be derived from the idea that if two convex bodies are separate, then one body does not touch the Minkowski sum of the two convex bodies with base on the other body.

One interesting thing is that we can effectively weave the two algorithms so that the hybrid algorithm outperforms by far any of the "pure-minded" ones. The reason is obvious: if there is a group of pairs of intersecting convex bodies, then the separation-searching algorithm has to go through all the separation cases before it reaches the conclusion that all pairs are not separate. Likewise, for pairs of separate bodies, the intersection-searching algorithm will perform much worse than the separation-searching algorithm. The hybrid algorithm does intersection search at one step, and does separation search at the next step in a staircase way. This feature makes it more efficient.

Another interesting thing is that the signs of some brackets of the homogeneous coordinates of the vertices of the two convex bodies are all we need in carrying out the judgement. By definition, in an n-dimensional affine space A^n, the homogeneous coordinates of a point $\mathbf{a} = (a_1, \ldots, a_n)$ is $(a_1, \ldots, a_n, 1)$, and the *bracket* of a sequence of $n+1$ points $\mathbf{a}_1, \ldots, \mathbf{a}_{n+1}$ is

$$[\mathbf{a}_1 \cdots \mathbf{a}_{n+1}] = \begin{vmatrix} a_{11} & \cdots & a_{(n+1)1} \\ \vdots & \ddots & \vdots \\ a_{1n} & \cdots & a_{(n+1)n} \\ 1 & \cdots & 1 \end{vmatrix} . \tag{1.1}$$

The $n+1$ points are *affinely dependent* if their bracket equals zero. When the bracket is nonzero, its sign is completely determined by the sequence of points and the orientation of the coordinate system. In other words, a change of the coordinate system does not change the sign of the bracket as long as the new coordinate system has the same orientation with the old one.

2. Brackets, Intersections and Some Notations

An n-dimensional affine space A^n can be taken as a hyperplane in an $(n+1)$-dimensional vector space V^{n+1} away from the origin. Its points are thus represented by vectors from the origin of V^{n+1} to the points in the hyperplane. When $n+1$ points $\mathbf{a}_1, \ldots, \mathbf{a}_{n+1}$ are affinely independent, the first n points generate a hyperplane in A^n, which is the intersection of the vector space spanned by the vectors representing the n points with the hyperplane representing A^n.

The sequence $\mathbf{a}_1, \ldots, \mathbf{a}_n$ determines an orientation of the hyperplane they generate. While an even permutation of the sequence does not change the orientation, an odd one reverses the orientation. Point $\mathbf{a}_{n+1}$ is said to be on the *positive* side of the oriented hyperplane if $[\mathbf{a}_1 \cdots \mathbf{a}_{n+1}] > 0$, and on the *negative* side if $[\mathbf{a}_1 \cdots \mathbf{a}_{n+1}] < 0$.

In application, it often occurs that brackets of different size are needed. For example, for points $\mathbf{a}_1, \mathbf{a}_2$ in a plane, the bracket $[\mathbf{a}_1\mathbf{a}_2\mathbf{a}_3]$ is meaningful for any point $\mathbf{a}_3$ in the plane, while the bracket $[\mathbf{a}_1\mathbf{a}_2]$ is meaningful on the line passing through the two points. The sign of the latter bracket equals that of $[\mathbf{a}_1\mathbf{a}_2\mathbf{a}_3]$, for any point $\mathbf{a}_3$ outside the line such that the bracket $[\mathbf{e}_1\mathbf{e}_2\mathbf{a}_3]$ formed by the coordinate system $\{\mathbf{e}_1, \mathbf{e}_2\}$ of the line and $\mathbf{a}_3$ is positive.

In the plane, the intersection of lines $\mathbf{12}, \mathbf{1'2'}$ is

$$\begin{aligned}\mathbf{12} \cap \mathbf{1'2'} &= c^{-1}([\mathbf{11'2'}]\mathbf{2} - [\mathbf{21'2'}]\mathbf{1}) \\ &= c^{-1}([\mathbf{122'}]\mathbf{1'} - [\mathbf{121'}]\mathbf{2'}),\end{aligned} \tag{2.1}$$

where $c = [\mathbf{11'2'}] - [\mathbf{21'2'}] = [\mathbf{122'}] - [\mathbf{121'}]$. The two lines are collinear if and only if $[\mathbf{11'2'}] = [\mathbf{21'2'}] = 0$. They are parallel if and only if $[\mathbf{11'2'}] = [\mathbf{21'2'}] \neq 0$.

In space, the intersection of the line $\mathbf{12}$ and the plane $\mathbf{1'2'3'}$ is

$$\begin{aligned}\mathbf{1'2'3'} &= d^{-1}([\mathbf{121'2'}]\mathbf{3'} + [\mathbf{122'3'}]\mathbf{1'} + [\mathbf{123'1'}]\mathbf{2'}) \\ &= d^{-1}([\mathbf{11'2'3'}]\mathbf{2} - [\mathbf{21'2'3'}]\mathbf{1}).\end{aligned}$$

where $d = [\mathbf{121'2'}] + [\mathbf{122'3'}] + [\mathbf{123'1'}] = [\mathbf{11'2'3'}] - [\mathbf{21'2'3'}]$. The line is on the plane if and only if $[\mathbf{11'2'3'}] = [\mathbf{21'2'3'}] = 0$. It is parallel to the plane if and only if $[\mathbf{11'2'3'}] = [\mathbf{21'2'3'}] \neq 0$.

The *sign function* $s : R \longrightarrow \{1, 0, -1\}$ is a mapping returning the sign a^s of a real number a. For a real matrix M, its *sign matrix* M^s is the matrix composed of the signs of the components of M.

Let $\mathbf{1'2'3'}$ be a triangle. By this we assume that points $\mathbf{1'}, \mathbf{2'}, \mathbf{3'}$ are not collinear. A point $\mathbf{1}$ is inside the triangle if and only if the following brackets have the same sign:

$$[\mathbf{11'2'}],\ [\mathbf{12'3'}],\ [\mathbf{13'1'}]. \tag{2.2}$$

Point $\mathbf{1}$ is on the border of the triangle if and only if the three brackets are either all ≥ 0 or all ≤ 0, and at least one bracket equals zero. Point $\mathbf{1}$ is *covered by the triangle* if it is in the closure of the solid triangle.

In this paper, for a vertex $\mathbf{i}$ of a polytope $\mathbf{1} \cdots \mathbf{n}$, the two vertices joined to $\mathbf{i}$ are denoted by $\mathbf{i} - \mathbf{1}$ and $\mathbf{i} + \mathbf{1}$ respectively. Such indices are always modulo $\mathbf{n}$. Often, we drop the bold face notation for points.

3. Intersection-Searching Criteria in the 2D Case

Two Triangles

They intersect if and only if either one vertex of a triangle is covered by the other triangle, or one side of a triangle intersects a side of the other triangle.

Proposition 3.1. *Let* **123** *and* **1′2′3′** *be two triangles in the plane. Let*

$$M = \begin{pmatrix} [\mathbf{12'3'}] & [\mathbf{13'1'}] & [\mathbf{11'2'}] \\ [\mathbf{22'3'}] & [\mathbf{23'1'}] & [\mathbf{21'2'}] \\ [\mathbf{32'3'}] & [\mathbf{33'1'}] & [\mathbf{31'2'}] \end{pmatrix}, \quad M' = \begin{pmatrix} [\mathbf{1'23}] & [\mathbf{1'31}] & [\mathbf{1'12}] \\ [\mathbf{2'23}] & [\mathbf{2'31}] & [\mathbf{2'12}] \\ [\mathbf{3'23}] & [\mathbf{3'31}] & [\mathbf{3'12}] \end{pmatrix}. \tag{3.2}$$

The two solid triangles do not intersect if and only if ± 1 *are in every row of* M^s *and* M'^s, *and*

$$\begin{aligned} &\max\big([\mathbf{11'2'}]^s[\mathbf{21'2'}]^s,\ [\mathbf{1'12}]^s[\mathbf{2'12}]^s \big) \\ =\ &\max\big([\mathbf{13'1'}]^s[\mathbf{23'1'}]^s,\ [\mathbf{1'12}]^s[\mathbf{3'12}]^s \big) \\ =\ &\max\big([\mathbf{11'2'}]^s[\mathbf{31'2'}]^s,\ [\mathbf{1'31}]^s[\mathbf{2'31}]^s \big) \\ =\ &\max\big([\mathbf{13'1'}]^s[\mathbf{33'1'}]^s,\ [\mathbf{1'31}]^s[\mathbf{3'31}]^s \big) \\ =\ &1. \end{aligned} \tag{3.3}$$

Proof. We only need to prove the sufficiency of the conditions for the two triangles not to intersect. Since every row of M^s and M'^s contains 1 and -1, no vertex of a triangle is covered by the other triangle.

When sides $(\mathbf{i}-\mathbf{1})(\mathbf{i}+\mathbf{1})$ and $(\mathbf{j}-\mathbf{1})'(\mathbf{j}+\mathbf{1})'$ are collinear, which is equivalent to the condition that the j-th column of M^s has only the i-th element being nonzero, the two sides do not intersect. When points $\mathbf{i}, \mathbf{j}'$ are on different sides of line $(\mathbf{i}-\mathbf{1})(\mathbf{i}+\mathbf{1})(\mathbf{j}-\mathbf{1})'(\mathbf{j}+\mathbf{1})'$, the two triangles do not intersect. When they are on the same side of the line, it can be easily proved that the two triangles intersect if and only if each of $(\mathbf{i}-\mathbf{1})\mathbf{i}, (\mathbf{i}+\mathbf{1})\mathbf{i}$ intersects $(\mathbf{j}-\mathbf{1})'\mathbf{j}'$ and $(\mathbf{j}+\mathbf{1})'\mathbf{j}'$. In both cases, the two triangles do not intersect if and only if at least one of the following pairs of sides do not intersect:

$$\begin{aligned} &\{(i-1)i,\ (j-1)'j'\}, \\ &\{(i+1)i,\ (j-1)'j'\}, \\ &\{(i-1)i,\ (j+1)'j'\}, \\ &\{(i+1)i,\ (j+1)'j'\}. \end{aligned} \tag{3.4}$$

Observe that the set $\{\mathbf{12}, \mathbf{1'2'}\}, \{\mathbf{12}, \mathbf{1'3'}\}, \{\mathbf{13}, \mathbf{1'2'}\}, \{\mathbf{13}, \mathbf{1'3'}\}$ always has non-empty intersection with the set (3.4) for any $1 \le i, j \le 3$. As a consequence, if sides **12**, **13** and sides **1′2′**, **1′3′** do not intersect, then at least one of the pairs of sides in (3.4) do not intersect, and the condition is also sufficient and necessary for the two triangles not to intersect.

When no two sides from different triangles are collinear, it can be easily proved that if the two triangles intersect, then there are two possibilities: either each side of each triangle intersects two sides of the other triangle, or two sides of each triangle each intersects two sides of the other triangle, and the third side of each triangle does not intersect any side of the other triangle. Thus the two triangles do not intersect if and only if sides **12**, **13** do not intersect sides **1′2′**, **1′3′**. □

A Triangle and a Convex Quadrilateral

Lemma 3.5. *Triangle* **123** *and convex quadrilateral* **1′2′3′4′** *do not intersect if and only if no vertex of either is covered by the other, and line segments* **12**, **13** *do not intersect line segments* **1′3′**, **2′4′**.

Proposition 3.6. *For triangle* **123** *and convex quadrilateral* **1′2′3′4′** *in the plane, define*

$$M = \begin{pmatrix} [\mathbf{11'2'}] & [\mathbf{12'3'}] & [\mathbf{13'4'}] & [\mathbf{14'1'}] \\ [\mathbf{21'2'}] & [\mathbf{22'3'}] & [\mathbf{23'4'}] & [\mathbf{24'1'}] \\ [\mathbf{31'2'}] & [\mathbf{32'3'}] & [\mathbf{33'4'}] & [\mathbf{34'1'}] \end{pmatrix},$$

$$M' = \begin{pmatrix} [\mathbf{1'12}] & [\mathbf{1'23}] & [\mathbf{1'31}] \\ [\mathbf{2'12}] & [\mathbf{2'23}] & [\mathbf{2'31}] \\ [\mathbf{3'12}] & [\mathbf{3'23}] & [\mathbf{3'31}] \\ [\mathbf{4'12}] & [\mathbf{4'23}] & [\mathbf{4'31}] \end{pmatrix}. \tag{3.7}$$

They do not intersect if and only if the following conditions are all satisfied:

(1) ± 1 are in every row of M^s and M'^s.

(2) $\max([\mathbf{11'3'}]^s[\mathbf{21'3'}]^s,\ [\mathbf{1'12}]^s[\mathbf{3'12}]^s) + \max([\mathbf{12'4'}]^s[\mathbf{22'4'}]^s,\ [\mathbf{2'12}]^s[\mathbf{4'12}]^s) \geq 1.$

(3) $\max([\mathbf{11'3'}]^s[\mathbf{31'3'}]^s,\ [\mathbf{1'13}]^s[\mathbf{3'13}]^s) + \max([\mathbf{12'4'}]^s[\mathbf{32'4'}]^s,\ [\mathbf{2'13}]^s[\mathbf{4'13}]^s) \geq 1.$

Two Convex Quadrilaterals

Lemma 3.8. *For two solid convex quadrilateral* **1234** *and* **1′2′3′4′**, *they do not intersect if and only if no vertex of either is covered by the other, and line segments* **13**, **24** *do not intersect line segments* **1′3′**, **2′4′**.

Proposition 3.9. *For two convex quadrilateral* **1234** *and* **1′2′3′4′**, *let*

$$M = \begin{pmatrix} [\mathbf{11'2'}] & [\mathbf{12'3'}] & [\mathbf{13'4'}] & [\mathbf{14'1'}] \\ [\mathbf{21'2'}] & [\mathbf{22'3'}] & [\mathbf{23'4'}] & [\mathbf{24'1'}] \\ [\mathbf{31'2'}] & [\mathbf{32'3'}] & [\mathbf{33'4'}] & [\mathbf{34'1'}] \\ [\mathbf{41'2'}] & [\mathbf{42'3'}] & [\mathbf{43'4'}] & [\mathbf{44'1'}] \end{pmatrix},$$

$$M' = \begin{pmatrix} [\mathbf{1'12}] & [\mathbf{1'23}] & [\mathbf{1'34}] & [\mathbf{1'41}] \\ [\mathbf{2'12}] & [\mathbf{2'23}] & [\mathbf{2'34}] & [\mathbf{2'41}] \\ [\mathbf{3'12}] & [\mathbf{3'23}] & [\mathbf{3'34}] & [\mathbf{3'41}] \\ [\mathbf{4'12}] & [\mathbf{4'23}] & [\mathbf{4'34}] & [\mathbf{4'41}] \end{pmatrix}. \tag{3.10}$$

They do not intersect if and only if the following conditions are all satisfied:

(*1*) ± 1 *are in every row of* M^s *and* M'^s.

(*2*) $\max\big([\mathbf{11'3'}]^s[\mathbf{31'3'}]^s,\ [\mathbf{1'13}]^s[\mathbf{3'13}]^s\big) + \max\big([\mathbf{12'4'}]^s[\mathbf{32'4'}]^s,\ [\mathbf{2'13}]^s[\mathbf{4'13}]^s\big) \geq 1.$

(*3*) $\max\big([\mathbf{21'3'}]^s[\mathbf{41'3'}]^s,\ [\mathbf{1'24}]^s[\mathbf{3'24}]^s\big) + \max\big([\mathbf{22'4'}]^s[\mathbf{42'4'}]^s,\ [\mathbf{2'24}]^s[\mathbf{4'24}]^s\big) \geq 1.$

Two Convex Polygons

An oriented convex polygon is represented by the sequence of its vertices: $\mathbf{a}_1 \cdots \mathbf{a}_m$. Let $\mathbf{1} \cdots \mathbf{n}$ and $\mathbf{1'} \cdots \mathbf{m'}$ be two convex polygons. Let

$$M = \begin{pmatrix} [\mathbf{11'2'}] & [\mathbf{12'3'}] & \cdots & [\mathbf{1m'1'}] \\ [\mathbf{21'2'}] & [\mathbf{22'3'}] & \cdots & [\mathbf{2m'1'}] \\ \vdots & \vdots & \vdots & \vdots \\ [\mathbf{n1'2'}] & [\mathbf{n2'3'}] & \cdots & [\mathbf{nm'1'}] \end{pmatrix},$$

$$M' = \begin{pmatrix} [\mathbf{1'12}] & [\mathbf{1'23}] & \cdots & [\mathbf{1'n1}] \\ [\mathbf{2'12}] & [\mathbf{2'23}] & \cdots & [\mathbf{2'n1}] \\ \vdots & \vdots & \vdots & \vdots \\ [\mathbf{m12}] & [\mathbf{m23}] & \cdots & [\mathbf{mn1}] \end{pmatrix}. \tag{3.11}$$

Case 1. When m, n are both even, the two polygons do not intersect if and only if the vertices of any of them are not covered by the other one, and line segments $\mathbf{i(i+n/2)}$ for $i = 1, \ldots, n/2$ do not intersect line segments $\mathbf{j'(j+m/2)'}$ for $j = 1, \ldots, m/2$.

Case 2. When one of m, n, say n, is even, the two polygons do not intersect if and only if the vertices of any of them are not covered by the other one, and line segments $\mathbf{i(i+n/2)}$ for $i = 1, \ldots, n/2$ do not intersect line segments $\mathbf{j'(j+(m-1)/2)'}$ and $\mathbf{(m-1)'m'}$ for $j = 1, \ldots, (m-1)/2$.

Case 3. When m, n are both odd, the two polygons do not intersect if and only if the vertices of any of them are not covered by the other one, and line segments $\mathbf{i(i+(n-1)/2)}$ and $\mathbf{(n-1)n}$ for $i = 1, \ldots, (n-1)/2$ do not intersect line segments $\mathbf{j'(j+(n-1)/2)'}$ and $\mathbf{(m-1)m}$ for $j = 1, \ldots, (m-1)/2$.

Proposition 3.12. *Two convex polygons* $\mathbf{1 \cdots n}$ *and* $\mathbf{1' \cdots m'}$ *do not intersect if and only if* ± 1 *are in every row of* M^s *and* M'^s, *and*

(a) *if both* m, n *are even, then for any* $1 \le i \le \frac{n}{2}$,

$$\sum_{j=1}^{\frac{m}{2}} \max\Big([ij'(j+\frac{m}{2})']^s [(i+\frac{n}{2})j'(j+\frac{m}{2})']^s, \\ [j'i(i+\frac{n}{2})]^s [(j+\frac{m}{2})'i(i+\frac{n}{2})]^s \Big) \ge \frac{m}{2} - 1, \tag{3.13}$$

(b) *if one of* m, n, *say* n, *is even, then for any* $1 \le i \le \frac{n}{2}$,

$$\max\Big([i(m-1)'m']^s [(i+\frac{n}{2})(m-1)'m']^s, \\ [(m-1)'i(i+\frac{n}{2})]^s [m'i(i+\frac{n}{2})]^s \Big) \\ + \sum_{j=1}^{[\frac{m}{2}]} \max\Big([ij'(j+[\frac{m}{2}])']^s [(i+\frac{n}{2})j'(j+[\frac{m}{2}])']^s, \\ [j'i(i+\frac{n}{2})]^s [(j+[\frac{m}{2}])'i(i+\frac{n}{2})]^s \Big) \ge [\frac{m}{2}] + 1, \tag{3.14}$$

(c) *if both* m, n *are odd, then for any* $1 \le i \le [\frac{n}{2}]$

$$\max\Big([i(m-1)'m']^s [(i+[\frac{n}{2}])(m-1)'m']^s, \\ [(m-1)'i(i+[\frac{n}{2}])]^s [m'i(i+[\frac{n}{2}])]^s \Big) \\ + \sum_{j=1}^{[\frac{m}{2}]} \max\Big([ij'(j+[\frac{m}{2}])']^s [(i+[\frac{n}{2}])j'(j+[\frac{m}{2}])']^s, \\ [j'i(i+[\frac{n}{2}])]^s [(j+[\frac{m}{2}])'i(i+[\frac{n}{2}])]^s \Big) \ge [\frac{m}{2}] + 1 \tag{3.15}$$

and

$$\max\Big([(n-1)(m-1)'m']^s[n(m-1)'m']^s,$$
$$[(m-1)'(n-1)n]^s[m'(n-1)n]^s\Big)$$
$$+\sum_{j=1}^{[\frac{m}{2}]}\max\Big([(n-1)j'(j+[\frac{m}{2}]')]^s[nj'(j+[\frac{m}{2}])']^s, \tag{3.16}$$
$$[j'(n-1)n]^s[(j+[\frac{m}{2}])'n(n-1))]^s\Big) \ \geq \ [\frac{m}{2}]+1.$$

4. Separation-Searching Criterion in the 2D Case

Proposition 4.1. *For two convex polygons, they do not intersect if and only if there exists a line passing through one edge of a polygon such that the two polygons are on different sides of the line and one polygon does not touch the line at all.*

Proof. We only need to prove the necessity of the condition. If two convex polygons A, B do not intersect, then their distance is nonzero. Let a, b be two points on the boundaries of the two polygons respectively such that their distance equals that of the two polygons. Such two points obviously exists. Pushing polygon A along the vector from a to b towards B until a, b are identified, we get a polygon A'. There are three cases.

Case 1. a, b are at the interiors of two edges E_a, E_b of the two polygons respectively.

There exists a neighborhood of b in E_b which is also in the edge E'_a of A' corresponding to E_a. So E_a, E_b are parallel, and the two polygons are outside the region between the two parallel lines. Obviously they do not intersect.

Case 2. a is a vertex of A while b is at the interior of edge E_b of B.

If there is any other point c other than b which is common to both A' and B, then by convexity, line segment bc belongs to both polygons. This reduces to Case 1. So we can assume that b is the unique common point between A', B. Then A, B are on different sides of line E_b, and A does not touch the line.

Case 3. a, b are both vertices.

We simply reflect the angle of A' at vertex b with respect to the vertex. The union of the reflected angle with the angle of B at b is an angle $\angle_b(A, B)$, called the *forbidden angle* of the two polygons at b. It has the property that no matter how A' is translated nearby the vertex b of B, it simply cannot enter the angle, and vice versa when A' is fixed and B is translated. Let ℓ be a side of the forbidden angle which does not pass through point a. Then the edge corresponding to ℓ separates A, B in the sense that the line passing through the edge separates A, B, and one polygon does not touch the line. □

Corollary 4.2. *For two convex polygons* $\mathbf{12}\cdots\mathbf{n}$ *and* $\mathbf{1'2'}\cdots\mathbf{m'}$, *let their corresponding matrices* M, M' *be defined in (3.11), and assume that their orientations are both positive. Then they do not intersect if and only if at least one column of* M^s *or* M'^s *is composed of* -1*'s.*

5. Intersection-Searching Criteria in the 3D Case

Two Convex Polygons

Proposition 5.1. *For two non-coplanar convex polygons* $\mathbf{12}\cdots\mathbf{n}$ *and* $\mathbf{1'2'}\cdots\mathbf{m'}$, *let*

$$M_1 = \big([\mathbf{11'2'3'}]\cdots[\mathbf{n1'2'3'}]\big), \quad M_2 = \big([\mathbf{1'123}]\cdots[\mathbf{m'123}]\big). \tag{5.2}$$

Let m_j *be the sum of the elements in* M_j^s. *The two polygons do not intersect if and only if one of the following conditions is satisfied.*

(1) $|m_1| = n$ *or* $|m_2| = m$.
(2) $|m_1| = n-1$, *and if the* i*-th element in* M_1^s *is zero, then* ± 1 *are in* $\{[i(i+1)1'2']^s,\ [i(i+1)2'3']^s, \cdots,\ [i(i+1)m'1')]^s\}$.
(3) $|m_2| = m-1$, *and if the* j*-th element in* M_2^s *is zero, then* ± 1 *are in* $\{[12j'(j+1)']^s, [23j'(j+1)']^s, \cdots, [n1j'(j+1)')]^s\}$.
(4) *In other cases, let* i_1 *be the first point in* $\mathbf{1}, \mathbf{2}, \cdots, \mathbf{n}$ *such that*

$$[i_1\mathbf{1'2'3'}]^s \notin \{\mathbf{0}, [(i_1+1)\mathbf{1'2'3'}]^s\},$$

and let i_2 *be the first point in* $\mathbf{n}, \cdots, \mathbf{2}, \mathbf{1}$ *such that*

$$[i_2\mathbf{1'2'3'}]^s \notin \{\mathbf{0}, [(i_2-1)\mathbf{1'2'3'}]^s, -[i_1\mathbf{1'2'3'}]^s\}.$$

Points j_1' *and* j_2' *in* $\mathbf{1'}, \cdots, \mathbf{m'}$ *can be found similarly. Then*

$$\begin{aligned} \big|\, &[i_1(i_1+1)j_1'(j_1+1)']^s + [i_1(i_1+1)j_2'(j_2-1)']^s + \\ &[i_2(i_2-1)j_1'(j_1+1)']^s + [i_2(i_2-1)j_2'(j_2-1)']^s \,\big| = 4. \end{aligned} \tag{5.3}$$

A Convex Polygon and a Convex Polytope

For a convex polytope A in the space, we use the symbol $\mathbf{a}_{ij}$ to denote the j-th vertex of the i-th face. The orientation of the face defined by the sequence $\mathbf{a}_{i1}, \ldots, \mathbf{a}_{ik_i}$ is induced from the orientation of the polytope such that $[\mathbf{a}_{i(j-1)}\mathbf{a}_{ij}\mathbf{a}_{i(j+1)}\mathbf{a}_{pq}] > 0$ for any $1 \leq j, 1 \leq k_i$ and any vertex $\mathbf{a}_{pq}$ outside the face. This orientation is said to be *positive.*

It is obvious that a polytope A intersects a polygon $B = \mathbf{b}_1 \cdots \mathbf{b}_m$ if and only if one of the vertices of B is covered by A, or B intersects one face of A. Let f be the number of faces of A. Let

$$M = \begin{pmatrix} [\mathbf{a}_{11}\mathbf{a}_{12}\mathbf{a}_{13}\mathbf{b}_1] & [\mathbf{a}_{21}\mathbf{a}_{22}\mathbf{a}_{23}\mathbf{b}_1] & \cdots & [\mathbf{a}_{f1}\mathbf{a}_{f2}\mathbf{a}_{f3}\mathbf{b}_1] \\ [\mathbf{a}_{11}\mathbf{a}_{12}\mathbf{a}_{13}\mathbf{b}_2] & [\mathbf{a}_{21}\mathbf{a}_{22}\mathbf{a}_{23}\mathbf{b}_2] & \cdots & [\mathbf{a}_{f1}\mathbf{a}_{f2}\mathbf{a}_{f3}\mathbf{b}_2] \\ \vdots & \vdots & \vdots & \vdots \\ [\mathbf{a}_{11}\mathbf{a}_{12}\mathbf{a}_{13}\mathbf{b}_m] & [\mathbf{a}_{21}\mathbf{a}_{22}\mathbf{a}_{23}\mathbf{b}_m] & \cdots & [\mathbf{a}_{f1}\mathbf{a}_{f2}\mathbf{a}_{f3}\mathbf{b}_m] \end{pmatrix}.$$

Then there exists a vertex of B which is covered by A if and only if one line of M^s has all its elements being nonnegative. And B intersect one face of A, as discussed in the previous section.

Two Convex Polytopes

Two convex polytopes intersect if and only if either one vertex of a polytope is covered by the other polytope, or one face of a polytope intersects a face of the other polytope.

6. Separation-Searching Criteria in the 3D Case

Proposition 6.1. *Two convex polytopes A and B do not intersect if and only if there exists a plane C having one of the following properties:*

(1) *C contains one face of a polytope, the two polytopes are on different sides of C and one polytope does not touch C.*
(2) *C passes through one edge of a polytope and is parallel to one edge of the other polytope, the two polytopes are on different sides of C and one polytope does not touch C.*

Corollary 6.2. *For two convex polytopes A and B, let their positive-oriented vertex sequences on the i-th and j-th faces respectively be*

$\mathbf{a}_{i1}, \dots \mathbf{a}_{ik_i}$ *and* $\mathbf{b}_{j1}, \dots, \mathbf{b}_{jl_j}$. *Let*

$$M_A = \begin{pmatrix} [\mathbf{a}_{11}\mathbf{a}_{12}\mathbf{a}_{13}\mathbf{b}_1] & [\mathbf{a}_{21}\mathbf{a}_{22}\mathbf{a}_{23}\mathbf{b}_1] & \cdots & [\mathbf{a}_{i1}\mathbf{a}_{i2}\mathbf{a}_{i3}\mathbf{b}_1] \\ [\mathbf{a}_{11}\mathbf{a}_{12}\mathbf{a}_{13}\mathbf{b}_2] & [\mathbf{a}_{21}\mathbf{a}_{22}\mathbf{a}_{23}\mathbf{b}_2] & \cdots & [\mathbf{a}_{i1}\mathbf{a}_{i2}\mathbf{a}_{i3}\mathbf{b}_2] \\ \vdots & \vdots & & \vdots \\ [\mathbf{a}_{11}\mathbf{a}_{12}\mathbf{a}_{13}\mathbf{b}_m] & [\mathbf{a}_{21}\mathbf{a}_{22}\mathbf{a}_{23}\mathbf{b}_m] & \cdots & [\mathbf{a}_{i1}\mathbf{a}_{i2}\mathbf{a}_{i3}\mathbf{b}_m] \end{pmatrix},$$

$$M_B = \begin{pmatrix} [\mathbf{b}_{11}\mathbf{b}_{12}\mathbf{b}_{13}\mathbf{a}_1] & [\mathbf{b}_{21}\mathbf{b}_{22}\mathbf{b}_{23}\mathbf{a}_1] & \cdots & [\mathbf{b}_{j1}\mathbf{b}_{j2}\mathbf{b}_{j3}\mathbf{a}_1] \\ [\mathbf{b}_{11}\mathbf{b}_{12}\mathbf{b}_{13}\mathbf{a}_2] & [\mathbf{b}_{21}\mathbf{b}_{22}\mathbf{b}_{23}\mathbf{a}_2] & \cdots & [\mathbf{b}_{j1}\mathbf{b}_{j2}\mathbf{b}_{j3}\mathbf{a}_2] \\ \vdots & & & \vdots \\ [\mathbf{b}_{11}\mathbf{b}_{12}\mathbf{b}_{13}\mathbf{a}_n] & [\mathbf{b}_{21}\mathbf{b}_{22}\mathbf{b}_{23}\mathbf{a}_n] & \cdots & [\mathbf{b}_{j1}\mathbf{b}_{j2}\mathbf{b}_{j3}\mathbf{a}_n] \end{pmatrix}.$$

The two polytopes do not intersect if and only if either one column of ${M_A}^s$ *is composed of* -1*'s, or one column of* ${M_B}^s$ *is composed of* -1*'s, or there exist an edge* $\mathbf{12}$ *of* A *and an edge* $\mathbf{1'2'}$ *of* B *such that*

(1) $[\mathbf{121'2'}]^s \neq \mathbf{0}$,

(2) *for two vertices* $i_1, i_2 \notin \{\mathbf{1}, \mathbf{2}\}$ *of* A, $i_1\mathbf{12}$ *and* $i_2\mathbf{12}$ *are different faces of* A *satisfying* $([\mathbf{12}i_1\mathbf{1'}] - [\mathbf{12}i_1\mathbf{2'}])^s = [\mathbf{121'2'}]^s$ *and* $([\mathbf{12}i_2\mathbf{1'}] - [\mathbf{12}i_2\mathbf{2'}])^s = [\mathbf{121'2'}]^s$,

(3) *for similar* $j_1', j_2' \notin \{\mathbf{1'}, \mathbf{2'}\}$ *of* B, $j_1'\mathbf{1'2'}$ *and* $j_2'\mathbf{1'2'}$ *are different faces of* B *satisfying* $([\mathbf{1'2'}j_1'\mathbf{1}] - [\mathbf{1'2'}j_1'\mathbf{2}])^s = [\mathbf{121'2'}]^s$ *and* $([\mathbf{1'2'}j_2'\mathbf{1}] - [\mathbf{1'2'}j_2'\mathbf{2}])^s = [\mathbf{121'2'}]^s$.

7. Hybrid Search and Simulations

For two coplanar and positively-oriented convex polygons $\mathbf{12}\cdots\mathbf{n}$ and $\mathbf{1'2'}\cdots\mathbf{m'}$, let their corresponding matrices M, M' be defined in (3.11). In the following cases, (1) and (2) mean that either one vertex is covered by the other, that there is an intersection. (3) and (4) mean that they do not intersect, and there is no other case in which they do not intersect.

(1) A row of M^s has no -1.
(2) A row of M'^s has no -1.
(3) A column of M^s are all -1's.
(4) A column of M'^s are all -1's.

The above list suggests a hybrid search strategy: we can scan the matrices M^s, M'^s in a staircase way and skip a lot of redundant searching.

For pairs of polygons with no intersection at all, the intersection-searching criterion is obviously less efficient than the separation-searching

Algorithm: Intersection Detection (Hybrid Search)

Input: Matrices M^s, M'^s of size $n \times m$ and $m \times n$ respectively.

Output: "Intersection" or "No Intersection".

Step 1. Set $\texttt{Mat} = M^s$, and set $\texttt{Mat}_1 = \texttt{Mat}_2 = \phi$ where ϕ is the empty set.

Step 2. Set $\texttt{col} = \texttt{row} = 1$.

Step 3. Scan the col-th column of Mat to find an element not equal to -1.
If no element is found, then
    output "No Intersection" and exit.
Let $i \geq \texttt{row}$ be the first row such that $\texttt{Mat}(i, \texttt{col}) \neq -1$.
If i does not exist, then
    if $\texttt{Mat} = M^s$, set $\texttt{Mat}_1 = \texttt{Mat}(1..n, \texttt{col} + 1..m)$,
        set $\texttt{Mat} = M'^s$ and go back to Step 2
    else set $\texttt{Mat}_2 = \texttt{Mat}(1..m, \texttt{col} + 1..n)$ and go to Step 5.
else set $\texttt{row} = i$.

Step 4. Scan the row-th row of Mat to find an element equal to -1.
If no element is found, then output "Intersection" and exit.
Let $j \geq \texttt{col}$ be the first column such that $\texttt{Mat}(\texttt{row}, j) = -1$
If j does not exist, then
    if $\texttt{Mat} = M^s$, set $\texttt{Mat} = M'^s$ and go back to Step 2,
    else go to Step 5.
else set $\texttt{col} = j$.

Step 5. If $\texttt{Mat}_1 = \texttt{Mat}_2 = \phi$, then output "Intersection" and exit.
Scan the columns of $\texttt{Mat}_1$ and $\texttt{Mat}_2$ to check if there is any column which are all -1's.
If there is any, output "No intersection" and exit,
else output "Intersection" and exit.

one. For intersecting polygons, the situation is reversed. The hybrid search algorithm (see next page) intertwines the column scan and the row scan of the sign matrices, and takes into consideration that only the column scan is complete in the judgement. In our simulation experiments, it performs much better than the algorithms based on the previous two criteria separately.

In our simulation, we take one convex polygon as inscribed in a circle and the other convex polygon as inscribed in a branch of a hyperbola. The total number of tests is q, and each time we randomly choose n and m points sequentially from the circle and the hyperbola respectively to form the vertices of our convex polygons. Figure 1 shows the performance curves of the three searching methods respectively. Four groups of tests are carried out for the three methods, each group composed of $q = 90$ tests. In each group, the percentage of intersection cases is used to tick the x-axis, and the total time consumed in the q tests is used to tick the y-axis. The superiority of the hybrid search is obvious.

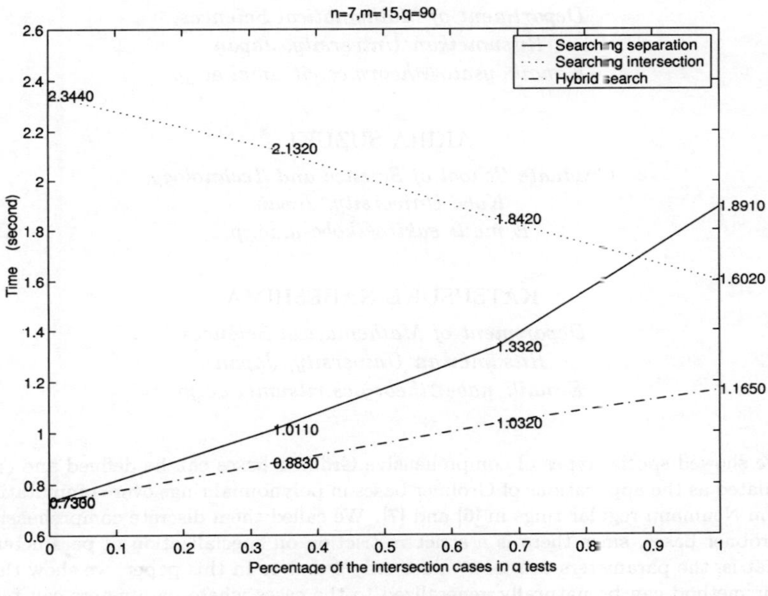

Figure 1. Performance curves of the criteria for two coplanar convex polygons.

In the 3D case, we have similar results and we omit them because of the space.

References

1. J. Goodman, J. O'Rourke eds., *Handbook of Discrete and Computational Geometry.* CRC Press, Boca Raton, New York, 1997.
2. D. Hestenes, H. Li, A. Rockwood. New algebraic tools for classical geometry. In G. Sommer ed., *Geometric Computing with Clifford Algebra.* Springer, 1999.
3. D. Hestenes, G. Sobczyk. *Clifford Algebra to Geometric Calculus.* D. Reidel, Dordrecht, Boston, 1984.
4. N. White. The bracket ring of combinatorial geometry, I. *Trans. Amer. Math. Soc.* **202** (1975), 79–103.

DISCRETE COMPREHENSIVE GRÖBNER BASES, II

YOSUKE SATO

Department of Mathematical Sciences,
Ritsumeikan University, Japan
E-mail: ysato@theory.cs.ritsumei.ac.jp

AKIRA SUZUKI

Graduate School of Science and Technology,
Kobe University, Japan
E-mail: sakira@kobe-u.ac.jp

KATSUSUKE NABESHIMA

Department of Mathematical Sciences,
Ritsumeikan University, Japan
E-mail: nabe@theory.cs.ritsumei.ac.jp

We showed special types of comprehensive Gröbner bases can be defined and calculated as the applications of Gröbner bases in polynomial rings over commutative Von Neumann regular rings in [6] and [7]. We called them discrete comprehensive Gröbner bases, since there is a strict restriction on specialization of parameters, that is, the parameters can take values only 0 and 1. In this paper, we show that our method can be naturally generalized to the cases where parameters can take any value from a given finite set.

1. Introduction

In [6] and [7], we proposed special types of comprehensive Gröbner bases called discrete comprehensive Gröbner bases using Weispfenning's theory of Gröbner bases in polynomial rings over commutative Von Neumann regular rings [9]. Roughly speaking, discrete comprehensive Gröbner bases are comprehensive Gröbner bases with parameters the specializations of which are restricted to only 0 and 1. One of the key facts for constructing discrete comprehensive Gröbner bases is that the quotient ring $R[X]/(X^2 - X)$ for a given Von Neumann regular ring R also becomes a Von Neumann regular ring. We gave an elementary direct proof of this fact in [7]. However, this fact essentially follows from the Chinese Remainder Theorem since

$R[X]/(X^2 - X) \cong R[X]/(X) \times R[X]/(X-1)$ (direct product). This observation leads us to generalize discrete comprehensive Gröbner bases as follows.

Let K be a field and $S_1, \ldots, S_n$ be non-empty finite subsets of K. Let $A_1, \ldots, A_n$ be indeterminates and for each $i = 1, \ldots, n$, let $p_i(A_i)$ be the polynomial $\prod_{k \in S_i}(A_i - k)$. Then the quotient ring $K[A_1, \ldots, A_n]/(p_1(A_1), \ldots, p_n(A_n))$ becomes a commutative Von Neumann regular ring. Let F be a finite set of polynomials in $K[A_1, \ldots, A_n, \overline{X}]$, where $\overline{X}$ are indeterminates distinct from $A_1, \ldots, A_n$. Considering F to be a finite set of polynomials in $(K[A_1, \ldots, A_n]/(p_1(A_1), \ldots, p_n(A_n)))[\overline{X}]$, construct a stratified Gröbner basis G of the ideal (F). Then G becomes a discrete comprehensive Gröbner basis of (F) in the following sense. For each $i = 1, \ldots, n$, let a_i be an element of S_i. Then the set of polynomials $G(a_1, \ldots, a_n) = \{g(a_1, \ldots, a_n, \overline{X}) \mid g \in G\}$ is the reduced Gröbner basis of the ideal generated by the set of polynomials $F(a_1, \ldots, a_n) = \{f(a_1, \ldots, a_n, \overline{X}) | f \in F\}$ in $K[\overline{X}]$.

We made an implementation to compute the above revised version of discrete comprehensive Gröbner bases for the case that K is the field of rational numbers. Through our computation experiments, we found that they are sufficiently practical.

The rest of the paper is organized as follows. In Section 2, we describe some mathematical facts which play important roles for the construction of our revised discrete comprehensive Gröbner bases. Our main results are shown in Section 3. In Section 4, we give some computation examples of our implementation. The reader is assumed to be familiar with the theory of Gröbner bases in polynomial rings over commutative Von Neumann regular rings. We refer the reader to [9], [5] or [7].

2. Some Basic Facts

In this section, we show some mathematical facts which are easy consequences of the Chinese Remainder Theorem.

Lemma 2.1. *Let K be a field and $a_1, a_2, \ldots, a_\ell$ be distinct elements of K. Let $p(X)$ be a polynomial defined by $p(X) = (X - a_1)(X - a_2) \cdots (X - a_\ell)$. Let R be a commutative ring which extends K. Then $R[X]/(p(X))$ is isomorphic to R^ℓ. Actually the mapping Φ from $R[X]/(p(X))$ to R^ℓ defined by $\Phi(h(X)) = (h(a_1), h(a_2), \ldots, h(a_\ell))$ is an isomorphism.*

Proof. The ideals $(X - a_1), (X - a_2), \ldots, (X - a_\ell)$ are clearly co-maximal in $K[X]$. Hence, they are also co-maximal in $R[X]$. By the Chinese Remainder Theorem, we have an isomorphism Φ from $R[X]/(p(X))$ to $\prod_{i=1\ldots\ell} R[X]/(X - a_i)$ defined by $\Phi(h(X)) = (h(a_1), h(a_2), \ldots, h(a_\ell))$. $R[X]/(X - a_i)$ is clearly isomorphic to R for each i. □

Using the above lemma we have the following.

Lemma 2.2. *Let K be a field and $S_1, S_2, \ldots, S_n$ be non-empty finite subsets of K. Let $A_1, \ldots, A_n$ be indeterminates and $p_i(A_i)$ be the polynomial $\prod_{k \in S_i}(A_i - k)$ for each $i = 1, \ldots, n$. Then $K[A_1, \ldots, A_n]/(p_1(A_1), \ldots, p_n(A_n))$ is isomorphic to K^M, where $M = |S_1||S_2|\ldots|S_n|$ and $|S_i|$ denotes the cardinality of S_i.*

Proof. We prove by induction on n. When n is 1, it follows directly from Lemma 2.1. Note that $K[A_1, \ldots, A_n]/(p_1(A_1), \ldots, p_n(A_n))$ is isomorphic to $R[A_n]/(p_n(A_n))$ with $R = K[A_1, \ldots, A_{n-1}]/(p_1(A_1), \ldots, p_{n-1}(A_{n-1}))$. By the induction hypothesis, R is isomorphic to $K^{M'}$, where $M' = |S_1||S_2|\ldots|S_{n-1}|$. Since R clearly includes K, we can apply Lemma 2.1 to have an isomorphism between $R[A_n]/(p_n(A_n))$ and $R^{|S_n|}$ which is isomorphic to K^M. □

By this lemma, we can see $K[A_1, \ldots, A_n]/(p_1(A_1), \ldots, p_n(A_n))$ is a commutative Von Neumann regular ring. This is the key fact in this paper. In order to have our discrete comprehensive Gröbner bases, we need to describe the isomorphism explicitly.

Lemma 2.3. *With the same notations in Lemma 2.2, let $\overline{\alpha}_1, \overline{\alpha}_2, \ldots, \overline{\alpha}_M$ be an enumeration of the set $\{(a_1, a_2, \ldots, a_n) | a_i \in S_i \text{ for each } i\}$. For each $j = 1, 2, \ldots, M$, let $\overline{\alpha}_j = (\alpha_1^j, \alpha_2^j, \ldots, \alpha_n^j)$. The mapping*

$$\Phi : K[A_1, \ldots, A_n]/(p_1(A_1), \ldots, p_n(A_n)) \rightarrow \prod_{j=1,\ldots,M} K[A_1, \ldots, A_n]/(A_1 - \alpha_1^j, \ldots, A_n - \alpha_n^j)$$

defined by

$$\Phi(h(A_1, A_2, \ldots, A_n)) = (h(\overline{\alpha}_1), h(\overline{\alpha}_2), \ldots, h(\overline{\alpha}_M))$$

is an isomorphism.

Proof. We actually showed this fact in the proof of Lemma 2.2 by applying Lemma 2.1 iteratively. □

3. Discrete Comprehensive Gröbner Bases

For any polynomial h of $R[\overline{X}]$, let h^i denotes the polynomial in $K[\overline{X}]$ obtained from h by replacing each coefficient c in h by the ith coordinate of c, which belongs to K^M after identifying R with K^M. The following lemma is a directly consequence of Theorem 2.3 of [9].

Lemma 3.1. *Let K be a field and R be a commutative Von Neumann regular ring defined as a finite direct product K^M of K for some natural number M. Fix a term order for the terms in the indeterminates $\overline{X}$ and let $G = \{g_1, \ldots, g_k\}$ be the stratified reduced Gröbner basis of an ideal $(f_1, \ldots, f_\ell)$ in a polynomial ring $R[\overline{X}]$. Then, $\{g_1^i, \ldots, g_k^i\}$ becomes the reduced Gröbner basis of the ideal $(f_1^i, \ldots, f_\ell^i)$ in the polynomial ring $K[\overline{X}]$ for each $i = 1, 2, \ldots, M$.*

We also have the following lemma.

Lemma 3.2. *With the same notations and conditions in Lemma 3.1, let $G_i = \{g_1^i, \ldots, g_k^i\}$ for each i. Then for any polynomial h in $R[\overline{X}]$, we have $(h\downarrow_G)^i = h^i \downarrow_{G_i}$ for each i. Here, $h\downarrow_G$ denotes the normal form of h with respect to the Gröbner basis G.*

Proof. The proof is essentially same as the proof for Property (2) of Theorem 3.3 [7] or the proof for Property (2) of Theorem 3.2 [8]. □

Now we are ready to state our revised discrete comprehensive Gröbner bases.

Theorem 3.3. *Let K be a field and $S_1, \ldots, S_n$ be non-empty finite subsets of K. Let $A_1, \ldots, A_n$ be indeterminates and let $p_i(A_i)$ be the polynomial $\prod_{k \in S_i}(A_i - k)$ for each $i = 1, \ldots, n$. Then, the quotient ring*

$$\mathcal{R} = K[A_1, \ldots, A_n]/(p_1(A_1), \ldots, p_n(A_n))$$

becomes a commutative Von Neumann regular ring as is shown in Lemma 2.2. Let F be a finite set of polynomials in $K[A_1, \ldots, A_n, \overline{X}]$, where $\overline{X}$ are indeterminates distinct from $A_1, \ldots, A_n$.

Fix a term order of terms of indeterminates $\overline{X}$. Considering F to be a finite set of polynomials in

$$\mathcal{R}[\overline{X}] = (K[A_1, \ldots, A_n]/(p_1(A_1), \ldots, p_n(A_n)))[\overline{X}],$$

construct the stratified Gröbner basis G of the ideal (F) in this polynomial ring. Then we have the following properties.

(1) *For any n-tuple $(a_1, a_2, \ldots, a_n)$ of elements of K such that $a_i \in S_i$ for each i, the set of polynomials*

$$G(a_1, \ldots, a_n) = \{g(a_1, \ldots, a_n, \overline{X}) \mid g \in G\}$$

is the reduced Gröbner basis of the ideal generated by the set of polynomials

$$F(a_1, \ldots, a_m) = \{f(a_1, \ldots, a_n, \overline{X}) \mid f \in F\}$$

in $K[\overline{X}]$.

(2) *For any $h(A_1, \ldots, A_n, \overline{X})$ in $\mathcal{R}[\overline{X}]$, we have*

$$(h(A_1, \ldots, A_n, \overline{X}) \downarrow_G)(a_1, \ldots, a_n, \overline{X}) = h(a_1, \ldots, a_n, \overline{X}) \downarrow_{G(a_1, \ldots, a_n)}.$$

Here, $r(a_1, \ldots, a_n, \overline{X})$ denotes the polynomial in $K[\overline{X}]$ obtained from a polynomial $r(A_1, \ldots, A_n, \overline{X})$ in $\mathcal{R}[\overline{X}]$ by substituting each A_i with a_i.

Proof. The first property follows from Lemma 2.3 and Lemma 3.1, the second property follows from Lemma 2.3 and Lemma 3.2. □

Let G be as in Theorem 3.3. Then we call G a *discrete comprehensive Gröbner basis.* Remember that G is nothing but our original discrete comprehensive Gröbner basis, when each set S_i is $\{0, 1\}$.

4. Computation Examples

We made an implementation to compute the revised version of discrete comprehensive Gröbner bases for the case that the coefficient field is the field of rational numbers. Though our program is very naive and written in prolog, it is sufficiently practical. The following are examples of our computation experiments.

Example 4.1. Let F be the set of polynomials

$$A_1^2A_2X_1^2X_2 + A_3X_1 + A_2,$$
$$A_1A_3X_1X_2^2 + A_2X_1X_2X_3^2 + X_1 + A_1,$$
$$A_1^2X_1X_3 + A_1X_2 + A_3X_3$$

with parameters A_1, A_2, A_3. Let

$$S_1 = \{-1, 0, 2\}, \quad S_2 = \{-1, 0, 1\}, \quad S_3 = \{-1, 1, 3\}.$$

Our program calculated the following discrete comprehensive Gröbner basis with the graded reverse lex order $>$ such that $X_1 > X_2 > X_3$.

```
-A1^2*A2^2+1/2*A1^2+A1*A2^2-1/2*A1+A2^2,
(1/2*A1^2*A2^2-1/2*A1^2-1/2*A1*A2^2+1/2*A1-A2^2+1)*X3,
(1/2*A1^2*A2^2-1/2*A1^2-1/2*A1*A2^2+1/2*A1-A2^2+1)*X1,
(1/2*A1^2*A2^2-1/2*A1*A2^2)*X3*X1+(-1/4*A1^2*A2^2+3/4*A1*A2^2)
*X2+(3/8*A1^2*A2^2*A3-5/8*A1*A2^2*A3)*X3,

   ⋮

(1/24*A1^2*A2^2*A3^2-1/12*A1^2*A2^2*A3+3/8*A1^2*A2^2+1/24
*A1^2*A2*A3^2-1/12*A1^2*A2*A3-1/8*A1^2*A2-1/12*A1*A2^2*A3^2+
1/6*A1*A2^2*A3-1/4*A1*A2^2-1/12*A1*A2*A3^2+1/6*A1*A2*A3+1/4
*A1*A2)*X2^2*X1+(-1/24*A1^2*A2^2*A3^2-7/24*A1^2*A2^2*A3+1/8
*A1^2*A2^2-1/24*A1^2*A2*A3^2+1/12*A1^2*A2*A3+1/8*A1^2*A2+
1/12*A1*A2^2*A3^2+11/24*A1*A2^2*A3-1/4*A1*A2^2+1/12*A1*A2
*A3^2-1/6*A1*A2*A3-1/4*A1*A2)*X2^2+(17/48*A1^2*A2^2*A3^2-
1/12*A1^2*A2^2*A3-1/8*A1^2*A2^2+1/24*A1^2*A2*A3^2-1/12*A1^2
*A2*A3-1/8*A1^2*A2-37/48*A1*A2^2*A3^2+1/6*A1*A2^2*A3+1/4*A1
*A2^2-1/12*A1*A2*A3^2+1/6*A1*A2*A3+1/4*A1*A2)*X3*X2+(1/24
*A1^2*A2^2*A3^2-1/12*A1^2*A2^2*A3-1/8*A1^2*A2^2+1/24*A1^2
*A2*A3^2+7/24*A1^2*A2*A3-1/8*A1^2*A2-1/12*A1*A2^2*A3^2+1/6
*A1*A2^2*A3+1/4*A1*A2^2-1/12*A1*A2*A3^2-11/24*A1*A2*A3+1/4
*A1*A2)*X2+(1/4*A1^2*A2^2-5/16*A1^2*A2*A3^2-3/4*A1*A2^2+
11/16*A1*A2*A3^2)*X3
```

The computation time was a few seconds by a personal computer with a CPU of Pentium III 1200 MHZ. We can of course get a similar result by calculating a full comprehensive Gröbner basis of

$$F \cup \{(A_1+1)A_1(A_1-2), (A_2+1)A_2(A_2-1), (A_3+1)(A_3-1)(A_3-3)\}.$$

However, `cgb` of CGB [1] and `dispgb` of DisPGB [2], which are the only available existing comprehensive Gröbner bases computation packages, did not terminate within one hour.

Example 4.2. Let F be the same set of polynomials as the above example. Let

$$S_1 = \{-3, -1, 0, 2, 5\}, S_2 = \{-3, -1, 0, 1, 5\}, S_3 = \{-7, -1, 1, 3, 6\}.$$

Our program calculated the discrete comprehensive Gröbner basis within 10 seconds and produced the following polynomial that consists of only parameters.

```
-1/225*A1^4*A2^4+2/225*A1^4*A2^3+16/225*A1^4*A2^2-2/225*
A1^4*A2-1/30*A1^4+1/75*A1^3*A2^4-2/75*A1^3*A2^3-16/75*
A1^3*A2^2+2/75*A1^3*A2+1/10*A1^3+1/15*A1^2*A2^4-2/15*A1^2*
A2^3-16/15*A1^2*A2^2+2/15*A1^2*A2+1/2*A1^2-19/225*A1*A2^4+
38/225*A1*A2^3+304/225*A1*A2^2-38/225*A1*A2-19/30*A1-1/15*
A2^4+2/15*A2^3+16/15*A2^2-2/15*A2
```

We can also get information on the parameters by calculating a Gröbner basis of

$$\begin{aligned} F\cup\ \{&(A_1+3)(A_1+1)A_1(A_1-2)(A_1-5),\\ &(A_2+3)(A_2+1)A_2(A_2-1)(A_2-5),\\ &(A_3+7)(A_3+1)(A_3-1)(A_3-3)(A_3-6)\}\end{aligned}$$

in the polynomial ring $\mathbf{Q}[X_1, X_2, X_3, A_1, A_2, A_3]$ with the block term order such that $[X_1, X_2, X_3] > [A_1, A_2, A_3]$. We, again, were not able to compute the Gröbner basis even by using RISA/ASIR [3] that has a very fast and sophisticated Gröbner bases computation package.

5. Conclusion and Remarks

Although we do not give a description in this paper, we can generalize Theorem 3.3 for an arbitrary polynomial $p_i(A_i)$. In order to construct discrete comprehensive Gröbner bases for such cases, we further need factorizations in polynomial rings over algebraically extended fields and have to handle fields which are represented as quotient rings of some polynomial rings. Since we have not made an implementation for such cases at this point we do not know if they are feasible.

References

1. A. Dolzmann, T. Sturm, W. Neun. CGB: Computing comprehensive Gröbner bases. **http://www.fmi.uni-passau.de/~redlog/cg/**, 1999.
2. A. Montes. A new algorithm for discussing Gröbner basis with parameters, *J. Symb. Comp.* **33**(1-2) (2002), 183–208.
3. M. Noro, T. Takeshima. Risa/Asir—A computer algebra system. *Proceedings of the 1992 International Symposium on Symbolic and Algebraic Computation* (*ISSAC 92*), 387–396. ACM Press, 1992.
4. D. Saracino, V. Weispfenning. On algebraic curves over commutative regular rings. In *Model Theory and Algebra*, a memorial tribute to A. Robinson, 307–387. *LNM* **498**, Springer, 1975.

5. Y. Sato. A new type of canonical Gröbner bases in polynomial rings over Von Neumann regular rings. *Proceedings of the 1998 International Symposium on Symbolic and Algebraic Computation (ISSAC 98)*, 317–321. ACM Press, 1998.
6. Y. Sato, A. Suzuki. Gröbner bases in polynomial rings over Von Neumann regular rings—their applications—(extended abstract). In X. Gao, D. Wang eds., *Proceedings of the Fourth Asian Symposium on Computer Mathematics (ASCM 2000)*, 59–63. *Lecture Notes Series on Computing, Vol. 8.* World Scientific, Singapore, 2000.
7. Y. Sato, A. Suzuki. Discrete comprehensive Gröbner bases. *Proceedings of the 2001 International Symposium on Symbolic and Algebraic Computation (ISSAC 2001)*, 292–296. ACM Press, 2001.
8. A. Suzuki, Y. Sato. An alternative approach to comprehensive Gröbner bases. *Proceedings of the 2002 International Symposium on Symbolic and Algebraic Computation (ISSAC 2002)*, 255–261. ACM Press, 2002.
9. V. Weispfenning. Gröbner bases in polynomial ideals over commutative regular rings. In J. H. Davenport ed., *Proceedings of EUROCAL '87*, 336–347. *LNCS* **378**. Springer, 1987.
10. V. Weispfenning. Comprehensive Gröbner bases. *J. Symb. Comp.* **14**(1) (1992), 1–29.

COMPUTATIONAL ASPECTS OF HYPERELLIPTIC CURVES

T. SHASKA
Department of Mathematics
University of California at Irvine
E-mail: tshaska@math.uci.edu

We introduce a new approach of computing the automorphism group and the field of moduli of points $\mathfrak{p} = [C]$ in the moduli space of hyperelliptic curves $\mathcal{H}_g$. Further, we show that for every moduli point $\mathfrak{p} \in \mathcal{H}_g(L)$ such that the reduced automorphism group of $\mathfrak{p}$ has at least two involutions, there exists a representative C of the isomorphism class $\mathfrak{p}$ which is defined over L.

1. Introduction

The purpose of this note is to introduce some new techniques of computing the automorphism group and the field of moduli of genus g hyperelliptic curves. Former results by many authors have focused on hyperelliptic curves of small genus, see [8,3,7,10,12], *et al.* We aim to find a method which would work for any genus.

Let C denote a genus g hyperelliptic curve defined over an algebraically closed field k of characteristic zero and $G := \mathrm{Aut}\,(C)$ its automorphism group. We denote by $\mathcal{H}_g$ the moduli space of genus g hyperelliptic curves and by $\mathcal{L}_g$ the locus in $\mathcal{H}_g$ of hyperelliptic curves with extra involutions. The locus $\mathcal{L}_g$ is a g-dimensional rational variety, see [6]. Equation (2.2) gives a normal form for curves in $\mathcal{L}_g$. This normal form depends on parameters $a_1, \dots, a_g \in k$, such that the discriminant of the right side $\Delta(a_1, \dots, a_g) \neq 0$. Dihedral invariants $(u_1, \dots, u_g)$ were introduced by Gutierrez and this author in [6]. The tuples $\mathfrak{u} = (u_1, \dots, u_g)$ (such that $\Delta_{\mathfrak{u}} \neq 0$) are in one-to-one correspondence with isomorphism classes of genus g hyperelliptic curves with automorphism group the Klein 4-group. Thus, dihedral invariants $u_1, \dots, u_g$ yield a birational parameterization of the locus $\mathcal{L}_g$. Computationally these invariants give an efficient way of determining a generic point of the moduli space $\mathcal{L}_g$. Normally, this is accomplished by invariants of $GL_2(k)$ acting on the space of binary forms of degree $2g+2$. These $GL_2(k)$-invariants are not known for $g \geq 3$. However, dihedral invariants are explicitly defined for all genera.

The full automorphism groups of hyperelliptic curves are determined in [2] and [1]. Most of these groups have non-hyperelliptic involutions (i.e., the corresponding curve is in $\mathcal{L}_g$). For each group G that occurs as a full automorphism group of genus g curves, one determines the G-locus in $\mathcal{L}_g$ in terms of the dihedral invariants. Given a genus g curve C we first determine if $C \in \mathcal{L}_g$. Then we compute its dihedral invariants and determine the locus $\mathcal{L}_G$ that they satisfy. This determines Aut (C). Present algorithms of computing the automorphism group of a hyperelliptic curve $Y^2 = F(X)$ are based on computing the roots of $F(X)$ and then finding fractional linear transformations that permute these roots. The algorithm we propose requires only determining the normal form of C (i.e., Eq. (2.2)). This requires solving a system of g-equations and four unknowns. For curves which have at least two involutions in their reduced automorphism group we find a nice condition on the dihedral invariants.

For $C \notin \mathcal{L}_g$ similar methods can be used. If $|\text{Aut}\,(C)| > 2$ and $C \notin \mathcal{L}_g$, then C has an automorphism of order N, where N is as in Lemma 3.5. For small genus these curves can be classified by ad-hoc methods. In general one needs to find invariants of such spaces for all $N > 2$ and implement similar methods as above. We intend this as the object of further research.

In Section 4, we introduce how to compute the field of moduli of genus g hyperelliptic curves with automorphism group of order > 4. Let $\mathcal{M}_g$ (resp., $\mathcal{H}_g$) be the moduli space of algebraic curves (resp., hyperelliptic curves) of genus g defined over k and L a subfield of k. It is well known that $\mathcal{M}_g$ (resp., $\mathcal{H}_g$) is a $3g-3$ (resp., $2g-1$) dimensional variety. If C is a genus g curve defined over L, then clearly $[C] \in \mathcal{M}_g(L)$. However, the converse is not true. In other words, the moduli space $\mathcal{M}_g$ of algebraic curves of genus g is a coarse moduli space.

The answer is not obvious if we restrict ourselves to the singular points of $\mathcal{M}_g$. Singular points of $\mathcal{M}_g$ (resp., $\mathcal{H}_g$) correspond to isomorphism classes of curves with nontrivial automorphism groups (resp., automorphism groups of order > 2). In general, we conjecture that for a singular point $\mathfrak{p} \in \mathcal{M}_g(L)$ (resp., $\mathfrak{p} \in \mathcal{H}_g(L)$) there is always a curve C defined over L which correspond to $\mathfrak{p}$. We focus on $\mathcal{H}_g$. A point $\mathfrak{p} = [C] \in \mathcal{H}_g$ is given by the g-tuple of dihedral invariants. We denote by Aut $(\mathfrak{p})$ the automorphism group of any representative C of $\mathfrak{p}$. More precisely, for hyperelliptic curves we conjecture the following:

Conjecture 1.1. *Let $\mathfrak{p} \in \mathcal{H}_g(L)$ such that $|Aut\,(\mathfrak{p})| > 2$. There exists a representative C of the isomorphism class $\mathfrak{p}$ which is defined over L.*

In this paper we show how dihedral invariants can be used to prove some special cases of this conjecture. A detailed discussion on this problem is intended in [11]. The condition $|\mathrm{Aut}\,(\mathfrak{p})| > 2$ in the above conjecture cannot be dropped. Determining exactly the points $\mathfrak{p} \in \mathcal{H}_g \setminus \mathcal{L}_g$ where such rational model C does not exist is still an open problem. For $g = 2$, Mestre [8] found an algorithm which determines such points. It is based on classical invariants of binary sextics.

Notation 1.2. Throughout this paper k denotes an algebraically closed field of characteristic zero, g an integer ≥ 2, and C a hyperelliptic curve of genus g. The moduli space of curves (resp., hyperelliptic curves) defined over k is denoted by $\mathcal{M}_g$ (resp., $\mathcal{H}_g$). Further, V_4 denotes the Klein 4-group and D_{2n} (resp., $\mathbb{Z}_n$) the dihedral group of order $2n$ (resp., cyclic group of order n).

2. Dihedral Invariants of Hyperelliptic Curves

Let k be an algebraically closed field of characteristic zero and C be a genus g hyperelliptic curve given by the equation $Y^2 = F(X)$, where $\deg(F) = 2g+2$. Denote the function field of C by $K := k(X,Y)$. Then, $k(X)$ is the unique degree 2 genus zero subfield of K. We identify the places of $k(X)$ with the points of $\mathbb{P}^1 = k \cup \{\infty\}$ in the natural way (the place $X = \alpha$ gets identified with the point $\alpha \in \mathbb{P}^1$). Then, K is a quadratic extension field of $k(X)$ ramified exactly at $n = 2g+2$ places $\alpha_1, \ldots, \alpha_n$ of $k(X)$. The corresponding places of K are called the *Weierstrass points* of K. Let $\mathcal{P} := \{\alpha_1, \ldots, \alpha_n\}$. Thus, $K = k(X,Y)$, where

$$Y^2 = \prod_{\alpha \in \mathcal{P},\ \alpha \neq \infty} (X - \alpha). \tag{2.1}$$

Let $G = \mathrm{Aut}\,(K/k)$. Since $k(X)$ is the only genus 0 subfield of degree 2 of K, then G fixes $k(X)$. Thus, $G_0 := Gal(K/k(X)) = \langle z_0 \rangle$, with $z_0^2 = 1$, is central in G. We define *the reduced automorphism group of K* to be the group $\overline{G} := G/G_0$. Then, $\overline{G}$ is naturally isomorphic to the subgroup of $\mathrm{Aut}\,(k(X)/k)$ induced by G. We have a natural isomorphism $\Gamma := PGL_2(k) \to \mathrm{Aut}\,(k(X)/k)$. The action of Γ on the places of $k(X)$ corresponds under the above identification to the usual action on $\mathbb{P}^1$ by fractional linear transformations $t \mapsto \frac{at+b}{ct+d}$. Further, G permutes $\alpha_1, \ldots, \alpha_n$. This yields an embedding $\overline{G} \hookrightarrow S_n$.

Because K is the unique degree 2 extension of $k(X)$ ramified exactly at $\alpha_1, \ldots, \alpha_n$, each automorphism of $k(X)$ permuting these n places extends to an automorphism of K. Thus, $\overline{G}$ is the stabilizer in Aut $(k(X)/k)$ of the set $\mathcal{P}$. Hence under the isomorphism $\Gamma \mapsto \text{Aut}\,(k(X)/k)$, $\overline{G}$ corresponds to the stabilizer $\Gamma_{\mathcal{P}}$ in Γ of the n-set $\mathcal{P}$.

An *extra involution of* K is an involution in G which is different from z_0 (the hyperelliptic involution). If z_1 is an extra involution and z_0 the hyperelliptic one, then $z_2 := z_0\, z_1$ is another extra involution. So the extra involutions come naturally in pairs. Suppose z_1 is an extra involution of K. Let $z_2 := z_1\, z_0$, where z_0 is the hyperelliptic involution. Then $K = k(X, Y)$ with equation

$$Y^2 = X^{2g+2} + a_g X^{2g} + \ldots + a_1 X^2 + 1, \tag{2.2}$$

see [6]. The dihedral group $H := D_{2g+2} = \langle \tau_1, \tau_2 \rangle$ acts on $k(a_1, \ldots, a_g)$ as follows:

$$\tau_1: \quad a_i \to \varepsilon^{2i} a_i, \qquad \text{for} \quad i = 1, \ldots, g;$$

$$\tau_2: \quad a_i \to a_{g+1-i}, \qquad \text{for} \quad i = 1, \ldots, [\frac{g+1}{2}].$$

The fixed field $k(a_1, \ldots, a_g)^H$ is the same as the function field of the variety $\mathcal{L}_g$. The invariants of such action are

$$u_i := a_1^{g-i+1}\, a_i + a_g^{g-i+1}\, a_{g-i+1}, \qquad \text{for} \quad 1 \leq i \leq g. \tag{2.3}$$

These are called the *dihedral invariants* for the genus g, and the tuple

$$\mathfrak{u} := (u_1, \ldots, u_g)$$

is called the *tuple of dihedral invariants*, see [6] for details.

It is easily seen that $\mathfrak{u} = 0$ if and only if $a_1 = a_g = 0$. In this case replacing a_1, a_g by a_2, a_{g-1} in the formula above would give new invariants. In [6] it is shown that $k(\mathcal{L}_g) = k(u_1, \ldots, u_g)$. The $(2g+2)$-degree field extension $k(a_1, \ldots, a_g)/k(u_1, \ldots, u_g)$ has equation

$$2^{g+1}\, a_g^{2g+2} - 2^{g+1}\, u_1\, a_g^{g+1} + u_g^{g+1} = 0 \tag{2.4}$$

and the map

$$\Phi : k \setminus \{\Delta \neq 0\} \to \mathcal{L}_g$$
$$\text{given by} \quad (a_1, \ldots, a_g) \mapsto (u_1, \ldots, u_g)$$

has Jacobian zero exactly on points which correspond to curves $C \in \mathcal{L}_g$ such that $V_4 \hookrightarrow \overline{G}$.

3. Automorphism Groups

In this section we suggest an algorithm for computing the full automorphism group of hyperelliptic curves. Let C be a genus g hyperelliptic curve with equation $Y^2 = F(X)$ where $\deg(F) = 2g+2$. Existing algorithms are based on finding all automorphisms of C. Instead, we search for only one automorphism (non-hyperelliptic) of C of order N. Most of the time $N = 2$ is enough since the majority of groups of order > 2 that occur as full automorphism groups have non-hyperelliptic involutions. It is well known that the order of a non-trivial automorphism of a hyperelliptic curve is $2 \leq N \leq 2(2g+1)$, where $2(2g+1)$ is known as the Wiman's bound.

If an automorphism of order $N = 2$ exists then $C \in \mathcal{L}_g$ and we use dihedral invariants to determine the automorphism group. We illustrate with curves of small genus.

The case $g = 2$ has been studied in [12]. Every point in $\mathcal{M}_2$ is a triple (i_1, i_2, i_3) of absolute invariants. We state the results of [12] without proofs.

Lemma 3.1. *Let C be a genus 2 curve such that $G := Aut(C)$ has an extra involution and $\mathfrak{u} = (u_1, u_2)$ its dihedral invariants. Then,*

(a) *$G \cong \mathbb{Z}_3 \rtimes D_8$ if and only if $(u_1, u_2) = (0, 0)$, or $(u_1, u_2) = (6750, 450)$;*
(b) *$G \cong GL_2(3)$ if and only if $(u_1, u_2) = (-250, 50)$;*
(c) *$G \cong D_{12}$ if and only if $u_2^2 - 220u_2 - 16u_1 + 4500 = 0$, for $u_2 \neq 18, 140 + 60\sqrt{5}, 50$;*
(d) *$G \cong D_8$ if and only if $2u_1^2 - u_2^3 = 0$, for $u_2 \neq 2, 18, 0, 50, 450$. Cases $u_2 = 0, 450$ and $u = 50$ are reduced to cases* (a) *and* (b) *respectively.*

The mapping $\Phi : (u_1, u_2) \to (i_1, i_2, i_3)$, parameterization of $\mathcal{L}_2$. The fibers of Φ of cardinality > 1 correspond to those curves C with $|\mathrm{Aut}(C)| > 4$. Dihedral invariants u_1, u_2 are given explicitly as rational functions of i_1, i_2, i_3. The curve $Y^2 = X^6 - X$ is the only genus 2 curve (up to isomorphism) which has extra automorphisms and is not in $\mathcal{L}_2$. The automorphism group in this case is $\mathbb{Z}_{10}$, see [12]. Thus, if $C \in \mathcal{L}_2$ we determine $\mathrm{Aut}(C)$ via Lemma 3.1; otherwise C is isomorphic to $Y^2 = X^6 - X$ or $\mathrm{Aut}(C) \cong \mathbb{Z}_2$.

The case $g = 3$ is given as an application in [6]. Let $C \in \mathcal{L}_3$ with equation as in Eq. (2.2). Dihedral invariants are

$$u_1 = a_1^4 + a_3^4, \quad u_2 = (a_1^2 + a_3^2)a_2, \quad u_3 = 2a_1a_3.$$

The analogue of Lemma 3.1 is proved in [6] for $g = 3$.

This technique can be used successfully for all g. We have implemented programs that determine $\mathrm{Aut}\,(C)$ for $C \in \mathcal{L}_g$ and for $g = 2, 3, 4, 5, 6$. In order to compute the automorphism group of a curve $C \in \mathcal{L}_g$ we transform this curve to its normal form (i.e., Eq. (2.2)) and then compute its dihedral invariants. If these invariants satisfy any locus $\mathcal{L}_G$ then the automorphism group is G, otherwise the automorphism group is V_4. The following lemma determines a nice condition for $\overline{G}$ to have at least two involutions.

Lemma 3.2. *For a curve $C \in \mathcal{L}_g$ the reduced automorphism group has at least two involutions if and only if*

$$(2^{g-1}u_1^2 + u_g^{g+1})\,(2^{g-1}u_1^2 - u_g^{g+1}) = 0 \tag{3.3}$$

Proof. Let $C \in \mathcal{L}_g$. Then, there is an involution $z_1 \in \overline{G}$ which fixes no Weierstrass points of C, see the proof of Lemma 1 in [6]. Thus, $z_1(X) = -X$. Let $z_2 \neq z_1$ be another involution in $\overline{G}$. Since, $z_2 \neq z_1$ then $z_2(X) = \frac{m}{X}$, where $m^2 = 1$. Then, $V_4 = \langle z_1, z_2 \rangle \hookrightarrow \overline{G}$ and z_2 or $z_1\, z_2$ is the transformation $X \mapsto \frac{1}{X}$, say $z_2(X) = \frac{1}{X}$. If g is odd, we have $\mathcal{P} = \{\pm\alpha_1, \pm\frac{1}{\alpha_1}, \dots, \pm\alpha_n, \pm\frac{1}{\alpha_n}\}$, where $n = [\frac{g+1}{2}]$; otherwise $\mathcal{P}$ contains also two points $\pm P$. Thus, $\pm P$ can be either fixed or permuted by $z_2(X) = \frac{1}{X}$. Hence, they are ± 1 or $\pm I$, where $I^2 = 1$. The equation of C is given by

$$Y^2 = \prod_{i=1}^{n}(X^4 - \lambda_i X^2 + 1), \quad \textit{if } g \textit{ is odd,}$$

$$Y^2 = (X^2 \pm 1)\prod_{i=1}^{n}(X^4 - \lambda_i X^2 + 1), \quad \textit{if } g \textit{ is even.}$$

Let $s := \lambda_1 + \ldots + \lambda_n$. If g is odd then $a_1 = a_g = -s$. Then, $u_1 = 2s^{g+1}$ and $u_g = 2s^2$ and they satisfy Eq. (3.3). If $z_2(X) = \frac{1}{X}$ fixes two points of $\mathcal{P}$ then one of the factors of the equation is $X^4 - 1$. Then, $a_1 = (-1)^{\frac{1}{g+1}}\, s$ and $a_g = (-1)^{\frac{1}{g+1}}\, s$. Hence, $a_g^{g+1} = a_g^{g+1} = -s^{g+1}$ and $u_1 = -2s^{g+1}$, $u_g = -2s^2$. Then, $2^{g-1}u_1^2 + u_g^{g+1} = 0$.

If g is even and $\{\pm 1\} \subset \mathcal{P}$ then $a_1 = a_g = s + 1$. If $\{\pm I\} \subset \mathcal{P}$ then $a_1 = a_g = 1 - s$. In both cases $2^{g-1}u_1^2 - u_g^{g+1} = 0$. The converse goes similarly. □

Remark 3.4. If $2^{g-1}u_1^2 + u_g^{g+1} = 0$, then one of the involutions z_2, $z_1 z_2$ of $\overline{G}$ lifts to an element of order 4 in G. If $2^{g-1}u_1^2 - u_g^{g+1} = 0$ both of them lift to involutions in G.

For $C \notin \mathcal{L}_g$ we check if C has automorphisms of order $3 \leq N \leq 2(2g+1)$, see Wiman [15]. The following lemma is a consequence of [2] and gives possible values for N. We only sketch the proof.

Lemma 3.5. *Let C be a genus g hyperelliptic curve with an automorphism of order $N > 2$. Then either $N = 3, 4$ or one of the following holds;*

(a) $N \mid (2g+1)$ *or* $N|2g$ *and* $N < g$ *(then* $Aut(C) \cong \mathbb{Z}_{2N}$*)*
(b) $N \mid 2g$ *and* N *is an even number such that* $6 \leq N \leq 2g-2$.
(c) $N = 4N'$ *such that* $N'|g$ *and* $N' < g$.

Proof. Let C be a genus g hyperelliptic curve with extra automorphisms such that $C \notin \mathcal{L}_g$. Then, the automorphism group of C is isomorphic to one of the following: $SL_2(3)$, $SL_2(5)$, W_3, $H_{N/2}$, $U_{N/2}$, $G_{N/2}$, $\mathbb{Z}_{2N}$ where $N \,|\, 2g+1$ or $N \,|\, 2g$ and $N < g$; see [2] for definitions of these groups. All other groups listed in Table 2 in [2] contain at least two involutions, hence they correspond to curves in $\mathcal{L}_g$. The only groups in the above list that might not contain an element of order 2, 3, or 4 are $U_{N/2}$, $G_{N/2}$. The group $G_{N/2}$ (resp., $U_{N/2}$) has an element of order N where N is as above. $\square$

To have a complete algorithm that works for any $g \geq 2$, one needs to classify (up to isomorphism) curves of genus g which are not in the locus $\mathcal{L}_g$. In order to do this, we need invariants which classify isomorphism classes of curves with an automorphism of order $N > 2$. However, for small genus ad-hoc methods can be used to identify such groups.

4. Field of Moduli

In this section we introduce a method to compute the field of moduli of hyperelliptic curves with extra automorphisms. Until recently this was an open problem even for $g = 2$. Further, we state some open questions for higher genus and prove Conjecture 1.1 for $\mathfrak{p} \in \mathcal{H}_g$ such that the reduced automorphism group of $\mathfrak{p}$ has at least two involutions.

Let C be a genus g hyperelliptic curve defined over k. We can write the equation of C as follows

$$Y^2 = X(X-1)(X^{2g-1} + c_{2g-2}X^{2g-2} + \ldots + c_1 X + c_0)$$

where the discriminant Δ of the right side is nonzero. Then, there is a map

$$\Phi_1 : \; k^{2g-1} \setminus \{\Delta \neq 0\} \to \mathcal{H}_g$$
$$\text{given by } \; (c_0, \ldots, c_{2g-2}) \mapsto \mathfrak{p} = [C]$$

of degree $d = 4g(g+1)(2g+1)$. We denote by J_Φ the Jacobian matrix of a map Φ. Then Conjecture 1.1 can be restated as follows:

Conjecture 4.1. *For each $\mathfrak{p}$ in the locus $\det(J_{\Phi_1}) = 0$ such that $\mathfrak{p} \in \mathcal{H}_g(L)$ there exists a representative C of the isomorphism class $\mathfrak{p}$ which is defined over L.*

For $g = 2$ this conjecture is a theorem as shown in [3]. The main result in [3] is to prove the case when the automorphism group is V_4. A method of Mestre is generalized which uses covariants of order 2 of binary sextics and a result of Clebsch. Such a method probably could be generalized to higher genus as claimed by Mestre [8] and Weber [14].

Remark 4.2. There is a mistake in the proof of Theorem 2 in [3]. In other words, the proof is incorrect when the Clebsch invariant $C_{10} = 0$. However, it can easily be fixed. A correct version of the algorithm has been implemented in *Magma* by P. van Wamelen.

For $g = 3$ the conjecture is proven by Gutierrez and this author [6] for all points $\mathfrak{p}$ with $|\mathrm{Aut}\,(\mathfrak{p})| > 4$. The proof uses dihedral invariants of hyperelliptic curves. A generalization of the method used in [8], [14] for $\mathfrak{p} \in \mathcal{H}_3$ such that $\mathrm{Aut}\,(\mathfrak{p}) \cong V_4$ would complete the case $g = 3$.

Next we focus on the locus $\mathcal{L}_g$. Let $C \in \mathcal{L}_g$. Then, C can be written in the normal form as in equation (2.2). The map

$$\Phi : k^g \setminus \{\Delta \neq 0\} \to \mathcal{L}_g, \quad \text{given by } (a_1, \dots, a_g) \mapsto (u_1, \dots, u_g)$$

has degree $d = 2g+2$. We ask a similar question as in Conjecture 4.1. Let $\mathfrak{p}$ be in the locus $\det(J_{\Phi_1}) = 0$ such that $\mathfrak{p} \in \mathcal{H}_g(L)$. Is there a representative C of the isomorphism class $\mathfrak{p}$ which is defined over L?

The determinant of the Jacobian matrix is

$$\det(J_\Phi) = (2^{g-1}u_1^2 + u_g^{g+1})\,(2^{g-1}u_1^2 - u_g^{g+1}).$$

The locus $\det(J_\Phi) = 0$ corresponds exactly to the hyperelliptic curves with $V_4 \hookrightarrow \overline{G}$ as shown by Lemma 3.2.

Theorem 4.3. *For each $\mathfrak{p}$ in the locus $\det(J_\Phi) = 0$ such that $\mathfrak{p} \in \mathcal{H}_g(L)$ there exists a representative C of the isomorphism class $\mathfrak{p}$ which is defined over L. Moreover, the equation of C over L is given by*

$$C: \quad Y^2 = u_1 X^{2g+2} + u_1 X^{2g} + u_2 X^{2g-2} + \dots \pm u_g X^2 + 2, \tag{4.4}$$

where the coefficient of X^2 is u_g (resp., $-u_g$) when $2^{g-1}u_1^2 - u_g^{g+1} = 0$ (resp., $2^{g-1}u_1^2 + u_g^{g+1} = 0$).

Proof. Let $\mathfrak{p} = (u_1, \ldots, u_g) \in \mathcal{L}_g(L)$ such that $2^{g-1}u_1^2 - u_g^{g+1} = 0$. All we need to show is that the dihedral invariants of C satisfy the locus $\det(J_\Phi) = 0$. By the appropriate transformation C can be written as

$$Y^2 = X^{2g+2} + \left(\frac{u_1}{2}\right)^{\frac{1}{g+1}} \cdot X^{2g} + \sum_{i=1}^{g-1} \frac{u_{g+1-i}}{u_1} \cdot \left(\frac{u_1}{2}\right)^{\frac{g+1-i}{g+1}} \cdot X^{2i} + 1.$$

Then, its dihedral invariants are

$$u_1(C) = \frac{u_1}{2} + \left(\frac{u_g}{u_1}\right)^{g+1} \cdot \left(\frac{u_1}{2}\right)^g = \frac{2^{g-1}u_1^2 + u_g^{g+1}}{2^g u_1}, \quad u_g(C) = u_g.$$

Substituting $u_g^{g+1} = 2^{g-1}u_1^2$ we get $u_1(C) = u_1$. Thus, C is in the isomorphism class determined by $\mathfrak{p}$ and defined over L.

Let $\mathfrak{p} = (u_1, \ldots, u_g) \in \mathcal{L}_g(L)$ such that $2^{g-1}u_1^2 + u_g^{g+1} = 0$. This case occurs only when g is odd, see the proof of Lemma 3.2. We transform C as above and have $u_1(C) = u_1$ and $u_g(C) = -u_g$. They are the other tuple $(u_1, ..., -u_g)$ corresponding to $\mathfrak{p}$. This completes the proof. □

The following is a consequence of Lemma 3.2. and Theorem 4.3.

Corollary 4.5. *Conjecture 1.1 holds for all $p \in \mathcal{L}_g$ such that the reduced automorphism group of $\mathfrak{p}$ has at least two involutions.*

5. Closing Remarks

Conjecture 1.1 was stated for the first time during a talk of the author in ANTS V, see [9]. It can be generalized to $\mathcal{M}_g$ instead of $\mathcal{H}_g$. However, little is known about the loci $\mathcal{M}_G$ (i.e., locus of curves in $\mathcal{M}_G$ with full automorphism group G). In [7] we introduce an algorithm that would classify such groups G for all g and give a complete list of "large" groups for $g \leq 10$. However, finding invariants that classify curves with automorphism group G is not an easy task, since the equations describing non-hyperelliptic curves are more complicated then the hyperelliptic case. A more theoretical approach on singular points of $\mathcal{M}_g$ probably would produce better results on Conjecture 1.1. At this time we are not aware of any such results.

Our approach would work (with necessary adjustments) even in positive characteristic. However, the goal of this note was to introduce such method rather than explore it to the full extent.

Computationally, dihedral invariants give an efficient way of determining a point of the moduli space $\mathcal{L}_g$. Using such invariants in positive characteristic could have applications in the arithmetic of hyperelliptic curves, including cryptography.

Acknowledgments

This paper was written during a visit at the University of Florida. I want to thank the Department of Mathematics at the University of Florida for their hospitality.

References

1. R. Brandt, H. Stichtenoch. Die Automorphismengrupen hyperelliptischer Kurven. *Manuscripta Math* **55** (1986), no. 1, 83–92.
2. E. Bujalance, J. M. Gamboa, G. Gromadzki. The full automorphism groups of hyperelliptic Riemann surfaces. *Manuscripta Math.* **79**(3-4) (1993), 267–282.
3. G. Cardona, J. Quer. Field of moduli and field of definition for curves of genus 2. Article math.NT/0207015.
4. A. Clebsch. *Theorie der Binären Algebraischen Formen*. Verlag von B.G. Teubner, Leipzig (1872).
5. P. Débes, M. Emsalem. On fields of moduli of curves. *J. Algebra* **211**(1) (1999), 42–56.
6. J. Gutierrez, T. Shaska. Hyperelliptic curves with extra involutions. 2002, (submitted).
7. K. Magaard, T. Shaska, S. Shpectorov, H. Völklein. The locus of curves with prescribed automorphism group. In H. Nakamura ed., *Communications in Arithmetic Fundamental Groups and Galois Theory*, 112–141. *RIMS Series* **6**, No. 1267. Kyoto University, 2002.
8. J. F. Mestre. Construction de courbes de genre 2 á partir de leurs modules. (French). In T. Mora, C. Traverso, eds., *Effective Methods in Algebraic Geometry* (*Castiglioncello, 1990*), 313–334. *Prog. Math.* **94**. Birkhäuser, Boston, 1991.
9. T. Shaska. Genus 2 curves with (3,3)-split Jacobian and large automorphism group, LNCS, **2369** (2002), 205–218.
10. T. Shaska. Genus 2 fields with degree 3 elliptic subfields, Forum Math., 2002, (in press).
11. T. Shaska. Field of moduli of hyperelliptic curves (in preparation).
12. T. Shaska, H. Völklein. Elliptic subfields and automorphisms of genus 2 function fields. *Algebra and Algebraic Geometry with Applications*, LNCS, (2002), (in press).
13. T. Shioda. Constructing curves with high rank via symmetry. *Amer. J. Math.* **120**(3) (1998), 551–566.
14. H. J. Weber. Hyperelliptic simple factors of $J_0(N)$ with dimension at least 3. *Experiment. Math* **6**(4) (1997), 273–287.
15. A. Wiman. Über die hyperelliptischen Curven vom den Geschlechte $p = 4, 5,$ und 6, welche eindeutige Transformationen in sich besitzen. *Bihang Kongl. Svenska Vetenskaps-Akademiens Handlingar* **21**(3) (1985), 1–41.

APPLICATION OF THE WU-RITT DIFFERENTIAL ELIMINATION METHOD TO THE PAINLEVÉ TEST*

FUDING XIE, HONGQING ZHANG, YONG CHEN, BIAO LI

Department of Applied Mathematics,
Dalian University of Technology,
Dalian 116024, P. R. China

The Painlevé property and Painlevé test are interesting topics for nonlinear differential equations arising from physics. In general, a formal Laurent series solution is supposed and recursion relations of the coefficients are derived. Whether a given equation possesses the Painlevé property or not can be determined by analyzing the resonance equations. In this paper, a new constructive method is proposed to judge whether a given equation passes the Painlevé test by expanding only finitely many terms and using the Wu-Ritt differential elimination method.

1. Introduction

The singularities of solutions of ordinary differential equation (ODE) are classified according to their nature (pole or zero, branch point, essential singularity) and their type (fixed or movable). A given ODE is said to *possess the Painlevé property* when its solutions have only a movable singular pole [1]. The Laurent series of its solution at the singular point z_0 is as follows:

$$u(z) = \frac{a_{-N}}{(z-z_0)^N} + \frac{a_{-N+1}}{(z-z_0)^{N-1}} + \ldots + \frac{a_{-1}}{z-z_0} + \sum_{i=0}^{\infty} a_n (z-z_0)^n.$$

When an (ODE) system possesses the Painlevé property, the system will be "integrable." M. J. Albowitz, A. Ramani and H. Segur [1] have proven that when a partial differential equation is solvable by the inverse scattering transform and a system of ordinary differential equations is obtained from this PDE by an exact similarity reduction, then the solution (of this system

*This work has been supported by the National Key Basic Research Development Project Foundation of China (Grant No:G1998030600) and the National Natural Science Foundation of China (Grant No:10072013).

of ODE's) associated with the Gel'fand-Levitan-Marchenko equation will possess the Painlevé property. Furthermore, they conjectured that, when all the ODE's obtained by exact similarity transforms from a given PDE have the Painlevé property, perhaps after a change of variables, then the PDE will be "integral" [1]. The extension to PDE of the Painlevé property from the ODE was done by J. Weiss, M. Tabor and G. Carnevale [12] in 1983. This method is generally called the WTC-method. Based on the WTC method, many results have been obtained [2,3,5,6,8,9,10,11].

The WTC-method is briefly described as follows. Let a given PDE be

$$A(u_{z_1}, u_{z_2}, \ldots) = 0, \tag{1.1}$$

and assume that

$$u(z) = \phi^{-p}(z) \sum_{j=0}^{\infty} u_j(z)\phi^j(z), \tag{1.2}$$

where $u_j(z)$ and $\phi(z)$ are analytic functions of $z = (z_1, z_2, \ldots, z_n)$ in a neighborhood on a movable manifold $\phi(z) = 0$, $u_0(z) \neq 0$, and p is an integer.

Substituting Eqn. (1.2) into Eqn. (1.1) and balancing the powers of ϕ, the value of p can be determined. The recursion relations for u_j for $j = 0, 1, 2, \cdots$ can be defined as follows:

$$bG\cdot(j+1)(j-\alpha_1)(j-\alpha_2)\ldots(j-\alpha_k)u_j - F(\phi, u_0, u_1, \ldots\ u_{j-1}) = 0, \tag{1.3}$$

where b is the coefficient of the highest order derivation in Eqn. (1.1), G is a polynomial in the partial derivatives of ϕ, and $-1, \alpha_1, \alpha_2, \ldots, \alpha_k$ are "resonance points." In general, $j = -1$ corresponds to the arbitrary singularity manifold ($\phi = 0$). On the other hand, it is possible to introduce an arbitrary function u_j for every positive resonance point and the compatibility conditions on the functions $(\phi, u_0, u_1, \ldots, u_{j-1})$ are required. Eqn. (1.1) is said to *possess the Painlevé property* if all of the compatibility conditions are satisfied [9,10,11]. It is difficult to find the recursion relations and the compatibility conditions at α_j since both involve extensive calculations.

The basic idea in this article is to expand a finite number of terms from Eqn. (1.2) and to prove the system is auto-compatible at each of the resonance points with the aid of the Wu-Ritt elimination method. The scheme of the algorithm is given in Section 2. In Section 3, we apply this new algorithm to the KdV equation and show that it is an efficient method. In the last section, some conclusions are suggested.

2. A New Truncation Method in Painlevé Test

For convenience, we set $u = u(x,t), u_j = u_j(x,t), \phi = \phi(x,t)$ and introduce

$$u = \phi^{-p} \sum_{j=0}^{n} u_j \phi^j, \tag{2.1}$$

where n is an integer.

Let Ord(PDE) denote the order of the PDE in Eqn. (1.1). We generally set $n \geq$ Ord(PDE)+1 so that the equations for resonance points are over-determined. If the Ord(PDE) is k and the PDE has the Painlevé property, there are exactly k resonance points. When there are h resonance points in $\{1, \ldots, k+1\}$, $k-h$ other resonance points will have to be determined. The number of equations for the resonance points is $k-h+1$. So there is sufficient information to determine the rest.

We substitute Eqn. (2.1) into Eqn. (1.1) and multiply the result by a suitable power of ϕ so that the lowest power of ϕ is zero in the final expression. So we have

$$P_0 + P_1\phi + P_2\phi^2 + \ldots + P_n\phi^n + \ldots + P_m\phi^m = 0 \tag{2.2}$$

where m is some integer, and for $i = 0, 1, 2, \ldots, m$, each P_i is a polynomial function of $u_0, u_1, \ldots, u_j$, with j no bigger than the maximum of n and i, and of the derivatives $\phi_x, \phi_t, \ldots$ of ϕ.

We can generally solve for u_0 by setting $P_0 = 0$ and substitute it into P_j for $j > 0$. The set of P_i's, $i = 1, \ldots, m$, may be divided into two parts:

$$C_1 = \{P_j \mid u_j \text{ occurs effectively in } P_j\},$$
$$C_2 = \{P_j \mid u_j \text{ does not occur in } P_j\}.$$

Obviously, the set C_2 is a set of compatibility conditions.

Let $\alpha_1, \alpha_2, \ldots, \alpha_k$ be the set of subscripts j for which P_j in the set C_2. Let $R = \{-1, \alpha_1, \ldots, \alpha_k\}$. Obviously, j is a resonance point. In fact, the set R will include all the positive resonance points if n is big enough. The question is how to find all the resonance points for a fixed n (for example $n = 10$).

Let $j = j_1, \ldots, j_s$ be all the non-zero subscripts for which $P_j \in C_1$, and let β_i be the integer coefficient of u_{j_i} in P_{j_i}. Let

$$A = \{(\beta_1, j_1), (\beta_2, j_2), \ldots, (\beta_s, j_s)\}.$$

We set

$$r_i = \frac{\beta_i}{b(j_i+1)(j_i-\alpha_1)\ldots(j_i-\alpha_k)} \quad (i = 1, 2, \ldots, s). \tag{2.3}$$

If every $r_i = 1$ for $i = 1, \ldots, s$, the set R would cover all the resonance points because β_i would equal to the coefficient of the u_{j_i} in Eqn. (1.3), which includes all the resonance points. We know that the Eqn. (1.1) possesses the Painlevé property if the set C_2 can be reduced to zero w.r.t. C_1. The resonance at j (where j is one of $j_1, \ldots, j_k$) introduces an arbitrary function u_j and a compatibility condition $P_j = 0$ on the functions $(\phi, u_1, \ldots, u_{j-1})$ that requires P_j to vanish identically [12]. This will be true when P_j can be reduced to zero w.r.t. $(P_1, \ldots, P_{j-1})$ with aid of differential remainder formula [12]. Wu-Ritt differential elimination method is just the powerful tool to deal with the problem. Here, we omit the definitions and formulas related to reduction in differential algebra (for details, see [7,4,13]). So the question of whether Eqn. (1.1) possesses the Painlevé property or not is now transformed to one of whether the set C_2 is reduced to zero w.r.t C_1. We use the order called II-type in [13]. The ranking on the variables is such that $u_n \succ u_{n-1} \succ \ldots \succ u_1 \succ \phi$.

Suppose now for some i, $r_i \neq 1$. Let $|R|$ be the cardinality of the set R. If $|R|$=Ord(PDE), it is shown that the PDE does not have the Painlevé property. If $|R| <$ Ord(PDE), the set R does not include all resonance points. Obviously, the value of any omitted resonance point is less than -1 or greater than n. Let $\gamma_1, \ldots, \gamma_q$ be these points and let

$$r_i' = (j_i - \gamma_1)(j_i - \gamma_2)\ldots(j_i - \gamma_q), \quad (i = 1, 2, \ldots, s).$$

The following is true from [2]: Since the j_i and γ_d are integers, we have

$$|\gamma_d| - |j_i| \leq |\gamma_d - j_i| \leq |r_i'|.$$

This means that the absolute value of γ_d is not greater than $|r_i'| + j_i$. It is easy to find the γ_d in this interval.

3. An Example

Consider the KdV equation:

$$u_t + uu_x + u_{xxx} = 0. \tag{3.1}$$

We get $p = 2$ by a leading-term analysis or the homogenous balance method. Let its formal solution be

$$u = \phi^{-2} \sum_{j=0}^{10} u_j \phi^j. \tag{3.2}$$

Substituting Eqn. (3.2) into Eqn. (3.1) and multiplying by ϕ^5, we get with the aid of *Maple 7*,

$$P_0 + P_1\phi + P_2\phi^2 + \ldots + P_{10}\phi^{10} + \ldots + P_{21}\phi^{21} = 0, \qquad (3.3)$$

where

$$\begin{aligned}
P_0 &= -2u_0^2\phi_x - 24u_0\phi_x^3,\\
P_1 &= 30u_1\phi_x^3 - 360\phi_x^3\phi_{xx},\\
P_2 &= 24u_2\phi_x^3 - 6\phi_x^2u_{1x} - 18u_1\phi_x\phi_{xx} - u_1^2\phi_x +\\
&\quad 24\phi_x^2\phi_t + 288\phi_x\phi_{xx}^2 + 168\phi_x^2\phi_{xxx},\\
P_3 &= 12u_3\phi_x^3 - 72\phi_{xx}\phi_{xxx} - 24\phi_x\phi_{xxxx} - u_1\phi_t - 12\phi_x^2u_{2x} - u_1\phi_{xxx} -\\
&\quad 3u_{1x}\phi_{xx} - 24u_2\phi_x\phi_{xx} + u_1u_{1x} - 3u_{1xx}\phi_x - u_2u_1\phi_x - 24\phi_x\phi_{xt},\\
P_4 &= -12\phi_x^2u_{3x} + u_1u_{2x} + -24\phi_x\phi_{xx}u_3 + u_{1t} + u_{1xxx} + u_2u_{1x},\\
P_5 &= -6u_5\phi_x^3 + u_{2t} + 6u_4\phi_x\phi_{xx} + u_2u_{2x} + u_{2xxx} + 6u_{4x}\phi_x^2 +\\
&\quad 3u_{3xx}\phi_x + u_3\phi_{xxx} - 24u_4\phi_x\phi_{xx} + 3u_{3x}\phi_{xx} + u_3\phi_t -\\
&\quad 12\phi_x^2u_{4x} + u_3u_{1x} + u_1u_{3x} + u_1u_4\phi_x + u_2u_3\phi_x,\\
P_6 &= 2u_4\phi_t + u_3^2\phi_x + 2u_4\phi_{xxx} + u_2u_{3x} + 6u_{4x}\phi_{xx} + u_1u_{4x} +\\
&\quad 2u_5u_1\phi_x + u_{3xxx} - 12u_{5x}\phi_x^2 - 6u_5\phi_x\phi_{xx} + 18u_{5x}\phi_x^2 + u_3u_{2x} +\\
&\quad u_4u_{1x} + 6u_{4xx}\phi_x + u_{3t} + 2u_2u_4\phi_x,\\
P_7 &= 24u_7\phi_x^3 + \ldots + 24u_7\phi_{xxx},\\
P_8 &= 72u_8\phi_x^3 + \ldots + 4u_6\phi_{xxx},\\
P_9 &= 150u_9\phi_x^3 + \ldots + 5u_7\phi_{xxx},\\
P_{10} &= 264u_{10}\phi_x^3 + \ldots + 3u_5^2\phi_x.
\end{aligned}$$

The set C_1, C_2, R, A are found to be

$$\begin{aligned}
C_1 &= \{P_1, P_2, P_3, P_5, P_7, P_8, P_9, P_{10}\},\\
C_2 &= \{P_4, P_6\},\\
A &= \{(30,1), (24,2), (12,3), (-6,5), (24,7), (72,8), (150,9), (264,10)\},\\
R &= \{-1, 4, 6\}.
\end{aligned}$$

We can easily examine that $r_i = 1$ for $i = 1, 2, \ldots, 7$. So R covers all the resonance points of Eqn. (3.1).

Obviously, C_2 is already reduced w.r.t $\{P_7, P_8, P_9, P_{10}\}$. Making use of *Maple 7* to compute the remainders, we can see whether C_2 can be reduced to zero w.r.t. $\{P_1, P_2, P_3, P_5\}$ or not. Solving for u_0 in P_0, we substitute it to the other P_i's. The reduction process of P_6 and P_4 are given by

$$\begin{aligned}
R_{43} &= \text{Remainder}(P_4, P_3) \\
&= -\phi_x u_1 \phi_{xt} - 3\phi_x u_{1xx}\phi_{xx} - 4\phi_x u_{1x}\phi_{xxx} \\
&\quad -72\phi_x\phi_{xx}\phi_{xxxx} - \phi_x u_1 \phi_{xxxx} + \phi_{xx}u_1\phi_t + \phi_{xx}u_1\phi_{xxx} \\
&\quad -\phi_{xx}u_1u_{1x} - 12\phi_x^3 u_2 xx - 24\phi_x^2\phi_{xxx} - 24\phi_x^2\phi_{xxxxx} - 72\phi_x\phi_{xxx}^2 \\
&\quad +\phi_x u_{1x}^2 - 2u_{1xxx}\phi_x^2 + \phi_x^2 u_{1t} + 72\phi_{xx}^2\ phi_{xxx} + 3u_{1x}\phi_{xx}^2 \\
&\quad -36\phi_x^2 u_{2x}\phi_{xx} - \phi_x u_{1x}\phi_t - 24u_2\phi_x^2\phi_{xxx} + \phi_x u_1 u_{1xx}, \\
R_{42} &= \text{Remainder}(R_{43}, P_2) \\
&= 12\phi_x^3\phi_t\phi_{xxx} + 300\phi_x^3\phi_{xx}\phi_{xxxx} + 12\phi_x^3\phi_{xx}\phi_{xt} + \phi_x^2\phi_{xx}u_1\phi_t \\
&\quad -12\phi_x^4\phi_{xxt} - 5u_{1xxx}\phi_x^4 + \phi_x^4 u_{1t} + 60\phi_x^4\phi_{xxxxx} + 300\phi_x^3\phi_{xxx}^2 \\
&\quad +10\phi_x^2\phi_{xx}u_1\phi_{xxx} - 300\phi_x^2\phi_{xx}^2\phi_{xxx} + 15\phi_x^2 u_{1x}\phi_{xx}^2 - \phi_x^3 u_1\phi_{xt} \\
&\quad -10\phi_x^3 u_1\phi_{xxxx} - 15\phi_x^3 u_{1xx}\phi_{xx} - 12\phi_x^2\phi_{xx}^2\phi_t - \phi_x^3 u_{1x}\phi_t \\
&\quad -25\phi_x^3 u_{1x}\phi_{xxx}, \\
R_{41} &= \text{Remainder}(R_{42}, P_1) = 0, \\
R_{65} &= \text{Remainder}(P_6, P_5) \\
&= u_1 u_3 u_{1x} - 12\phi_{xx}u_3\phi_{xxx} - 12\phi_{xx}u_2u_{2x} + 3\phi_x u_3 u_{1xx} + u_1u_2u_{2x} \\
&\quad + \cdots (28 \text{ terms}), \\
R_{63} &= \text{Remainder}(R_{65}, P_3) \\
&= 12\phi_x^2\phi_t u_{1x}^2 + 636\phi_x^3 u_1\phi_{xxxt} + 756\phi_x^3 u_{1xx}\phi_{xxxx} \\
&\quad +120\phi_x^2 u_1^2\phi_{xxxxx} - 936\phi_x^3 u_{1xxxx}\phi_{xx} + \cdots (163 \text{ terms}), \\
R_{62} &= \text{Remainder}(R_{63}, P_2) \\
&= 627056640\phi_x^2\phi_{xx}^4\phi_{xxx} - 39813120\phi_x^2\phi_{xx}^4 u_{1x} \\
&\quad -456192\phi_x^4\phi_{xx}^2 u_{1t} - 540\phi_x u_1^4\phi_{xx}^2 + \cdots (111 \text{ terms}), \\
R_{61} &= \text{Remainder}(R_{62}, P_1) = 0,
\end{aligned}$$

where some expressions are too long and are omitted. The set C_2 can be reduced to zero w.r.t C_1. This shows that the system of partial differential equations $\{P_0 = 0, P_1 = 0, \ldots, P_{10} = 0\}$ is compatible, so Eqn. (3.1) possesses the Painlevé property. This result agrees with the one that appeared in [12].

4. Conclusion

The WTC method employs expensive computation and ingenious skills. In this paper, we give another truncation method to decide whether a given equation possesses the Painlevé property without using the recursion relations. It can be implemented on a computer since it is a constructive. This method can also be applied to PDEs other than the KdV equation, such as the KP equation, the MKdV equation, the Bousinesq equation, and Burgers equation.

References

1. M. J. Ablowitz, A. Ramani, H. Segur. A connection between nonlinear evolution equations and ordinary differential equations of P-type I. *J. Math. Phys.* **21** (1980), 715–721.
2. R. Conte. Invariant Painlevé analysis of partial differential equations. *Phys. Lett. A.* **140**(7–8) (1989), 383–390.
3. M. V. Hulstman, W. D. Halford. Exact solutions to KdV equations with variable coefficients and/or nonuniformities. *Comput. Math. Appl.* **29**(1) (1995), 39–47.
4. E. R. Kolchin. *Differential algebra and algebra groups.* Academic Press, New York, 1973.
5. A. Pickering. A new truncation in Painlevé analysis. *J. Phys. A:Math. Gen.* **26** (1993), 4395–4405.
6. A. Pickering. The singular manifold method revisited. *J. Math. Phys.* **37**(4) (1996), 1984–1927.
7. J. F. Ritt. *Differential Algebra.* Dover Publications Inc., New York, 1950.
8. B. Tian, Y. T. Gao, W. Hong. On Backlund transformation for a generalized Burgers equation and solitonic solutions. it Phys. Lett. A. **286**(3) (2000), 81–84.
9. J. Wiess. On classes of integrable systems and the Painlevé property. *J. Math. Phys.* **25**(1) (1984), 13–24.
10. J. Wiess. The Sine-Gordon equations: complete and partial integrability. *J. Math. Phys.* **25**(7) (1984), 2226–2235.
11. J. Wiess. The Painlevé property and Backlund transformations for the sequence of Boussinesq equations. *J. Math. Phys.* **26**(2) (1985), 258–269.
12. J. Weiss, M. Tabor, G. Carnevale. The Painlevé property for partial differential equations. *J. Math. Phys.* **24**(3) (1983), 522–526.
13. W. T. Wu. On the foundation of algebra differential geometry. *MM Research Preprints* **3** (1989).

DEGREE REDUCTION OF RATIONAL CURVES BY μ-BASES*

FANGLING ZENG,† FALAI CHEN

Department of Mathematics
University of Science and Technology of China
Hefei, Anhui 230026, People's Republic of China
†*E-mail: zella@mail.ustc.edu.cn*

The μ-basis is an intermediate representation between the parametric form and the implicit form of planar rational curves (Cox *et al.*, [6]). In this paper, degree reduction of rational curves using μ-basis is considered. We resort to optimal approximation based on Chebyshev polynomials, and find that in certain cases, the degree reduction approach using the μ-basis method results in smaller errors than the direct approach does. Furthermore, the μ-basis of the degree-reduced curve is directly obtained by μ-basis method.

Keywords: rational curves, degree reduction, μ-basis, optimal approximation

1. Introduction

In Computer Aided Geometric Design and Geometric Modelling, there is considerable interest to approximate curves (resp. surfaces) using simpler forms of curves (resp. surfaces). An example is the approximation of rational curves (resp. surfaces) using polynomial curves (resp. surfaces), or of high degree polynomial curves (resp. surfaces) using low degree polynomial ones. This interest is due to practical needs such as the communication of data between diverse CAD/CAM systems. In the past twenty years, there is a wealth of literature focusing on the problem of degree reduction, that is, using low degree curves to approximate high degree curves. We don't review those methods here. The interested reader is referred to [2,9] and the references therein. In this paper, we propose a new approach to solve the degree reduction problem based on the μ-basis method.

*This work is supported by NKBRSF on Mathematical Mechanics (no. G1998030600), Outstanding Youth Grant of NSF of China (no. 60225002), NSF of China (no.19971087), the TRAPOYT in Higher Education Institute of MOE of China, and the Doctoral Program of MOE of China (no. 20010358003).

The μ-basis of a planar rational curve was introduced in (Cox *et al.* [6]) as a new method to implicitize the rational curve. One nice property of the μ-basis is that the parametric equation can be recovered from the μ-basis. Thus, the μ-basis serves as a compact and useful representation of a planar rational curve—connecting its implicit equation and parametric equation, and facilitating the study of many properties of the curve.

In this paper, we apply the μ-basis to the degree reduction of planar rational curves based on an optimal approximation method. Generally, degree reduction by a direct optimal approximation method produces better results than by the μ-basis method. However, when the denominator of the rational curve has a minimal value close to zero, the direct method may behave badly at this point. In this case, a μ-basis can be applied to reduce the error. Furthermore, the μ-basis of the degree-reduced curve by the μ-basis method is directly obtainable without further calculation.

The paper is organized as follows. In the next section, we briefly review the concept of μ-basis. In Section 3, we discuss degree reduction using the direct optimized approximation method. In Section 4, an optimal approximation method is combined with the μ-basis method to present a new algorithm for the degree reduction problem. Error bounds are given for both the direct approximation method and the μ-basis method. Finally in the last section, we provide two examples to compare the two methods. The examples show that in certain cases where the direct approximation method behaves badly, the μ-basis method provides an alternative to reduce the degree of the rational curves with smaller errors. Besides, the μ-basis and the implicit equation of the degree-reduced curve can be directly obtained.

2. The μ-Basis of a Planar Rational Curve

A planar rational curve in homogenous form is given by the following parametric equation

$$\mathbf{P}(t) = (a(t), b(t), c(t)), \qquad t \in [0,1] \tag{2.1}$$

where

$$a(t) = \sum_{i=0}^{n} a_i t^i, \qquad b(t) = \sum_{i=0}^{n} b_i t^i, \qquad c(t) = \sum_{i=0}^{n} c_i t^i \tag{2.2}$$

and n is the degree of the rational curve. We assume a, b, c are relatively prime, and $c(t) > 0$ for $t \in [0,1]$.

Let

$$f = c(t)x - a(t), \quad g = c(t)y - b(t)$$

and consider the ideal I generated by f and g:

$$I := \langle f, g\rangle := \{h_1 f + h_2 g \mid h_1, h_2 \in k[x,y,t]\} \subset k[x,y,t],$$

where $k[x,y,t]$ is the polynomial ring defined over the field of real numbers. Denote by I_{jk} the set of polynomials in I whose total degree is at most j in x, y and degree at most k in t, and let μ be the smallest integer for which $I_{1,\mu}$ is nonzero. It is shown in (Cox *et al.* [6]) that $\mu \le [n/2]$ and

Proposition 2.3. *There exist two polynomials $p \in I_{1,\mu}$ and $q \in I_{1,n-\mu}$ such that p and q form a basis for the ideal $I = \langle p, q\rangle$.*

The pair of polynomials p and q are called the *μ-basis* of the rational curve (2.1). The equations $p = 0$ and $q = 0$ define two pencils of lines which intersect in the curve. If $\mu < n - \mu$, p is unique up to a scalar multiple, and q is unique up to a scalar multiple plus a multiple of p by an element of $k[t]$. If $\mu = n - \mu = n/2$, p and q can be any basis of the two-dimensional vector space $I_{1,n/2}$. Write

$$p = p_x x + p_y y + p_w, \qquad q = q_x x + q_y y + q_w, \tag{2.4}$$

and $\mathbf{p} = (p_x, p_y, p_w)$, $\mathbf{q} = (q_x, q_y, q_w)$. Sometimes, we also call $\mathbf{p}$ and $\mathbf{q}$ the μ-basis of the rational curves (2.1). One nice property of μ-basis is that it can recover the parametric equation of the rational curve. In fact, we have

Proposition 2.5. (*Cox et al.* [6]) *Let $\mathbf{p}, \mathbf{q}$ be a μ-basis of the rational curve $\mathbf{P}(t)$. Then*

$$\mathbf{P}(t) = \mathbf{p}(t) \times \mathbf{q}(t). \tag{2.6}$$

The μ-basis can be computed efficiently using the module computation. The details can be found in (Zheng *et al.* [10]) or (Chen *et al.* [3]).

3. Degree Reduction by Direct Approximation

In this section, we review the degree reduction method by an optimal approximation based on Chebyshev polynomials and derive an error bound for the direct degree reduction.

Given a planar rational curve as defined in (2.1) of degree n, we hope to find a planar rational curve $\widetilde{\mathbf{P}}(t) = (\widetilde{a}(t), \widetilde{b}(t), \widetilde{c}(t))$ of degree $n-1$ such that the norm of the error function $\epsilon(t)$ is minimized. Here

$$\varepsilon(t) := \sqrt{\left(\frac{a(t)}{c(t)} - \frac{\widetilde{a}(t)}{\widetilde{c}(t)}\right)^2 + \left(\frac{b(t)}{c(t)} - \frac{\widetilde{b}(t)}{\widetilde{c}(t)}\right)^2} \tag{3.1}$$

and $||\varepsilon(t)||_\infty = \max_{0\le t\le 1} |\varepsilon(t)|$.

This problem can be solved by Chebyshev polynomials:

$$T_n(t) = \cos(n \arccos(2t - 1)). \tag{3.2}$$

There is a recursive formula for $T_n(t)$:

$$\begin{aligned} T_0(t) = 1, \quad T_1(t) = 2t - 1, \\ T_{n+1}(t) = 2(2t-1)T_n(t) - T_{n-1}(t), \quad n = 1, 2, ... \end{aligned} \tag{3.3}$$

One of the important properties of Chebyshev polynomials $T_n(t)$ is:

Proposition 3.4. *Let π_{n-1} be the set of polynomials whose total degree is less than or equal to $n-1$. Then the best uniform approximation from π_{n-1} to a polynomial $h(t)$ of degree n is*

$$\widetilde{h}(t) = h(t) - \ell(h)T_n(t)/2^{2n-1}, \tag{3.5}$$

where $\ell(h)$ denotes the leading coefficient of h.

The proof of this proposition can be found in [2]. Based upon the above proposition, if we set

$$\begin{aligned} \widetilde{\mathbf{P}}(t) &= (\widetilde{a}(t), \widetilde{b}(t), \widetilde{c}(t)) \\ &= \left(a(t) - a_nT_n(t)/2^{2n-1}, b(t) - b_nT_n(t)/2^{2n-1}, c(t) - c_nT_n(t)/2^{2n-1}\right) \end{aligned} \tag{3.6}$$

then $\widetilde{\mathbf{P}}(t)$ is a good degree reduction curve to the rational curve (2.1). $\widetilde{\mathbf{P}}(t)$ is called the *degree reduction curve by direct method.*

To ensure the degree reduction curve $\widetilde{\mathbf{P}}(t)$ is well defined, $\widetilde{c}(t)$ should satisfy $\widetilde{c}(t) > 0$ for $t \in [0, 1]$. A sufficient condition for this situation is

$$\min_{t \in [0,1]} |c(t)| - |c_n|/2^{2n-1} > 0, \tag{3.7}$$

and we will make the above assumption in the following.

Theorem 3.8. *The degree reduction error by the direct approximation method satisfies*

$$\| \varepsilon(t) \|_\infty \leq \frac{\| (a, b, c) \|_\infty \cdot \| (a_n, b_n, c_n) \|_\infty}{2^{2n-1} \min |c| (\min |c| - |c_n|/2^{2n-1})}, \tag{3.9}$$

where $\min |c| = \min\limits_{0 \leq t \leq 1} |c(t)|$.

Proof. From (3.6), one has

$$\varepsilon(t) = \sqrt{\left(\frac{a(t)}{c(t)} - \frac{\widetilde{a}(t)}{\widetilde{c}(t)}\right)^2 + \left(\frac{b(t)}{c(t)} - \frac{\widetilde{b}(t)}{\widetilde{c}(t)}\right)^2}$$
$$= \frac{|T_n(t)|}{2^{2n-1}|c||c - \frac{c_nT_n(t)}{2^{2n-1}}|}\sqrt{(a_nc - ac_n)^2 + (b_nc - bc_n)^2},$$

$$\sqrt{(a_nc - ac_n)^2 + (b_nc - bc_n)^2} \leq \| (a,b,c) \times (a_n, b_n, c_n) \|_\infty$$
$$\leq \| (a,b,c) \|_\infty \| (a_n, b_n, c_n) \|_\infty,$$

where $\| (x,y,z) \|_\infty = \max_{t\in[0,1]} \sqrt{x^2+y^2+z^2}$, and $|T_n(t)| \leq 1$ for $t \in [0,1]$. Thus (3.9) holds and the theorem is proved. □

Degree reduction by the direct method generally produces good results because of the property of Chebyshev polynomials. However, from the error bound in (3.9) one can see that, at the points where $c(t)$ is sufficiently small, the error may be large. The examples in the last section illustrate that this could happen in practice. In the next section, we introduce the μ-basis method as another technique for degree reduction, which can to some extent help deal with this situation.

4. Degree Reduction by μ-Basis

Write the μ-basis of $\mathbf{p}$ and $\mathbf{q}$ in the form:

$$\mathbf{p} = (p_x, p_y, p_\omega) = \mathbf{p}_0 + \mathbf{p}_1 t + \ldots + \mathbf{p}_\mu t^\mu,$$
$$\mathbf{p}_i = (p_{xi}, p_{yi}, p_{\omega i}), \quad i = 0,1,\ldots\mu,$$
$$\mathbf{q} = (q_x, q_y, q_\omega) = \mathbf{q}_0 + \mathbf{q}_1 t + \ldots + \mathbf{q}_{n-\mu} t^{r-\mu},$$
$$\mathbf{q}_i = (q_{xi}, q_{yi}, q_{\omega i}), \quad i = 0,1,\ldots n-\mu.$$

We use an optimal approximation method to reduce the degree of $\mathbf{q}$ by setting

$$\widetilde{\mathbf{q}} = \mathbf{q} - \mathbf{q}_{n-\mu} T_{n-\mu}(t)/2^{2(n-\mu)-1}. \tag{4.1}$$

Now let

$$\widetilde{\mathbf{P}}(t) = (\widetilde{a}, \widetilde{b}, \widetilde{c}) = \mathbf{p} \times \widetilde{\mathbf{q}}. \tag{4.2}$$

Then $\widetilde{\mathbf{P}}(t)$ also gives a degree reduction curve to (2.1) and is called the *degree reduction curve by μ-basis method.* Note that the μ-basis of $\widetilde{\mathbf{P}}(t)$ is exactly $\mathbf{p}$ and $\widetilde{\mathbf{q}}$. Thus an advantage of μ-basis method is that the μ-basis of the degree-reduced curve is directly obtained without further computation.

Theorem 4.3. *Suppose*

$$\min_{0\le t\le 1}|c| - \max_{0\le t\le 1}|p_x q_{y,n-\mu} - p_y q_{x,n-\mu}|/2^{2(n-\mu)-1} > 0. \tag{4.4}$$

Then the degree reduction error by the μ-basis method satisfies

$$\begin{aligned} &\| \varepsilon(t) \|_\infty \\ &\leqslant \frac{\| (a,b,c) \|_\infty \| \mathbf{p}\times \mathbf{q}_{n-\mu} \|_\infty}{\min|c|\,(2^{2(n-\mu)-1}\min|c| - \max_{0\le t\le 1}|p_x q_{y,n-\mu} - p_y q_{x,n-\mu}|)} \end{aligned} \tag{4.5}$$

where $\min|c| = \min\limits_{0\le t\le 1}|c|$.

Proof. From

$$\begin{aligned} \widetilde{\mathbf{P}}(t) &= (\widetilde{a},\widetilde{b},\widetilde{c}) = \mathbf{p}\times\widetilde{\mathbf{q}} \\ &= \mathbf{p}\times(\mathbf{q} - \mathbf{q}_{n-\mu}T_{n-\mu}(t)/2^{2(n-\mu)-1}) \\ &= \mathbf{P}(t) - \mathbf{p}\times\mathbf{q}_{n-\mu}T_{n-\mu}(t)/2^{2(n-\mu)-1}, \end{aligned}$$

one has

$$\begin{aligned} \varepsilon(t) &\le \frac{1}{|c\widetilde{c}|}\,\| (a,b,c)\times(\widetilde{a},\widetilde{b},\widetilde{c}) \|_\infty \\ &= \frac{|T_{n-\mu}(t)|}{2^{2(n-\mu)-1}|c\widetilde{c}|}\,\| (a,b,c)\times(\mathbf{p}\times\mathbf{q}_{n-\mu}) \|_\infty \\ &\le \frac{|T_{n-\mu}(t)|}{2^{2(n-\mu)-1}|c\widetilde{c}|}\,\| (a,b,c) \|_\infty \| \mathbf{p}\times\mathbf{q}_{n-\mu} \|_\infty . \end{aligned}$$

When condition (4.4) is satisfied, $\widetilde{c}(t) > 0$ for $t\in[0,1]$, and (4.5) follows immediately. □

We especially note that a_n, b_n and c_n are the leading coefficients of the three components of polynomial vector $\mathbf{p}\times\mathbf{q}_{n-\mu}$. Thus if

$$|c_n| > 2^{2\mu}\|p_x q_{y,n-\mu} - p_y q_{x,n-\mu}\|_\infty \tag{4.6}$$

and

$$\|(a_n,b_n,c_n)\|_\infty > 2^{2\mu}\|\mathbf{p}\times\mathbf{q}_{n-\mu}\|_\infty, \tag{4.7}$$

then the upper bound for the μ-basis method is smaller than the upper bound for the direct method. Thus, in this case, it is possible that the μ-basis method produces smaller errors. The examples in next section show this is true even if these sufficient conditions are not satisfied.

Remark 4.8. Since the μ-basis element $\mathbf{q}$ is not unique, it is natural to ask which choice of $\mathbf{q}$ gives the best result. To answer this question, let $\overline{\mathbf{q}}$ be another choice of $\mathbf{q}$. Then $\overline{\mathbf{q}} = \mathbf{q}+h(t)\mathbf{p}$ with $h(t)$ being a polynomial whose degree is less than or equal to $n-2\mu$. By the degree reduction formula (3.5), we need to find a $\overline{\mathbf{q}}$ such that the norm of the leading coefficient vector of $\overline{\mathbf{q}}$ is as small as possible. Let the degree of $h(t)$ be $n-2\mu$ and the leading coefficient of $h(t)$ be α. Then the leading coefficient vector of $\overline{\mathbf{q}}$ is

$$\overline{\mathbf{q}}_{n-\mu} = \mathbf{q}_{n-\mu} + \alpha \mathbf{p}_{\mu},$$

the sqaure of the norm of which is thus

$$\|\overline{\mathbf{q}}_{n-\mu}\|^2 = \|\mathbf{q}_{n-\mu} + \alpha \mathbf{p}_{\mu}\|^2 = \|\mathbf{p}_{\mu}\|^2\alpha^2 + 2\mathbf{p}_{\mu}\cdot \mathbf{q}_{n-\mu}\alpha + \|\mathbf{q}_{n-\mu}\|^2.$$

Set $\alpha = -\mathbf{p}_{\mu}\cdot\mathbf{q}_{n-\mu}/\|\mathbf{p}_{\mu}\|^2$, and the norm of the leading coefficient vector of $\overline{\mathbf{q}}$ is minimized.

Remark 4.9. If we require that $\tilde{\mathbf{q}}$ interpolates the two end points of $\mathbf{q}$, by (2.6), then the degree reduced curve $\widetilde{\mathbf{P}}(t)$ will interpolate the end points of the original curve $\mathbf{P}(t)$.

5. Examples and Comparisons

In this section, we compare the two degree reduction methods through illustrated examples. The examples show that the μ-basis method is another choice to reduce the degree of a rational curve when the polynomial $c(t)$ is sufficiently small at some parameter values.

Example 5.1. (Illustrations in Fig. 1) Consider a rational curve of degree nine whose parametric equation is as follows

$$\begin{aligned} a(t) &= -3.7 + 15.2\,t + 40.7\,t^2 - 179.58\,t^3 - 15.6\,t^4 - 27.18\,t^5 \\ &\quad +470.2\,t^6 - 240\,t^7 + 237.8\,t^8 - 302.4\,t^9, \\ b(t) &= 3.7 - 15.2\,t - 8.08\,t^3 + 15.6\,t^4 - 0.68\,t^5 + 132.2\,t^6 \\ &\quad -160.2\,t^7 + 162.6\,t^8 - 129.0\,t^9, \\ c(t) &= 10.23 - 44.44\,t^2 - 200.2\,t^3 + 200.1\,t^4 + 922.84\,t^5 \\ &\quad -881.1\,t^6 - 563.0\,t^8 + 560.4\,t^9. \end{aligned}$$

Then μ-basis can be computed as

$$\begin{aligned} \mathbf{p} &= (4.04 - 81.3\,t^3 + 80.1\,t^4, -6.19 + 118.9\,t^3 \\ &\quad -120\,t^4, 3.7 - 15.2\,t + 15.6\,t^4), \\ \mathbf{q} &= (1 + 2\,t^5, 1 - 11\,t^2 + 4\,t^5, 2(\,t^3 + \,t^5)). \end{aligned}$$

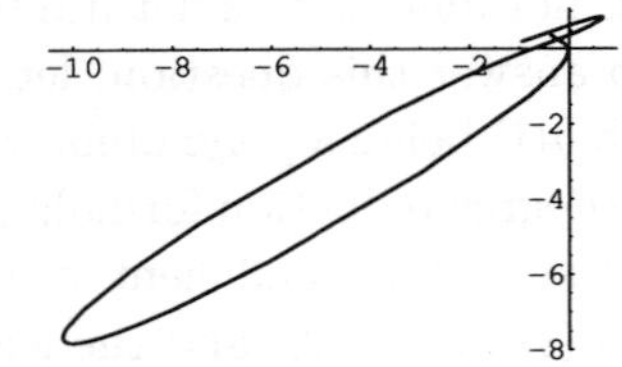

(a) The curve segment $\mathbf{P}(t)$

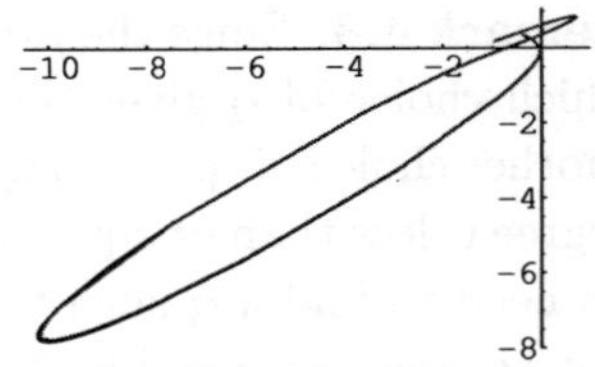

(b) Direct degree reduction curve and $\mathbf{P}(t)$

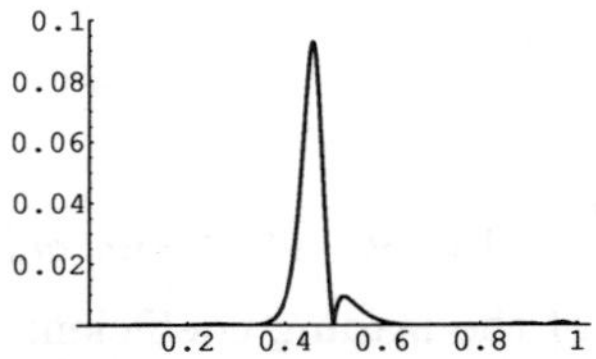

(c) The error function by direct degree reduction method

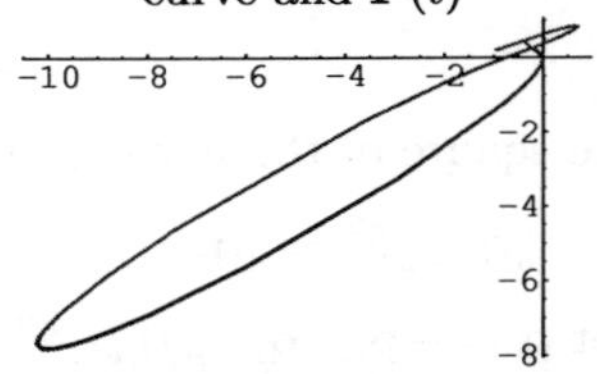

(d)Degree reduction curve by μ-basis method and $\mathbf{P}(t)$

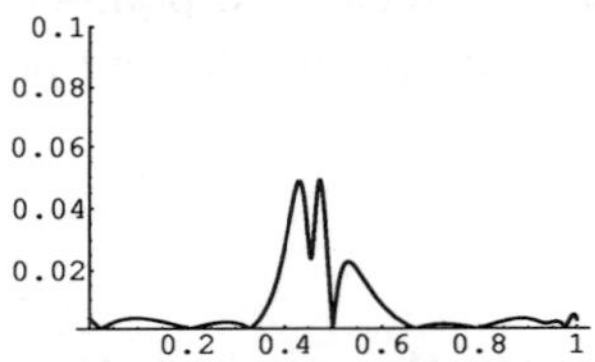

(e) The error function by μ-basis method

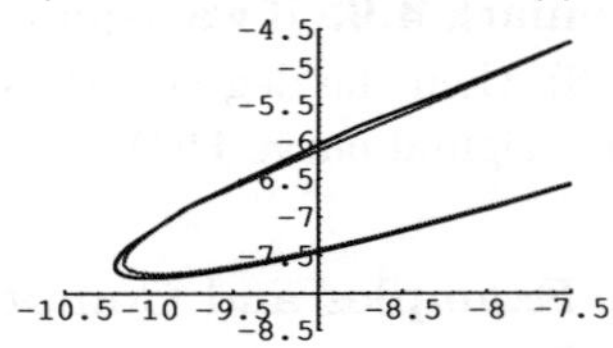

(f) Local amplification ($\mathbf{P}(t)$ and the degree reduced curve by direct method)

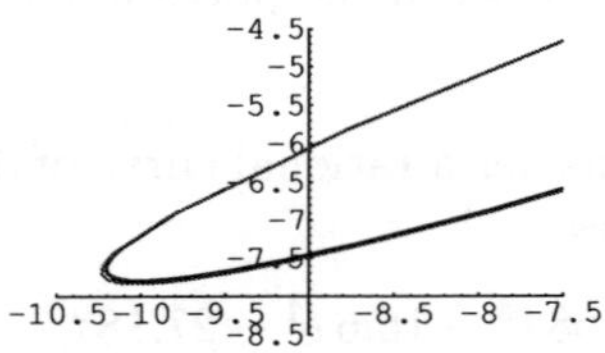

(g) Local amplification ($\mathbf{P}(t)$ and the degree reduced curve by μ-basis method)

Figure 1. Comparison of the degree reduction error functions and the degree reduction curves by the two methods in Example 5.1.

Thus

$$\begin{aligned}
\widetilde{\mathbf{q}} &= \frac{1}{256}(257-50\,t + 400\,t^2 - 1120\,t^3 + 1280\,t^4, 258-100\,t \\
&\quad -2016\,t^2 - 2240\,t^3 + 2560\,t^4, 1-50\,t + 400\,t^2 - 608\,t^3 + 1280\,t^4), \\
\widetilde{\mathbf{P}}(t) &= \mathbf{p} \times \widetilde{\mathbf{q}} \\
&= (-3.75309 + 17.973\,t + 13.5281\,t^2 - 72.1593\,t^3 - 240.363\,t^4 \\
&\quad +367.313\,t^5 - 347.038\,t^6 + 1016\,t^7 - 756\,t^8,
\end{aligned}$$

$$3.69867 - 15.193\,t + 2.4375\,t^2 - 30.0249\,t^3 + 64.2691\,t^4$$
$$+63.6289\,t^5 - 293.869\,t^6 + 528.487\,t^7 - 3225\,t^8,$$
$$10.2857 - 2.78711\,t - 22.1431\,t^2 - 263.731\,t^3 + 327.525\,t^4$$
$$+399.73\,t^5 + 788.275\,t^6 - 2633.38\,t^7 + 1401\,t^8).$$

Example 5.2. (Illustrations in Fig. 2) Let the rational curve be

$$a = -4 + 30\,t + 14\,t^2 - 198\,t^3 + 492\,t^5 - 324\,t^6 + 58\,t^7 - 132\,t^8,$$
$$b = 2 - 15\,t + t^3 + 303\,t^5 - 415\,t^6 + 41\,t^7 + 95\,t^8,$$
$$c = 13 - 154\,t^2 + 137\,t^3 + 287\,t^4 - 189\,t^5 - 727\,t^7 + 713\,t^8.$$

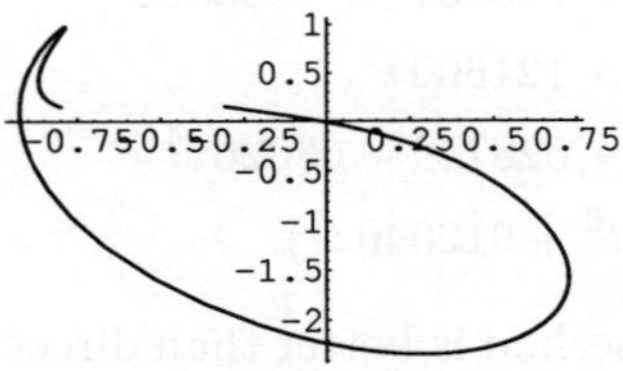

(a) The curve segment $\mathbf{P}(t)$

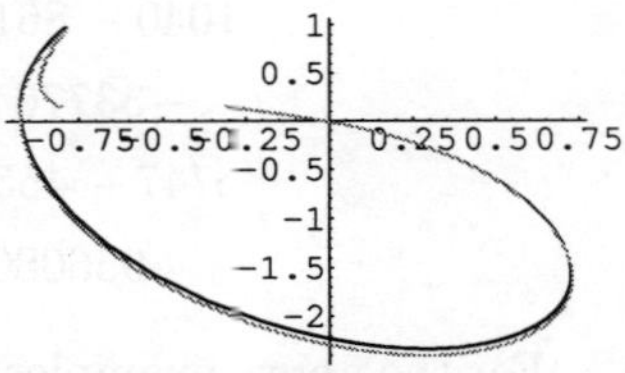

(b) Direct degree reduction curve and $\mathbf{P}(t)$

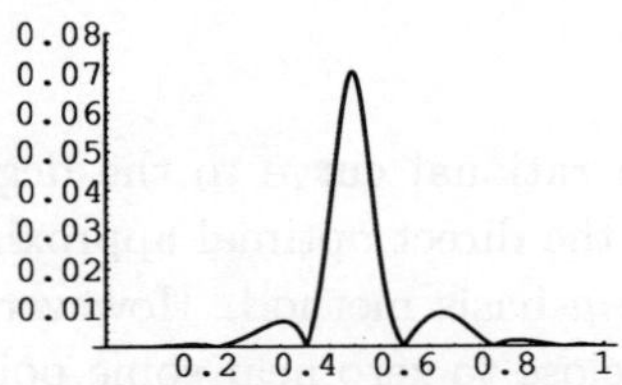

(c) Error function by the direct degree reduction method

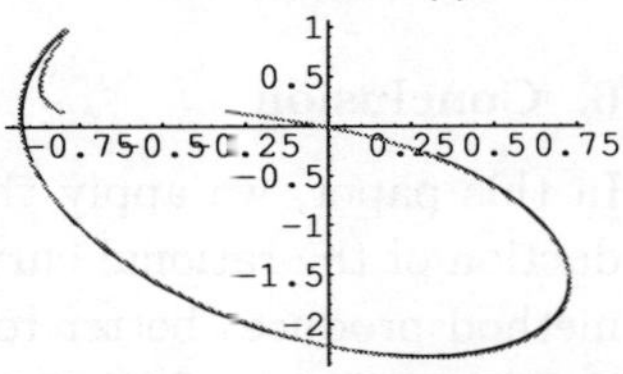

(d) Degree reduction curve by the μ-basis method and $\mathbf{P}(t)$

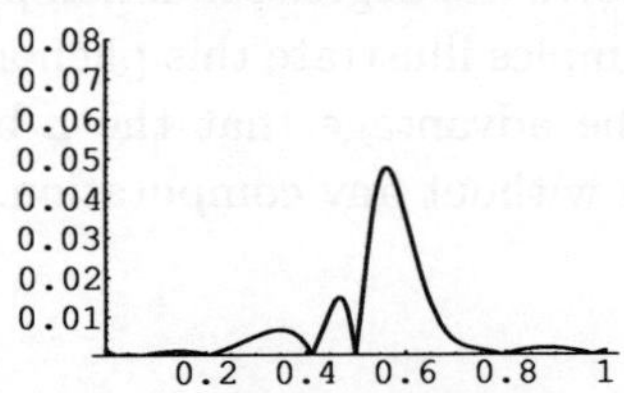

(e) The error function by the μ-basis method

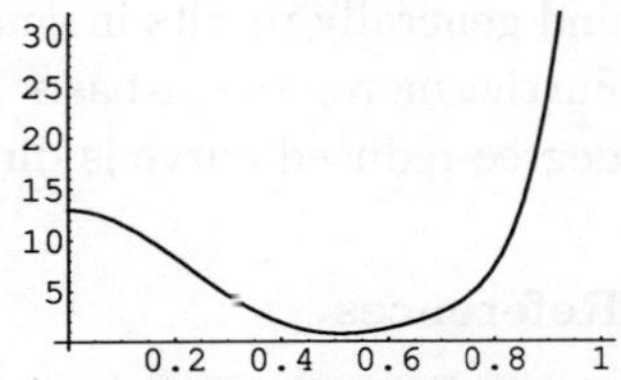

(f) The graphic of $c(t)$

Figure 2. Error functions and degree reduction curves for Example 5.2.

The μ-basis is

$$\mathbf{p} = (2-41\,t^2+40\,t^3, -9+58\,t^2-57\,t^3, 2-15\,t+15\,t^3),$$
$$\mathbf{q} = (1+9\,t^5, 2-7\,t^2+5\,t^5, 7\,t^3+t^5).$$

Thus

$$\widetilde{\mathbf{q}} = \frac{1}{512}(521-450\,t+3600\,t^2-10080\,t^3+11520\,t^4, 1029-250\,t$$
$$-1584\,t^2-5600\,t^3+6400\,t^4, 1-50\,t+400\,t^2+2464\,t^3+1280\,t^4),$$

and

$$\begin{aligned}\widetilde{\mathbf{P}}(t) &= \mathbf{p}\times\widetilde{\mathbf{q}}\\ &= \frac{1}{512}(-2067+16385\,t-4124\,t^2-53128\,t^3\\ &\quad -78520\,t^4+239872\,,t^5+17792\,t^6-168960\,t^7,\\ &\quad 1040-8615\,t+13191\,t^2-73363\,t^3+183330\,t^4\\ &\quad -33776\,t^5-197280\,t^6+121600\,t^7,\\ &\quad 6747-4550\,t-43175\,t^2+5287\,t^3-63026\,t^4\\ &\quad +956080\,t^5-1729120\,t^6+912640\,t^7).\end{aligned}$$

For the above examples, the μ-basis method is better than direct degree reduction method near the zero of $c(t)$.

6. Conclusion

In this paper, we apply the μ-basis of a rational curve to the degree reduction of the rational curve. Generally, the direct optimal approximation method produces better results than the μ-basis method. However, when the denominator of the rational curve is close to zero near some point, the direct optimal approximation method behaves badly. In this case, the μ-basis method provides another choice to solve the degree reduction problem and generally results in small errors. Examples illustrate this phenomenon. Furthermore, the μ-basis method has the advantage that the μ-basis of degree-reduced curve is directly obtained without any computation.

References

1. G. Brunnett, T. Schreiber, J. Braun. The geometry of optimal degree reduction of Bézier curves. *Computer Aided Geometric Design* **13**(1996), 773–788.
2. F. Chen, W. P. Lou. Degree reduction of interval Bézier curves. *Computer Aided Design* **32** (2000), 571–582.

3. F. Chen, W. Wang. μ-Basis of a planar rational curve—properties and computation, 2001, submitted.
4. F. Chen, J. Zheng, T. Sederberg. The mu-basis of a rational ruled surface. *Computer Aided Geometric Design* **18** (2001), 61–72.
5. D. Cox, L. Little, D. O'Shea. *Ideals, Varieties and Algorithms*, Springer-Verlag, New York, 1992.
6. D. Cox, L. Little, D. O'Shea. *Using Algebraic Geometry*, Springer-Verlag, New York, 1998.
7. D. Cox, T. Sederberg, F. Chen. The moving line ideal basis of planar rational curves. *Computer Aided Geometric Design* **15** (1998), 803–827.
8. M. Eck. Degree reduction of Bézier curves. *Computer Aided Geometric Design* **10** (1993), 237–51.
9. D. Lutterkort, J. Peters, U. Reif. Polynomial degree reduction in the L2-norm equals best Euclidean approximation of Bézier coefficients, *Computer Aided Geometric Design* **16**(1999), 607–612.
10. J. Zheng, T. W. Sederberg. A direct approach to the mu-basis of planar rational curves. *J. Symbolic Computation* **31** (2001), 619–629.

A MOLECULAR INVERSE KINEMATICS PROBLEM: AN APPROXIMATION APPROACH AND CHALLENGES

MING ZHANG, R. ALLEN WHITE

The University of Texas M.D. Anderson Cancer Center, Department of Biomathematics, Houston, TX 77030, USA

E-mail: {ming,allen}@mdanderson.org

The efficient computation of molecular conformations (3D-shapes) is critical for many applications in computational biology and chemistry. In this paper we study one such molecular inverse kinematics problem, where we are given an initial conformation of a molecule and the target positions of feature atoms in that molecule. We wish to compute new conformations of the molecule automatically where the feature atoms reach their target positions. A system of polynomial equations can be derived from these target positions, which are described as geometric constraints. Solving for the exact solutions of these equations is a very daunting, practically impossible task. We present an approximation approach based on the Gröbner basis method from algebraic geometry and develop a subdivision algorithm to approximate the solutions. This algorithm still suffers from the huge computational complexity of determining the Gröbner basis. A fast and reliable floating point arithmetic version of the Gröbner basis computation is needed to solve the molecular inverse kinematics problem in an acceptable manner.

1. Introduction

A complete description of the motion of a molecule that includes quantum mechanical effects, dynamic averaging, electronic interactions, vibrational and rotational motion is well beyond current technologies. Indeed a first approximation to the complete panoply raises important challenges, and yet, as described below, still has important potential applications. In this paper, a molecule is modeled by atoms (spheres) and bonds (line segments) connecting the atoms. Some bonds can rotate—torsional angles—along their own directions, so that molecules may have different shapes in space as long as the atoms (spheres) do not collide with each other. Each valid shape of the molecule is called a *conformation* and the set of all possible conformations is called the *conformational space*. The efficient computation of molecular conformations with respect to predefined geometric or energy constraints is of critical importance in computational biology and chemistry.

A rapid solution of the conformational search problem would directly benefit a number of computer applications including drug design. A fundamental assumption in rational drug design is that drug activity is obtained through the molecular binding of one molecule (the ligand) to a specific site on another, usually larger, molecule (the receptor, often a protein). The question is how to place features of the ligand in predefined positions of the receptor [8,10]—thereby defining the inverse kinematics problem. Other applications of molecular inverse kinematics include the placement of side chains and loops in protein folding [17] and the computation of protein folding pathways [19].

In this paper we explore a specific conformational search problem which we call the molecular inverse kinematics problem. Here the initial conformation of a molecule is known and the target positions of some feature atoms of the molecule are user-specified. The torsional angles of the molecule are then computed so that the feature atoms can achieve their target positions.

Our terminology is borrowed from the robotics literature since molecules can be modeled as tiny robots [4,6] and hence we can talk about the kinematics for molecules as robots. In general, the position of (the center of) any atom in the molecule is represented as trigonometric or polynomial functions. When the feature atoms of the ligand molecule are set to the target positions, a system of trigonometric or polynomial equations is obtained [12,21]. Thus the solutions of a system of trigonometric or polynomial equations become the solutions of the inverse kinematics problem.

The molecular inverse kinematics problem in this paper is different from the inverse problems for serial robot manipulators [12,16]. A serial robot manipulator usually has a single chain with 6 links and the tip of a manipulator is determined by 6 parameters. The molecular inverse kinematics problem usually encounters molecules with multiple branches and each branch may have longer chains. Thus the resulting system of polynomial equations may be much bigger and more complex.

Theoretically the molecular inverse kinematics problem can be solved by examining all possible conformations with respect to the target positions of feature atoms. However, the number of degrees of freedom (DOF) of a molecule—the number of rotatable bonds here—are on the order of 10 to 30 even for small molecules [7,21]. Discretization of each DOF requires $\sim$50 values. As a result, any systematic enumeration of all the conformations (exponential in DOF) is prohibitively expensive. Recent work has investigated a variety of randomized and heuristic approaches [8,9,18]. However, these methods suffer from the lack of good starting values for the minimizations, thus leading to conformations trapped in local minima [9] and alternative methods are needed.

One such approach to the molecular inverse kinematics problem is to solve for the variables directly. In particular, algebraic geometry methods have been used to solve the systems of polynomial equations resulting from inverse kinematics problem [5,12,11,13,16]. However, finding the solutions of systems of multivariate polynomials is by no means an easy task. Currently, there are *no good, general solvers* to solve multivariate (non-linear) polynomial equations [11,15].

Realizing the computational impracticality of computing the exact solutions of the required polynomial equations, we naturally seek an approximation approach. More specifically we seek approximations of the solutions for use as initial values for a subsequent optimization procedure which will then find the solutions rapidly from these initial values [9]. In this paper, we introduce a new approximation method using a subdivision scheme.

This paper is organized as follows. Section 2 reviews the derivation of systems of polynomial equations from geometric constraints and formally defines the molecular inverse kinematics problem. Section 3 reviews a technique based on the Gröbner basis method from algebraic geometry for isolating and locating the real solutions of polynomial equations. In Section 4, we introduce a subdivision algorithm to approximate the real solutions of the molecular inverse kinematics problem. We conclude with a discussion of the approximation approach and future work in Section 5.

2. Molecular Inverse Kinematics Problem

In this section, we first review the derivation of the equations describing the positions of atoms in a molecule and then formally define the molecular inverse kinematics problem.

In order to speed up the computation and focus only on the critical rotatable bonds, we partition the atoms of a molecule into atomgroups. An *atomgroup* is a (maximum) set of connected atoms such that none of the bonds inside the atomgroup rotates. Thus only the bonds between the atomgroups rotate [21]. For simplicity, we also assume that there are no cycles of atomgroups in the molecule. When one atomgroup is chosen as the root, the molecule becomes a tree with the atomgroups at the nodes.

We first attach local frames to atomgroups to facilitate the calculation of atom positions. A local frame $F_i = \{Q_i; \mathbf{u}_i, \mathbf{v}_i, \mathbf{w}_i\}$ is attached to atomgroup g_i as follows (Figure 1. (a)): Q_i is the atom in g_i connected to g_{i-1}

by bond b_i; $\mathbf{w}_i$ is the unit vector along bond b_i pointing toward g_{i-1}; $\mathbf{u}_i$ is an arbitrary unit vector perpendicular to $\mathbf{w}_i$; $\mathbf{v}_i = \mathbf{w}_i \times \mathbf{u}_i$.

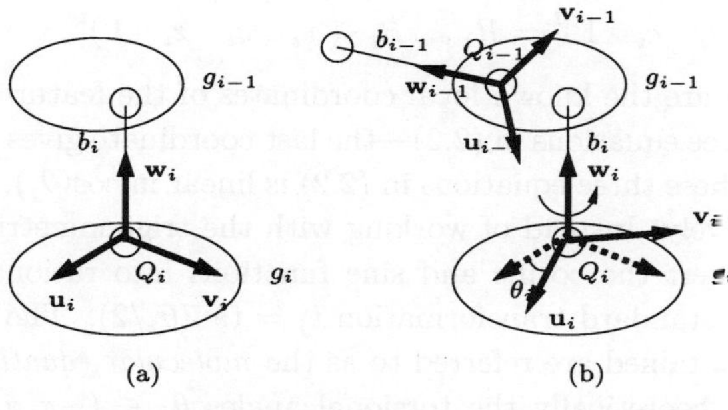

Figure 1. (a) Local frame $F_i = \{Q_i; \mathbf{u}_i, \mathbf{v}_i, \mathbf{w}_i\}$ at atomgroup g_i. (b) Local frames $F_i = \{Q_i; \mathbf{u}_i, \mathbf{v}_i, \mathbf{w}_i\}$ at atomgroup g_i and $F_{i-1} = \{Q_{i-1}; \mathbf{u}_{i-1}, \mathbf{v}_{i-1}, \mathbf{w}_{i-1}\}$ at g_{i-1}.

Next we determine the relations between neighbor local frames which will be used to calculate atom positions. Suppose the frame at atomgroup g_i is $F_i = \{Q_i; \mathbf{u}_i, \mathbf{v}_i, \mathbf{w}_i\}$ and the frame at its parent atomgroup g_{i-1} is $F_{i-1} = \{Q_{i-1}; \mathbf{u}_{i-1}, \mathbf{v}_{i-1}, \mathbf{w}_{i-1}\}$. Let the torsional angle of bond b_i be θ_i (Figure 1. (b)). For each atom A in atomgroup g_i, the coordinates (x_i, y_i, z_i) in F_i and the coordinates $(x_{i-1}, y_{i-1}, z_{i-1})$ in F_{i-1} are related by

$$(\, x_{i-1} \quad y_{i-1} \quad z_{i-1} \quad 1\,)^t = R_i \cdot (\, x_i \quad y_i \quad z_i \quad 1\,)^t,$$

where

$$R_i = \text{a } 4 \times 4 \text{ constant matrix} \cdot \begin{pmatrix} \cos\theta_i & -\sin\theta_i & 0 & 0 \\ \sin\theta_i & \cos\theta_i & 0 & 0 \\ 0 & 0 & 1 & 0 \\ 0 & 0 & 0 & 1 \end{pmatrix}.$$

Finally, the position of any atom A in atomgroup g_i can be calculated. Suppose $g_i, g_{i-1}, \ldots, g_0$ is a sequence of atomgroups, where g_j is the parent atomgroup of g_{j+1}, $0 \le j \le i-1$, and g_0 is the root atomgroup. Then the position of A is

$$(\, x \quad y \quad z \quad 1\,)^t = R_1 \ldots R_i \cdot (\, x_i \quad y_i \quad z_i \quad 1\,)^t, \tag{2.1}$$

where (x_i, y_i, z_i) are the coordinates of A in the local frame at atomgroup g_i. (Here we assume that the Euler angles are known and the Euler matrix [4] has been multiplied into the constant matrix in R_1.)

The molecular inverse kinematics problem can now be formally defined. Given (i) a molecule in an initial conformation, and (ii) the target positions

(a_i, b_i, c_i) of some feature atoms, solve for the values of torsional angles such that the feature atoms reach their target positions. That is, for each feature atom A_i, if the target position is (a_i, b_i, c_i), then

$$(\,a_i \quad b_i \quad c_i \quad 1\,)^t = R_1 \ldots R_i \cdot (\,x_i \quad y_i \quad z_i \quad 1\,)^t\,, \tag{2.2}$$

where (x_i, y_i, z_i) are the known local coordinates of the feature atom A_i.

There are three equations in (2.2)—the last coordinate gives the identity $1 = 1$. Each of these three equations in (2.2) is linear in $\cos(\theta_j), \sin(\theta_j), j = 1, \ldots, i$, respectively. Instead of working with the trigonometric equations directly, we convert the cosine and sine functions into rational functions by applying the standard transformation $t_j = \tan(\theta_j/2)$. The polynomial equations thus obtained are referred to as the *molecular equations.*

Notice that theoretically the torsional angles $\theta_j \in (-\pi, \pi)$, and thus $-\infty < t_j < \infty$. But for practical purposes, it is usually enough to assume $-\pi/2 < \theta_j < \pi/2$ and so $-1 < t_j < 1$. Also notice that each molecular equation is quadratic in each of the t_j involved. Since only real solutions are meaningful for our problem, we will target the real solutions only.

3. Locating Real Roots

In this section, we review some background algebraic geometry on locating real solutions of polynomial equations. For more information, see [3,14].

Throughout this section, we let k be a subfield of $\mathbb{R}$ (real numbers), usually $k = \mathbb{Q}$ (rational numbers), or a finite extension field of $\mathbb{Q}$. Let I be an ideal in the polynomial ring $k[t_1, \ldots, t_n]$ and we denote the set of real solutions of I by $V(I) = \{a \in \mathbb{R}^n \mid f(a) = 0, \forall f \in I\}$. Assume that $V(I)$ is finite.

Let $A = k[t_1, \ldots, t_n]/I$ be the quotient ring. Then A is a vector space (also an algebra) over k. For any $h \in k[t_1, \ldots, t_n]$, define the mapping

$$m_h : A \longrightarrow A, \qquad m_h([g]) = [h] * [g] = [h * g] \in A,$$

where the brackets represent the equivalent classes. Then m_h is a linear mapping from A to A. Fix a monomial order, and let B be a monomial basis of A as a vector space over k. Then we can represent m_h by its matrix with respect to the basis B.

For any $h \in k[t_1, \ldots, t_n]$, define the bilinear form

$$S_h : A \times A \longrightarrow k, \qquad S_h([f], [g]) = \mathrm{Trace}(m_{h*f*g}),$$

where the trace of a matrix is the sum of the diagonal entries, and m_{h*f*g} is a linear mapping defined by the polynomial $h * f * g$. It is easy to see that S_h is symmetric, i.e., $S_h([f],[g]) = S_h([g],[f])$. Let M_h be the matrix of S_h with respect to basis B. Then M_h is symmetric. Specifically, suppose $B = \{v_1, \ldots, v_d\}$, then M_h has dimension $d \times d$, and the (i, j) entry of M_h is Trace($m_{h*v_i*v_j}$).

Note that the Gröbner basis is involved here implicitly. The basis B consists of all the monomials which are not multiples of the leading monomials of the polynomials of a Gröbner basis (with respect to the fixed monomial order). We also use the Gröbner basis to compute both m_h and S_h.

Since we seek only the real solutions of polynomial equations, the search space for the solutions of the molecular inverse kinematics problem is $\mathbb{R}^n$. For any $h \in k[t_1, \ldots, t_n]$ (a decomposition function), define

$$H^+ = \{a \in \mathbb{R}^n \mid h(a) > 0\}, \quad H^- = \{a \in \mathbb{R}^n \mid h(a) < 0\},$$
$$H^0 = \{a \in \mathbb{R}^n \mid h(a) = 0\}.$$

Note that H^+, H^-, H^0 do not intersect with each other, and $\mathbb{R}^n = H^+ \cup H^- \cup H^0$. Our goal is to locate the solutions of the ideal I in H^+, H^-, H^0 by properly choosing the decomposition functions.

Recall that the signature of a matrix M, denoted by $\sigma(M)$, is the difference between the number of positive eigenvalues and the number of negative eigenvalues. We state two main theorems below. The second theorem holds the key to the algorithm we develop in Section 4.

Theorem 3.1. [3] *Assume that $V(I)$ is finite. Let S_h be the bilinear form defined by* Eqn. (2.1) *and M_h be its matrix. Then the signature of M_h is determined by*

$$\sigma(M_h) = \#\{a \in V(I) \mid h(a) > 0\} - \#\{a \in V(I) \mid h(a) < 0\},$$

where the # sign denotes the number of elements in a set.

Theorem 3.2. *Assume that $V(I)$ is finite. The numbers of solutions of I in each of H^+, H^-, H^0 are*

$$\#\{a \in V(I) \mid h(a) > 0\} = (\sigma(M_h) + \sigma(M_{h^2}))/2,$$
$$\#\{a \in V(I) \mid h(a) < 0\} = (\sigma(M_{h^2}) - \sigma(M_h))/2,$$
$$\#\{a \in V(I) \mid h(a) = 0\} = \sigma(M_1) - \sigma(M_{h^2}),$$

where M_1, M_h, M_{h^2} are the matrices of the bilinear forms defined by the polynomials $1, h, h^2$ respectively.

Proof. The theorem follows from these equalities:

$$\begin{aligned}\sigma(M_h) &= \#\{a \in V(I) \mid h(a) > 0\} - \#\{a \in V(I) \mid h(a) < 0\},\\ \sigma(M_{h^2}) &= \#\{a \in V(I) \mid h^2(a) > 0\}\\ &= \#\{a \in V(I) \mid h(a) > 0\} + \#\{a \in V(I) \mid h(a) < 0\},\\ \sigma(M_1) &= \#\{a \in V(I)\}.\end{aligned}$$

□

To obtain the solution counts in each of $V(I) \cap H^+$, $V(I) \cap H^-$, and $V(I) \cap H^0$, we need to compute the signatures $\sigma(M_h)$, $\sigma(M_{h^2})$, and $\sigma(M_1)$. We could solve for the eigenvalues of the matrix and obtain the signature. However this is unnecessary since we do not need the actual eigenvalues. A better way is to use the characteristic polynomials of symmetric matrices. The number of sign changes of the coefficient sequence of the characteristic polynomial of a matrix gives the number of the positive and negative eigenvalues and hence the signature [3]. But, the most reliable way to compute the signature of a symmetric matrix, that we are aware of, is to use the Bunch-Kaufman factorization [1]. A symmetric matrix M can be factored into $M = QDQ^T$ where D is symmetric and block diagonal with 1-by-1 or 2-by-2 diagonal blocks. The signature of M is the difference between the number of positive 1-by-1 blocks and number of negative 1-by-1 blocks of D.

4. Approximating Inverse Kinematics Solutions

In Section 2, we derived a system of molecular equations from the geometric constraints of the feature atoms. In Section 3, using Theorem 3.2, we count the numbers of real solutions in the regions (H^+, H^-, H^0) of the search space $\mathbb{R}^n$. In this section, we develop a subdivision algorithm to approximate the real solutions of the molecular equations, and hence the solutions of the molecular inverse kinematics problem. We first identify the intervals in which the coordinates of the real solutions lie. Then we identify the boxes (regions in the search space $\mathbb{R}^n$) where the real solutions reside. Both steps are accomplished by properly choosing the decomposition functions.

Suppose the molecular equations $f_1 = 0, \ldots, f_l = 0$ involve n variables $t_1, \ldots, t_n$. Let I be the ideal generated by $f_1, \ldots, f_l$ in $\mathbb{Q}[t_1, \ldots, t_n]$.

Identification of intervals

We give the following algorithm to identify the intervals in which the coordinates of the real solutions lie:

(1) For the variable t_1, choose a sequence of real numbers $\{a_1, \ldots, a_p\}$, $a_i < a_{i+1}$. For each $i = 1, \ldots, p-1$, let $h_i = (t_1 - a_i) * (t_1 - a_{i+1})$. Then the solution count, $\#\{H_i^- \cap V(I)\}$, is the number of solutions in $V(I)$ whose t_1 coordinates lie in (a_i, a_{i+1}). The sequence $\{a_1, \ldots, a_p\}$ should have a span wide enough to cover the t_1 coordinates of all the solutions in $V(I)$ such as $a_i = \tan(\theta_{1,i}/2)$, where $\theta_{1,i}$ samples the range of θ_1.

(2) Keep the intervals (a_i, a_{i+1}) with positive solution counts in Step 1. If the solution count is bigger than 1 or the length of the interval is bigger than a predefined threshold, subdivide the interval into two: $(a_i, (a_i + a_{i+1})/2)$ and $((a_i + a_{i+1})/2, a_{i+1})$. Keep the sub-intervals with positive solution counts. Repeat the process until the solution count of each interval is 1 and the length is below the threshold.

(3) Repeat Steps 1 and 2 for other variables $t_2, \ldots, t_n$.

Therefore, for each variable t_i, $i = 1, \ldots, n$, we obtain a sequence of intervals $\{(\ell_1, u_1), \ldots, (\ell_q, u_q)\}$ ("ℓ" and "u" for lower the upper bounds). For each interval, there exists a single real solution whose t_i coordinate lies between $= ell_j$ and u_j, $1 \le j \le q$.

Identification of Boxes

We now combine two sequences of intervals (for two variables) into rectangles and then form boxes which are the approximations of the solutions to the molecular inverse kinematics problem.

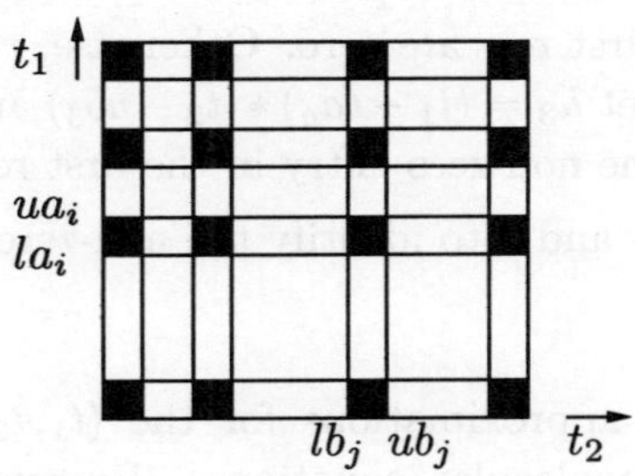

Figure 2. The interval sequence $\{(\ell a_1, ua_1), \ldots, (\ell a_q, ua_q)\}$ for t_1 and interval sequence $\{(\ell b_1, ub_1), \ldots, (\ell b_q, ub_q)\}$ for t_2. Each shaded rectangle contains one solution.

Let us first check the interval sequences $\{(\ell a_1, ua_1), \ldots, (\ell a_{q'}, ua_{q'})\}$ for t_1 and $\{(\ell b_1, ub_1), \ldots, (\ell b_{q''}, ub_{q''})\}$ for t_2. The number of solutions of the molecular equations is fixed, say q, no matter whether we count them with respect to t_1 or t_2, so $q' = q'' = q$. As illustrated in Figure 2, there is a solution in each interval $(\ell a_i, ua_i)$ and interval $(\ell b_j, ub_j)$. Therefore, each shaded rectangle contains exactly one solution.

We abstract Figure 2 into a matrix $E = (e_{ij})$ of dimension $q \times q$: the number of solutions in the shaded rectangle $(\ell a_i, ua_i) \times (\ell b_j, ub_j)$ is e_{ij}. Then the solution count problem can be rephrased using the matrix model. The entries e_{ij} of matrix $E_{q\times q}$ are either 1 or 0. The sum of each row is 1; the sum of each column is 1. Moreover, at any (i,j), $1 \le i,j \le q$, we break the matrix E into four sub-matrices

$$\begin{pmatrix} E_{11} & E_{12} \\ E_{21} & E_{22} \end{pmatrix}, \quad \begin{array}{ll} E_{11} = \{e_{k\ell} \mid k \ge i, \ell \le j\}, & E_{12} = \{e_{k\ell} \mid k \ge i, \ell > j\}, \\ E_{21} = \{e_{k\ell} \mid k < i, \ell \le j\}, & E_{22} = \{e_{k\ell} \mid k < i, \ell > j\}. \end{array}$$

It is easy to see that the sum of the entries of E_{11}, E_{22} and the sum of the entries of E_{12}, E_{21} can be computed. In fact, if we choose a decomposition function $h = (t_1 - \ell a_i) * (t_2 - ub_j)$, then all the real solutions which make h positive are counted in E_{12} and E_{21}; all the real solutions which make h negative are counted in E_{11} and E_{22}.

Next we show how to determine the non-zero entries of E and hence identify the rectangles where the molecular equations have a real solution.

(1) For row 1, choose $h_1 = (t_1 - \ell a_q) * (t_2 - ub_1)$. If the solution count r_1 in H_1^- equals q, then $e_{11} = 1$ and $e_{1j} = 0$, $j = 2, \ldots, q$.

(2) Otherwise, $e_{11} = 0$. Let $h_2 = (t_1 - \ell a_q) * (t_2 - ub_2)$ and compute the solution count r_2 in H_2^-. If $r_2 > r_1$, then $e_{12} = 1$ and all other entries in the first row are zero. Otherwise, $r_1 > r_2$ (since $r_1 \ne r_2$) and $e_{12} = 0$. Let $h_3 = (t_1 - \ell a_q) * (t_2 - ub_3)$ and continue the above process until the non-zero entry in the first row of E is identified.

(3) Repeat Steps 1 and 2 to identify the non-zero entries of other rows of matrix E.

Now we have the approximations for the (t_1, t_2) coordinates of the real solutions of the molecular equations. By examining the intervals $\{(\ell a_1, ua_1), \ldots, (\ell a_q, ua_q)\}$ for t_1 and $\{(\ell_{i,1}, u_{i,1}), \ldots, (\ell_{i,q}, u_{i,q})\}$ for t_i, we get approximations for the (t_1, t_i), $i = 2, \ldots, q$, coordinates of the real solutions. Combining all the (t_1, t_i) rectangular information together, we are able to approximate the real solutions of the molecular equations with the boxes in $\mathbb{R}^n$. The 3-dimensional case is illustrated in Figure 3.

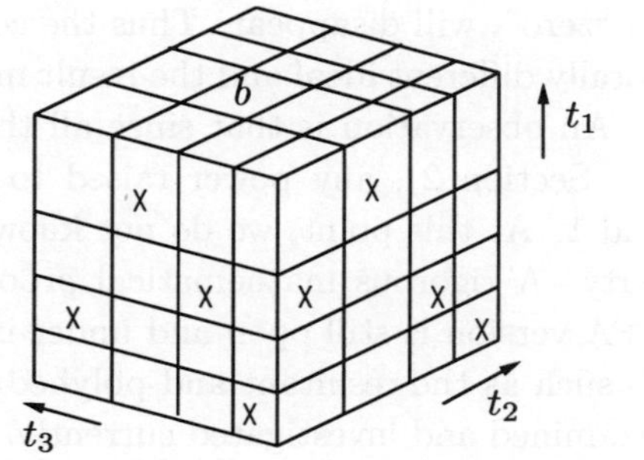

Figure 3. Identifying solutions in boxes. A "×" indicates a solution in the corresponding rectangle. Each level has exactly one solution; so each box, e.g. *b*, at the intersection along the axes perpendicular to the faces with "×" contains a solution.

5. Discussion

We have implemented the algorithm described in Section 4 using the libraries GRÖBNER [20] and SAC [2]. All computations are carried out using exact numbers (rationals). For small molecules with a very small number of atomgroups (3 equations and 3 variables), the algorithm successfully finds the approximations of the real solutions of the molecular inverse kinematics problem. It is much faster (not surprisingly) than Mathematica and Maple (several minutes versus several days). However, systems with 6 equations and 6 variables—the second simplest case as the number of equations increases by 3 when a new feature is added—are presently beyond the capability of the program. The initial coordinates of the target positions have 3 or 4 digits (rationals). In the case of 6 equations, the coefficients in the Gröbner basis computation soon grow to hundreds of digits and out of control.

The biggest challenge of the algorithm arises from the computation of of Gröbner basis. The Gröbner basis computation has double exponential complexity if the arithmetics are carried out using exact numbers. This has become the bottle-neck of the algorithm. A floating point arithmetic (FPA) version of the Gröbner basis computation, which should also be free of numeric error accumulation, will allow the algorithm to be practical in solving the molecular inverse kinematics problem for medium or even large size molecules.

A problem with the reliable FPA version of the Gröbner basis computation is that a term with very small coefficient (in absolute value), which is considered to be "zero", will disappear. Thus the computation may produce a basis of a totally different ideal and the result may not necessarily be reliable any more. An observation is that since all the variables $t_1, \ldots, t_n$ are in $(-1, 1)$ (c.f. Section 2), any power raised to the variables is still bounded by -1 and 1. At this point, we do not know how to take advantage of this property. A rigorous mathematical proof of the validity and efficiency of the FPA version is still open and under investigation.

Other methods such as the resultant and polyhedral homotopy continuation are being examined and investigated currently. It will be interesting to see how far we can push the limit of these methods or their combinations in solving the molecular inverse kinematics problem. Since solving the molecular inverse kinematics problem will have a direct impact in computational biology and chemistry, we want to bring it to the community of computational mathematicians.

Acknowledgments

This work is partially supported by a grant from National Cancer Institute of America, CA16672. We would like to thank Professors Ron Goldman, Lydia Kavraki, Dan Sorensen at Rice University, and Professor Bernd Sturmfels at University of California at Berkeley for insightful discussions. We are also very grateful to the reviewers for their suggestions.

References

1. E. Anderson, Z. Bai, C. Bischof, S. Blackford, J. Demmel, J. Dongarra, J. Du Croz, A. Greenbaum, S. Hammarling, A. Mckenney, D. Sorensen. *LAPACK User's Guide.* Third Edition, SIAM, Philadelphia, 1999.
2. B. Buchberger, G. E. Collins, M. J. Encarnación, H. Hong, J. R. Johnson, W. Krandick, R. Loos, A. M. Mondache, A. Neubacher, and H. Vielhaber. *SACLIB 1.1 User's Guide, RISC-Linz Report Series Technical Report* **93-19**, Johannes Kepler University, A-4040 Linz, Austria, 1993.
3. D. Cox, J. Little, D. O'Shea. *Using Algebraic Geometry.* Springer-Verlag, New York, 1998.
4. J. J. Craig. *Introduction to Robotics.* Addison-Wesley, Reading, MA, 1989.
5. I. Emiris, B. Mourrain. Computer algebra Methods for studying and computing molecular conformations, *Algorithmica.* **25**(2) (1999), 372–402.
6. P. W. Finn, L. E. Kavraki. Computational approaches to drug design. *Algorithmica* **25** (1999), 347–371.
7. D. R. Henry, A. G. Ozkabak. Conformational flexibility in 3D structure searching. In P. von R. Schleyer *et al.* eds., *Encyclopedia of Computational Chemistry*, Wiley, New York, 1998.

8. G. Jones, P. Willett, R. C. Glen, A. R. Leach, R. Taylor. Further development of a genetic algorithm for ligand docking and its application to screening combinatorial libraries. *ACS Symposium Series* (*Rational Drug Design: Novel Methodology and Practical Applications*) **719**(1999), 271–291.
9. S. M. Lavalle, P. W. Finn, L. E. Kavraki, J. C. Latombe. A randomized kinematics-based approach to pharmacophore-constrained conformational search and database screening. *Journal of Computational Chemistry* **21**(9) (2000), 731–747.
10. C. Lemmen, T. Lengauer, G. Klebe. FlexS: a method for fast flexible ligand superposition. *J. of Medicinal Chemistry* **41** (1998), 4502–4520.
11. D. Manocha. Numerical methods for solving polynomial equations. *Proceedings of Symposium in Applied Mathematics* **53** (1998), 41–66.
12. D. Manocha, Y. Zhu, W. Wright. Conformational analysis of molecular chains using Nano-Kinematics. *Computer Application of Biological Sciences (CABIOS)* **11**(1) (1995), 71–86.
13. T. G. Nikitopoulos, I. Z. Emiris. Molecular conformation search by matrix perturbations. Preprint at `http://www-sop.inria.fr/galaad/emiris/`, 2001.
14. P. Pedersen, M. F. Roy, A. Szpirglas. Counting real zeros in the multivariate case. In F. Eyssette and A. Galligo, eds., *Computational Algebraic Geometry*, 203–224. Birkhauser, Boston, 1993.
15. W. Press, B. Flannery, S. Teukolsky, W. Betterling. *Numerical Recipes: The Art of Scientific Computing.* Cambridge U. Press, Cambridge, 1990.
16. M. Raghavan, B. Roth. Solving Polynomial Systems for the Kinematic Analysis and Synthesis of Mechnisms and Robot Manipulators. *ASME J. Mechanical Design* **117**(2) (1995), 71–79.
17. R. Samudrala, E. S. Huang, P. Koehl, M. Levitt. Constructing side chains on near-native main chains for *ab initio* protein structure prediction. *Protein Eng.* **3** (2000), 453–457.
18. A. Smellie, S. D. Kahn, S. L. Teig. Analysis of conformational coverage, I: Validation and estimation of coverage. *Journal of Chemical Information and Computer Sciences,* **35** (1995), 285–294.
19. G. Song, N. M. Amato. Using motion planning to study protein folding pathways. *Proceedings of the Fifth Annual International Conference on Computational Biology* (2001), 287–296.
20. W. Windsteigner, B. Buchberger. *GRÖBNER: A Library for Computing Gröbner Bases based on SACLIB, Manual for Version 2.0. RISC-report* **93-72**, 1994.
21. M. Zhang, L. E. Kavraki. A new method for fast and accurate derivation of molecular conformations, *Journal of Chemical Information and Computer Sciences* **42**(1) (2002), 64–70.

DISPLACEMENT STRUCTURE IN COMPUTING APPROXIMATE GCD OF UNIVARIATE POLYNOMIALS

LIHONG ZHI
Mathematics Mechanization Research Center
Institute of Systems Science
Chinese Academy of Sciences
Beijing, China 100080
E-mail:lzhi@mmrc.iss.ac.cn

We propose a fast algorithm for the computing approximate GCD of univariate polynomials with coefficients that are given only to a finite accuracy. The algorithm is based on a stabilized version of the generalized Schur algorithm for Sylvester matrix and its embedding. All computations can be done in $O(n^2)$ operations, where n is the sum of the degrees of the polynomials. The stability of the algorithm is also discussed.

1. Introduction

Let $f(x)$ and $g(x)$ be given polynomials represented as

$$f(x) = f_n x^n + f_{n-1}x^{n-1} + \cdots + f_1 x + f_0,$$
$$g(x) = g_m x^m + g_{m-1}x^{m-1} + \cdots + g_1 x + g_0,$$

where $f_i, g_i \in \mathbb{R}$ and $||f||_2 = ||g||_2 = 1$. Many papers have already discussed the approximate GCD problem [8,18,19,21,22,23,26,24]. There are many different definitions about approximate GCDs. In the following text, we make use of the definition from [18]. For a given tolerance ϵ, we are going to find an approximate ϵ-GCD. In [9], we have already derived a backward stable method for computing the approximate GCD. The method is based on the classical QR factorization of the Sylvester matrix of f, g and their reversals. Utilizing the special structure of Sylvester matrix, we proposed a combined QR factoring algorithm using Givens rotations and Householder transformations. But the cost of the algorithm is still $O(n^3)$ (assuming $n \geq m$).

More recently, various results on matrices with a displacement structure have been reported in [2,3,4,5,6,7,13,17]. It is well-known that the Sylvester matrix is a quasi-Toeplitz matrix with displacement rank at most two. An algorithm based on fast QR factorization was suggested in [25], but stability was not guaranteed. In [6] a modified fast QR factorization for matrices with a shift structure (for example, Toeplitz or quasi-Topelitz matrices) was derived. The algorithm is provably both fast and backward stable for solving linear systems of equations, the coefficients matrices of which are structured. This motivates us to extend the stabilized version of the generalized Schur algorithm for computing the approximate GCD efficiently. In the following sections, we first introduce the displacement structure of the Sylvester matrix and its embedding. We then show a fast algorithm for computing the approximate GCD, giving an example that illustrates a good behavior of the algorithm. The backward error and primitiveness test are discussed briefly. All algorithms to be presented in Section 4 are also based on fast algorithms for structured matrix. We conclude with a short account on open problems about stability and structured perturbation.

2. Displacement Structure of a Sylvester Matrix and its Embedding

Let Z_i denote the $i \times i$ lower shift matrix with ones on the first subdiagonal and zeros elsewhere. The *displacement* of an $n \times n$ Hermitian matrix R was originally defined by Kailath, Kung, and Morf [16] as

$$\nabla R = R - Z_n R Z_n^T. \tag{2.1}$$

If ∇R has low rank r $(< n)$ independent of n, then R is said to be *structured with respect to the displacement defined by* (2.1) and r is referred to as the *displacement rank* of R.

The Sylvester matrix of $f(x), g(x)$ is:

$$S = S(f,g) = \begin{bmatrix} f_n & f_{n-1} & \cdots & f_1 & f_0 & & & \\ & f_n & f_{n-1} & \cdots & \ddots & \ddots & & \\ & & \ddots & \ddots & & \ddots & \ddots & \\ & & & f_n & f_{n-1} & \cdots & f_1 & f_0 \\ g_m & g_{m-1} & \cdots & g_1 & g_0 & & & \\ & g_m & g_{m-1} & & \ddots & \ddots & & \\ & & \ddots & \ddots & & \ddots & \ddots & \\ & & & g_m & g_{m-1} & \cdots & g_1 & g_0 \end{bmatrix}. \tag{2.2}$$

Let e_i denotes the i-th column of the $(n+m) \times (n+m)$ identity matrix.

Theorem 2.3. *The Sylvester matrix S is a quasi-Toeplitz matrix, that is, $S - Z_{n+m}SZ_{n+m}^T$ has displacement rank at most 2.*

Proof. It is trivial to note that $S - Z_{m+n}SZ_{m+n}^T = [e_1, e_{n+1}][u, v]^T$ with

$$u = [f_n, \ldots, f_0, 0, \ldots, 0]^T,$$
$$v = [g_m, \ldots, g_1, g_0 - f_n, -f_{n-1}, \ldots, -f_1]^T. \qquad \square$$

In order to compute the QR factorization of the structured matrix S, we need to apply the generalized Schur algorithm to the following properly defined embedding matrix M_5 of the Sylvester matrix.

Theorem 2.4. *The* $2(n+m) \times 2(n+m)$ *augmented matrix*

$$M_5 = \begin{bmatrix} S^TS & S^T \\ S & 0 \end{bmatrix} \tag{2.5}$$

has displacement rank at most 5.

Proof. We can verify that

$$M_5 - FM_5F^T = GJG^T, \tag{2.6}$$

where $J = (1 \oplus 1 \oplus -1 \oplus -1 \oplus -1)$ is a signature matrix, and

$$F = Z_{n+m} \oplus Z_{n+m},$$
$$G = \begin{bmatrix} x_0^T & y_0^T & x_1^T & y_1^T & 0 \\ e_1 & e_{m+1} & e_{m+1} & 0 & e_1 \end{bmatrix}, \tag{2.7}$$

and where

$$x_0 = \text{Row}(S, 1),$$
$$x_1 = \text{Row}(S, m)Z_{n+m}^T,$$
$$y_0 = \text{Row}(S, m+1),$$
$$y_1 = \text{Row}(S, n+m)Z_{n+m}^T. \qquad \square$$

It is clear that the generator G can be computed directly from the Sylvester matrix S instead of the embedding matrix M_5. As in [6], after applying the first $n+m$ steps of the generalized Schur algorithm to (F, G), we have the following partial triangulation:

$$M_5 = \begin{bmatrix} L \\ U \end{bmatrix} D^{-1} \begin{bmatrix} L^T & U^T \end{bmatrix} + \begin{bmatrix} 0 & 0 \\ 0 & -I \end{bmatrix}, \tag{2.8}$$

where L is $(n+m) \times (n+m)$ lower-triangular, U is an $(m+n) \times (m+n)$ matrix, and D is an $2(m+n) \times 2(m+n)$ diagonal matrix. By equating terms on both sides of the above equality we conclude that

$$S^T S = R^T R, \quad S = QR, \quad QQ^T = I, \tag{2.9}$$

where $Q = UD^{-1/2}$ and $R = (LD^{-1/2})^T$. The cost of the algorithm is $O(n^2)$. Notice that Q may not be an orthogonal matrix even in the well-conditioned case [6].

3. The Matrix M_5 for Computing the GCD of Univariate Approximate Polynomials

For exact (or infinite precision) computation, it is well-known that $S(f,g)$ is of rank $m+n-r$ if and only if $\deg(\text{GCD}(f,g)) = r$. Let R be an upper triangular factor of S. Then the last non-zero row of R gives the coefficients of GCD(f,g). See [20] for a proof.

For polynomials with coefficients that are given only to a finite accuracy, the above statement must be used carefully. A backward stable method such as QR factoring using Givens rotations or Householder transformations may not result in an R whose numeric rank equals to the degree of a numeric (approximate) GCD. Although QR factoring with column pivoting can possibly reveal the rank of a matrix, pivoting is forbidden in GCD computations. However, in [9], it has been proved that if all the common roots of f, g lie inside the unit circle, the computed R using QR factoring without pivoting will give the coefficients of an approximate GCD; otherwise, the last "non-zero" row of R will only be a factor of the approximate GCD which includes all common roots inside or close to the unit circle. Other common roots outside of the unit circle can be recovered from the QR factoring of the Sylvester matrix of the reversals of f, g. See [9] for details.

In [25], a fast QR factorization combined with an efficient rank estimator was applied to compute the approximate GCD of univariate polynomials. The method has two unsolved issues. One is that the stability of the algorithm is unknown. The other is that the rank estimator has difficulty in deciding the rank in the presence of perturbations. Moreover, according to [9], even if we can estimate the rank correctly, the computed R may still have a numeric rank different from the correctly estimated rank.

The second issue has been discussed in [9] extensively. So now let us concentrate on the stability problem. Chandrasekaran and Sayed derived a stable and fast solver for non-symmetric systems of linear equations with shift structured coefficient matrices. Can it be extended to solve the approximate GCD problem? We have derived an explicit formula for the generator of M_5 in the previous section. Now let us see if the stability problem of the fast algorithm can also be solved for the approximate GCD computation.

Two important properties follow from the Householder QR or the Givens QR factorization [12] [14], namely,

$$||S - QR||_2 = O(u\,||S||_2), \tag{3.1}$$

$$||Q^{-1}S - R||_2 = ||Q^T S - R||_2 = O(u\,||S||_2), \tag{3.2}$$

where u is the machine precision. The first property shows that the GCD of f, g can be written approximately as a linear combination of the polynomials formed by the rows of R. The second condition tells that any polynomials formed from the rows of R can be written approximately as a polynomial combination of f, g. Now let us check these properties for the R computed by fast the QR factoring of M_5.

Suppose $2(n+m)$ steps of the generalized Schur algorithm can be completed to give

$$\widehat{M} = \begin{bmatrix} \widehat{R}^T & 0 \\ \widehat{Q} & \Delta \end{bmatrix} \begin{bmatrix} \widehat{R} & \widehat{Q}^T \\ 0 & \Delta^T \end{bmatrix}. \tag{3.3}$$

Since the generalized Schur algorithm is backward stable, the first property can be easily derived. In the well-conditioned case, although $\widehat{Q}$ need not be orthogonal, it is still true that $\Delta^{-1}\widehat{Q}$ is numerically orthogonal and $||\Delta^{-1}||_2^2$ is bounded by $1/5$. So the second property can be derived from

$$||(\Delta^{-1}\widehat{Q})^T \Delta^{-1} S - \widehat{R}||_2 = O(u\,||S||_2).$$

In the ill-conditioned case, we cannot guarantee that Δ is well-conditioned or $\Delta^{-1}\widehat{Q}$ is numerically orthogonal. Nonetheless, if we restrict the perturbation β introduced in [6] to be 0, then the last $m+n$ negative steps not failing implies $\lambda_{min}(\widehat{Q}^T\widehat{Q}) > u$. So we have $||\widehat{Q}^{-1}S - \widehat{R}||_2 = O(\sqrt{u}\,||S||_2)$. Though the second property is not guaranteed, we may still obtain useful information from $\widehat{R}$, assuming that $\sqrt{u}$ is of the size of the tolerance.

In practice, we can always perturb the polynomials f, g within the tolerance of the coefficients to obtain a well-conditioned $\widehat{Q}$ even though the perturbed S may still be ill-conditioned. Since $||\widehat{Q}\widehat{Q}^T - \Delta\Delta^T||_2 = (O(u))$, the condition number of $\widehat{Q}$ is very close to the condition number of Δ, which is a triangular matrix, and so its condition number can be estimated efficiently. In the case when $\widehat{Q}$ or Δ is well-conditioned, we can guarantee that the two properties (3.1) and (3.2) hold.

Example 3.4. Let

$$f := 0.02077971692\,x^{13} + 0.09350872615\,x^{12} - 0.2246306892\,x^{11},$$
$$-0.4552056739\,x^{10} + \cdots - 0.09935302169\,x + 0.6520463574\,10^{-8}$$
$$g := -0.03804013712\,x^{11} - 0.1616705828\,x^{10} + 0.3870583952\,x^{9}$$
$$+0.2277653210\,x^{8} + \cdots + 0.03233411610\,x - 0.1940046900.$$

Setting Digits $=$ 15 in *Maple* 8, we get $\gcd(f,g) = 1$. Note that if we choose Digits= 10 in *Maple*, we will get $\gcd(f,g) = x^2 - 0.4623160489\,x - 0.5507384540$ instead. Both the QuasiGCD and the EpsilonGCD in the SNAP package in *Maple* 8 fail for this example. When we apply the generalized Schur algorithm to the rank 5 generator G and J, the algorithm breaks down at step 22. It is interesting to check that the 22-nd row of the computed R gives a polynomial very close to the polynomials c_1, c_2 given below. Since it is not clear how the algorithm works in such case, we would rather follow [6] to introduce a small perturbation to the matrix M_5 in order to avoiding the early breakdown of the algorithm. We add 10^{-8} to f, g to get $\widehat{f}, \widehat{g}$, then both the positive and negative steps of the generalized Schur algorithms succeed. We have

$$||S(\widehat{f},\widehat{g}) - \widehat{Q}\widehat{R}||_2 = 0.400670\,10^{-13}.$$

The condition number of Δ is 30.5154. The orthogonality of $\Delta^{-1}\widehat{Q}$ can be verified as

$$||\Delta^{-1}\widehat{Q}(\Delta^{-1}\widehat{Q})^T - I||_2 = 0.483682\,10^{-10}.$$

Consequently,

$$||(\Delta^{-1}\widehat{Q})^T\Delta^{-1}S - \widehat{R}||_2 = 0.123234\,10^{-11}.$$

So the two properties can be achieved even in the ill-conditioned case. The norms of the last two rows of $\widehat{R}$ are less than 10^{-8}. Forming the polynomial from the last third row, we will obtain a monic factor of the GCD of f, g as

$$c_1 := x^2 + 0.46231633\,x - 0.55073866\,76.$$

Now, if we apply the classical QR factoring to $S(f,g)$, the norms of the last two rows of R are less than 10^{-10}. The last third row of R gives

$$c_2 := x^2 + 0.4623160154\,x - 0.5507384196.$$

It is clear that c_1, c_2 are very close to each other.

4. Backward Error Analysis

Although it has been proved that the modified generalized Schur algorithm in[6] is backward stable and fast for solving system of linear equations, it is still not fully proved that the fast QR factorization for the approximate GCD computation is backward stable. So it is important to check the backward error after we have obtained a candidate for the approximate GCD. There are two main steps for checking the backward error:

(1) by an approximate polynomial division, and
(2) by a test for the primitiveness of the cofactors.

Approximate Polynomial Division

Let $f(x), c(x)$ be given polynomials having degree $n > n_1$ respectively, and suppose $||f||_2 = ||c||_2 = 1$:

$$f(x) = f_n x^n + f_{n-1}x^{n-1} + \cdots + f_1 x + f_0,$$
$$c(x) = c_{n_1} x^{n_1} + c_{n_1-1}x^{n_1-1} + \cdots + c_1 x + c_0. \tag{4.1}$$

We are trying to find the cofactor h which minimizes $||f - ch||_2$. There are many ways to solve this least square problem. We present one method based on the displacement structure of the following matrix. Define A as

$$A = \begin{bmatrix} c_{n_1} & & \\ \vdots & \ddots & \\ \vdots & & c_{n_1} \\ c_0 & & \\ & \ddots & \vdots \\ & & c_0 \end{bmatrix} \in \mathbb{R}^{(n+1)\times(n-n_1+1)}. \tag{4.2}$$

We can write the minimization problem in matrix form as $\min ||AV_h - V_f||_2$, where V_f and V_h are the coefficient vectors of the polynomials f and h respectively. Clearly, A is a generalized Toeplitz matrix. The minimization problem can be solved using the normal equations [12].

Theorem 4.3. *Let A be of full rank, and let $T = A^T A$ be a symmetric positive definite Toeplitz matrix. Then the difference*

$$T - Z_{n-n_1+1} T Z_{n-n_1+1}^T$$

has displacement rank at most 2.

Proof. We can verify that

$$T - Z_{n-n_1+1} T Z_{n-n_1+1}^T = \begin{bmatrix} 1 & 0 \\ t_1 & t_1 \\ \vdots & \vdots \\ t_{n-n_1} & t_{n-n_1} \end{bmatrix} \begin{bmatrix} 1 & 0 \\ 0 & -1 \end{bmatrix} \begin{bmatrix} 1 & 0 \\ t_1 & t_1 \\ \vdots & \vdots \\ t_{n-n_1} & t_{n-n_1} \end{bmatrix}^T$$
$$= GJG^T,$$

where $[1, t_1, \cdots, t_{n-n_1}]^T = A^T \text{Column}(A, 1)$. □

Remark 4.4. Applying the modified generalized Schur algorithm to the generator G, we obtain the Cholesky factorization using $O(n^2)$ operations:

$$T = A^T A = R^T R. \tag{4.5}$$

By solving $R^T y = AV_f$ and $RV_h = y$, we can find the cofactor h.

For Example 3.4, the above algorithm finds the backward error to be:

$$||f - c_1 \cdot f/c_1||_2 = 0.5566147 \cdot 10^{-7},$$
$$||g - c_1 \cdot g/c_1||_2 = 0.1643287 \cdot 10^{-6}.$$

Test for Primitiveness

After dividing f, g by the common divisor $c(x)$, it is necessary to check the primitiveness of the polynomials $f/c, g/c$ to guarantee that the computed approximate GCD $c(x)$ is of the highest possible degree. As stated in [1], it is equivalent to computing the condition number of the Sylvester matrix $S = S(f/c, g/c)$ by solving two systems of linear equations with S^T as coefficient matrix:

$$S^T x = b, \quad b \in \mathbb{R}^{m+n}. \tag{4.6}$$

One system corresponds to b being the coefficient vector of the polynomial 1, and the other system, of the polynomial x^{m+n-1}. Since S^T is also a quasi-Toeplitz matrix, we can directly apply the fast and stable solver to the following embedding of Sylvester matrix:

$$M_4 = \begin{bmatrix} SS^T & S \\ S^T & 0 \end{bmatrix}. \tag{4.7}$$

Theorem 4.8. *The matrix M_4 has displacement rank at most 4.*

Proof. We can verify that

$$R = M_4 - FM_4F^T = GJG^T, \tag{4.9}$$

where $F = Z_{n+m} \oplus Z_{n+m}$ and $J = (1 \oplus 1 \oplus -1 \oplus -1)$. The generator can be written as

$$G = [x_1, x_2, y_1, y_2], \tag{4.10}$$

where $x_k[i]$ and $y_k[i]$ denote the ith entry in the vectors x_k and y_k, respectively, for $k = 1, 2$ and they satisfy the following:

$$\begin{aligned} x_1 &= \text{Column}(R, 1), \quad \text{except} \quad x_1[m+1] = 0, \\ y_1 &= \text{Column}(R, 1), \quad \text{except} \quad y_1[1] = 0, y_1[m+1] = 0, \\ x_2 &= \text{Column}(R, m+1), \quad \text{except} \quad x_2[m+1] = 1/2, \\ y_2 &= \text{Column}(R, m+1), \quad \text{except} \quad y_2[m+1] = -1/2. \end{aligned}$$ □

Remark 4.11. Another different rank 4 generator was given by Claude-Pierre Jeannerod in [15]. It is still unknown which generator is better for computation.

Continuing with Example 3.4, the condition number of $S(f/c_1, g/c_1)$ is of order 10^9. This means that $f/c_1, g/c_1$ are not prime to each other. The classical QR factoring with pivoting running in *Matlab* tells us the numeric rank of $S(f, g)$ is 21, not 22, as showed by the above QR factoring. Actually, the missing common root of f, g is -5.787684. In order to find this common root, it is necessary to reapply the fast QR factoring algorithm to the Sylvester matrix of the reversals of f/c_1 and g/c_1. See [9] for details. The approximate GCD of f and g is:

$$c := x^3 + 6.250020509\, x^2 + 2.125011483\, x - 3.187512489.$$

The backward errors are:

$$\begin{aligned} ||f - c \cdot f/c||_2 &= 0.119003 \cdot 10^{-6}, \\ ||g - c \cdot g/c||_2 &= 0.285738 \cdot 10^{-6}. \end{aligned}$$

5. Concluding Remarks

This paper proposes a new fast algorithm for computing an approximate GCD of two univariate polynomials. The algorithm has been implemented in *Maple* 8. Some experimental results are included. The work reported here is just a first attempt to use displacement structure on the approximate

GCD computations. There are many interesting and important aspects that have not been explored yet. It would be interesting to compare this method with the `QuasiGCD` and the `EpsilonGCD` algorithms in `SNAP` since it has been proved that `QuasiGCD` and `EpsilonGCD` are weakly stable and fast (of $O(n^2)$ in general). There are two unsolved problems in our method: (1) Whether the fast QR factorization is backward stable for computing the approximate GCD; and (2) how to find a structured perturbation (of Sylvester type) to avoid the early breakdown of the fast algorithm. We will pursue these problems in a future paper.

Acknowledgments

Lihong Zhi would like to thank Professors Matu-Tarow Noda, Robert M. Corless, Stephen M. Watt, George Labahn, and Dr. Claude-Pierre Jeannerod for useful discussions.

References

1. B. Beckermann, G. Labahn. When are two polynomials relatively prime? In S. M. Watt, H. J. Stetter, eds. *Special issue of the JSC on Symbolic Numeric Algebra for Polynomials. Journal of Symbolic Computation* **26**(6) (1998), 677–689.
2. A.W. Bojanczyk, R. P. Brent, F. De Hoog. *QR* factorization of Toeplitz matrices. *Numerische Mathematik* **49** (1986) 81–94.
3. A. W. Bojanczyk, R. P. Brent, F. De Hoog, D. R. Sweet. On the stability of the Bareiss and related Toeplitz factorization algorithms. *SIAM J. Matrix Anal. Appl.* **16** (1995), 40–57.
4. R. P. Brent, A. W. Bojanczyk, F. R. de Hoog. Stability analysis of a general Toeplitz systems solver. *Numerical Algorithms* **10** (1995), 225–244.
5. S. Chandrasekaran, A. H. Sayed. Stailizing the generalized Schur algorithm. *SIAM J. Matrix Anal. Appl.* **17** (1996) 950–983.
6. S. Chandrasekaran, A. H. Sayed. A fast stable solver for nonsymmetric Toeplitz and quasi-Toeplitz systems of linear equations. *SIMAX* **19**(1) (1998), 107–139.
7. J. Chun. Fast array algorithms for structured matrices. PhD thesis, Stanford University, 1989.
8. R. Corless, P. Gianni, B. Trager, S. Watt. The Singular Value Decomposition for polynomial systems. In A. H. M. Levelt, ed. *Proceedings of International Symposium on Symbolic and Algebraic Computation*, Montréal, Canada, 195–207. ACM Press, 1995.
9. R. Corless, S. Watt, L. Zhi. QR factoring to compute the GCD of univariate approximate polynomials. submitted, 2002.
10. R. Corless, S. Watt, L. Zhi, A report on the SNAP package in Maple. *Tech. Rep.*, ORCCA, in preparation 2002.

11. J. Demmel. On condition numbers and the distance to the nearest ill-posed problem. *Numer. Math.* **51** (1987), 251–289.
12. G. Golub, C. Van Loan. *Matrix computations.* Johns Hopkins, 3rd edition, 1996.
13. M.Gu. Stable and efficient algorithms for structured systems of linear equations. *SIAM J. Matrix Anal. Appl.* **19** (1997), 279–306.
14. N. Higham. *Accuracy and Stability of Numerical Algorithms*, 1996.
15. C.-P. Jeannerod, L. Zhi. Computing low rank generators of Sylvester matrix embeddings. Manuscript, 2002.
16. T. Kailath, S. Y. Kung, M. Morf. Displacement ranks of a matrix. *Bull. Amer. Math. Soc.* **1** (1979), 769–773.
17. T. Kailath, A. Sayed. Displacement structure: theory and applications. *SIAM Review* **37**(3) (1995), 297–386.
18. N. Karmarkar, Y. N. Lakshman. Approximate polynomial greatest common divisors and nearest singular polynomials. *Proceedings of International Symposium on Symbolic and Algebraic Computation* (*Zürich, Switzerland 1996*), 35–42. ACM Press, 1996.
19. N. K. Karmarka, Y. N. Lakshman. On approximate GCDs of univariate polynomials. In S. M. Watt, H. J. Stetter, eds. *Special issue of the JSC on Symbolic Numeric Algebra for Polynomials. Journal of Symbolic Computation* **26**(6) (1998), 653–666.
20. M. A. Laidacker. Another theorem relating Sylvester's matrix and the greatest common divisor. *Mathematics Magazine* **42** (1969), 126–128.
21. M.-T. Noda, T. Sasaki. Approximate GCD and its application to ill-conditioned algebraic equations. *Journal of Computational and Applied Mathematics*, **38** (1991), 335–351.
22. A. Schönhage. The fundamental theorem of algebra in terms of computational complexity. *Tech. Rep.*, Math. Dept., University of Tübingen, 1982.
23. A. Schönhage. Quasi-gcd computations. *Journal of Complexity* **1** (1985), 118–137.
24. H. Stetter, The nearest polynomial with a given zero, and similar problems. *SIGSAM Bulletin: Communications on Computer Algebra* **33**(4) (1999), 2–4.
25. C. J. Zarowski, X. Ma, F. W. Fairman. A *QR*-factorization method for computing the greatest common divisor of polynomials with real-valued coefficients. *IEEE Trans. Signal Processing* **48** (2000), 3042–3051.
26. L. Zhi, M.-T. Noda. Approximate gcd of multivariate polynomials. In X.-S. Gao, D. Wang, eds. *Proceedings of the Fourth Asian Symposium in Computer Mathematics*, 9–18. World Scientific, Singapore, 2000.

Author Index

N. Aris, 40

L. Burlakova, 52

F. Chen, 64, 265
J. Cheng, 77
Yi. Chen, 227
Yo. Chen, 258
E. Chionh, 114
C. Ciliberto, 87
F. Cioffi, 87

L. Deng, 64

E. Fan, 103
M. Foo, 114

X. Gao, 129

K. Hazaveh, 145
W. Hereman, 163
M. Hickman, 163
C.M. Hoffmann, 1
V. Houseaux, 174
H. Huang, 189

G. Jacob, 174
D.J. Jeffrey, 145

R. Khanin, 204
I. Kotsireas, 217

E. Lau, 217
B. Li, 258
H. Li, 227
Z. Li, 189

R. Miranda, 87

K. Nabeshima, 240

T. Oaku, 23
F. Orecchia, 87
N.E. Oussous, 174

M. Petitot 174

A. Rahman, 40
G.J. Reid, 145

Y. Sato, 240
J. Schicho, 17
T. Shaska, 248
Y. Shiraki, 23
A. Suzuki, 240

N. Takayama, 23

D. Wang, 129
S.M. Watt, 145
R.A. White, 276
A.D. Wittkopf, 145

F. Xie, 258

W. Yang, 1

H. Zhang, 258
F. Zeng, 265
M. Zhang, 276
A. Zhou, 189
L. Zhi, 288